Principles of Vegetable Cultivation

PRINCIPLES
OF
VEGETABLE CULTIVATION

BY

L.H. BAILEY

DISCOVERY PUBLISHING HOUSE
NEW DELHI- 1100 02

Reprinted-1999
Reprint, 2010

ISBN 81-7141-489-3

Published by:
DISCOVERY PUBLISHING HOUSE
4831/24, Ansari Road, Prahlad Street,
Darya Ganj, New Delhi-110 002 (*INDIA*)
Phone: 3279245
Fax: 91-11-3253475

Printed at Mehra Offset Press, Delhi.

CONTENTS

CHAPTER I

CHAPTER II

CHAPTER III

CHAPTER IV

CHAPTER V

CHAPTER VII

CHAPTER VIII

CHAPTER XI

CHAPTER XII

CHAPTER XIII

CHAPTER XIV

CHAPTER XVIII

CHAPTER XIX

CHAPTER XX

CHAPTER I

THE SUBJECT

All plants are vegetables; yet by custom we designate the oleraceous or esculent herbs in a class by themselves, calling them "vegetables" in a restricted sense. The growing of these plants is known as *vegetable-gardening,* an awkward and ambiguous term, although now well restricted by usage. Sturtevant propounded the term *olericulture** (ol′ericulture, from Latin *olus,* kitchen herbs), but it is little used. Its etymology is good, but the word is rather formidable, and it naturally implies only the culture of vegetables, whereas the subject gathers to itself much fact and interest not closely connected with the manual practices. This term should properly cover the subjects

*E. Lewis Sturtevant, Proc. Western N. Y. Hort. Soc., 1886, 25: "As we indulge in special studies we often find a necessity for additional words to our language which shall express more definitely our subject than those in common use. . . . I suggest in addition *pomiculture,* or fruit-culture, and *olericulture,* or vegetable-culture."

associated with the *olerarium,* which is the olery or vegetable-garden. Effort has been taken to make a new term from plainer sources, as Day's *vegeculture,** but this is linguistically imperfect, although custom may eventually sanction it, or something like it, and in that event the justification will lie in considering it a contradiction of "vegetable-culture"; *vegeticulture* would be better.

Historically, the garden vegetables are specially those of edible herbage and root, eaten with meats or other foods rather than as desserts. They are cooked as pot-herbs, or eaten raw as salads. Cabbages and all their kind, spinach, lettuce, beet-root, onion tribes, are of this class. But now we add many fruits, and some of them are strictly desserts, as the melons, which may be treated in European books on fruit-culture, as strawberries may be treated in books on vegetable-gardening. But the vegetables, in current usage, are products of herbaceous plants and usually of annuals, whereas the fruits (if we conveniently forget the strawberry and do not define too closely with the banana and a few others) are products of woody plants. But although the definition may be difficult, my reader knows what a vegetable is; or if he does not know, he may more or less inform himself as he turns these pages.

The term vegetable-gardening, then, comprises a wide range of products limited by usage. Associated with the subject is also a large series of commercial questions in manufacture, transportation, refrigeration, marketing. This book deals primarily with the gardening phase of the subject, as its title indicates, for "horticulture ends at the

---

*Harry A. Day, F.R.H.S. Vegeculture: How to grow vegetables, salads, and herbs in town and country. London, 1917.

factory door," as it is written in the Annals of Horticulture for 1891; wherefore we may compile an inventory at the outset of vegetable-garden plants. The list is not nearly complete for the countries of the world, but it contains sufficient species for purposes of illustration, and it includes all those grown to any extent in the United States and Canada. The first obligation of the horticulturist is to know his plants and be able to grow them.

If the reader is curious to compare this inventory with a catalogue of species of fruits (as in Principles of Fruit-Growing, 20th and subsequent editions) he will note the marked dissimilarities in the representations of the famlies of plants. The great Rose family, so abundant in pomological material, is practically unrepresented, whereas the Pea family, barely represented in the fruits, is fertile in important species. The Rue family (Rutaceæ, yielding the citrus fruits) is unrepresented, as also Myrtaceæ, Vitaceæ, Palmaceæ, and the nut-yielding families as Juglandaceæ and Fagaceæ. On the other hand, certain families come prominently into this list that are absent from the other, as Cruciferæ, Chenopodiaceæ, Umbelliferæ, Labiatæ, Compositæ. The fruits and the vegetables represent unlike parts of the plant kingdom, showing that there is a real divergence between pomiculture and olericulture

Group I. The Herbage Vegetables

in which the leaves and growing parts are eaten.

Agaricaceæ, Mushroom Family.

Mushroom, *Agaricus campestris*.

Gramineæ, Grass Family.

Bamboo, *Phyllostachys pubescens*, and others.

Co-ba, hydropyrum, *Zizania latifolia*.

Liliaceæ, Lily Family.

Asparagus, *Asparagus officinalis.*

Chive, chives, *Allium Schœnoprasum.*

Iridaceæ, Iris Family.

Saffron, *Crocus sativus.*

Moraceæ, Mulberry Family.

Hop (young shoots), *Humulus Lupulus.*

Polygonaceæ, Buckwheat Family.

Rhubarb, pie-plant, *Rheum Rhaponticum.*

Spinach dock, herb patience, *Rumex Patientia.*

Spinach dock (of Chinese), *Rumex dentatus.*

Sorrel, *Rumex Acetosa.*

French sorrel, *Rumex scutatus.*

Chenopodiaceæ, Goosefoot Family.

Spinach, *Spinacia oleracea.*

Orach, *Atriplex hortensis.*

Mercury, Good King Henry, *Chenopodium Bonus-Henricus.*

Blite, *Chenopodium capitatum.*

Lamb's quarters, goosefoot, pigweed, *Chenopodium album* (taken in the fields, scarcely cultivated).

Beet, beet-root, *Beta vulgaris* (see also Group II).

Chard, *Beta vulgaris* var. *Cicla.*

Quinoa, *Chenopodium Quinoa* (see also Group III).

Amaranthaceæ, Amaranth Family.

Amaranth, *Amaranthus gangeticus*, and *A. Blitum* (*A. oleraceus*). Other species are collected for greens.

Phytolaccaceæ, Pokeweed Family.

Scoke, *Phytolacca americana, P. esculenta.*

Aizoaceæ, Carpetweed Family.

New Zealand spinach, *Tetragonia expansa.*

Ice-plant, *Mesembryanthemum crystallinum.*

Portulacaceæ, Purslane Family.

Purslane, *Portulaca oleracea.*

Winter purslane, *Montia perfoliata.*

Basellaceæ, Madeira-vine Family.

Basella, Malabar nightshade, *Basella rubra* and *B. alba.*

Madeira-vine, *Boussingaultia baselloides.*

Cruciferæ, Mustard Family.

Cabbage, *Brassica oleracea* var. *capitata*.

Cauliflower, broccoli, *Brassica oleracea* var. *botrytis*.

Brussels sprouts, *Brassica oleracea* var. *gemmifera*.

Kale, *Brassica oleracea* vars. *acephala* and *ramosa*.

Kohlrabi, *Brassica caulo-rapa*.

Pe-tsai, *Brassica pekinensis*.

Mustard, *Brassica alba*, *B. nigra*, *B. juncea*, *B. japonica*, *B. rugosa*, and others.

Water-cress, *Roripa Nasturtium-aquaticum*.

Tropical cress, *Roripa indica*.

Cress, *Lepidium sativum*.

Upland cress, *Barbarea vulgaris* and *B. verna*.

Sea-kale, *Crambe maritima*.

Rocket-salad, *Eruca sativa*.

Turkish or oriental rocket, *Bunias orientalis*.

Scurvy-grass, *Cochlearia officinalis*.

Shepherd's purse, *Capsella Bursa-pastoris*.

Rosaceæ, Rose Family.

Burnet, *Sanguisorba minor*.

Tropæolaceæ, Tropæolum Family.

Indian cress, *Tropæolum minus* and *T. majus*.

Tiliaceæ, Linden Family.

Jew's mallow, edible jute, *Corchorus olitorius*.

Malvaceæ, Mallow Family.

Roselle, *Hibiscus Sabdariffa*.

Araliaceæ, Ginseng Family.

Udo, *Aralia cordata*.

Umbelliferæ, Parsley Family.

Celery, *Apium graveolens*.

Parsley, *Petroselinum hortense*.

Lovage, *Levisticum officinale*.

Myrrh, sweet cicely, *Myrrhis odorata*.

Chervil, *Anthriscus Cerefolium*.

Angelica, *Archangelica officinalis*.

Fennel, *Fœniculum vulgare* and botanical varieties.

Alexanders, *Smyrnium Olustrum.*
Samphire, *Crithmum maritimum.*
Mitsuba (of Japan), *Cryptotænia canadensis.*

Convolvulaceæ, Morning-glory Family.
Kan-kun, young-tsai, *Ipomœa reptans.*
Alanga, moonflower (calyces), *Calonyction aculeatum.*

Labiatæ, Mint Family.
Sage, *Salvia officinalis.*
Clary, *Salvia Sclarea.*
Hyssop, *Hyssopus officinalis.*
Thyme, *Thymus vulgaris* and *T. Serpyllum.*
Lavender, *Lavandula vera* and *L. Spica.*
Rosemary, *Rosmarinus officinalis.*
Horehound, *Marrubium vulgare.*
Mint, *Mentha citrata, M. rotundifolia.*
Peppermint, *Mentha piperita.*
Spearmint, *Mentha spicata.*
Pennyroyal, *Mentha Pulegium.*
Basil, *Ocimum Basilicum, O. suave,* and *O. minimum.*
Balm, *Melissa officinalis.*
Marjoram, *Origanum Majorana, O. vulgare,* and *O. Onites.*
Savory, summer, *Satureia hortensis.*
Savory, winter, *Satureia montana.*
Catnip, *Nepeta Cataria.*

Valerianaceæ, Valerian Family.
Corn-salad, fetticus, *Valerianella Locusta.*
Italian corn-salad, *Valerianella eriocarpa.*
African valerian, *Fedia Cornucopiæ.*

Compositæ, Composite or Sunflower Family.
Lettuce, *Lactuca sativa* and botanical varieties.
Chicory, witloof, *Cichorium Intybus* (see also Group II).
Endive, *Cichorium Endivia.*
Artichoke, *Cynara Scolymus.*
Cardoon, *Cynara Cardunculus.*
Pot marigold, *Calendula officinalis.*
Chrysanthemum, *Chrysanthemum coronarium.*
Costmary, *Chrysanthemum Balsamita.*

Wormwood, *Artemisia Absinthium* and *A. pontica.*
Mugwort, *Artemisia vulgaris.*
Southernwood, *Artemisia Abrotanum.*
Tarragon, *Artemisia Dracunculus.*
Para cress, *Spilanthes oleracea.*
Dandelion, *Taraxacum officinale.*
Tansy, *Tanacetum vulgare.*

Group II. The Root Vegetables

in which certain underground parts are eaten.

Alismaceæ, Water-Plantain Family.
Arrow-head, *Sagittaria sagittifolia.*
Cyperaceæ, Sedge Family.
Water-chestnut (of the Orient), *Eleocharis dulcis.*
Chufa, *Cyperus esculentus.*
Araceæ, Arum Family.
Culcas, Egyptian taro, *Colocasia antiquorum.*
Taro, *Colocasia esculenta.*
Yautia, malanga, tanier, *Xanthosoma sagittifolium.*
Konjac, koniakum, mo-yii, *Amorphophallus Konjac.*
Zingiberaceæ, Ginger Family.
Ginger, *Zingiber officinale.*
Cannaceæ, Canna Family.
Canna, Queensland arrow-root, *Canna edulis.*
Arrowroot, *Maranta arundinacea.*
White topinambour, topee-tamboo, *Calathea Alluia.*
Liliaceæ, Lily Family.
Onion, *Allium Cepa.*
Welsh onion, *Allium fistulosum.*
Portuguese onion, *Allium lusitanicum.*
Shallot, *Allium ascalonicum.*
Garlic, *Allium sativum.*
Rocambole, *Allium Scorodoprasum.*
Leek, *Allium Porrum.*
Lily, *Lilium* species.

Dioscoreaceæ, Yam Family.

Yam (true), *Dioscorea Batatas, D. alata,* and others.

Air potato, *D. bulbifera* (tubers mostly aerial).

Chenopodiaceæ, Goosefoot Family.

Beet, beet-root, mangel, *Beta vulgaris* (see also Group I).

Olluco, *Ullucus tuberosus.*

Nymphæaceæ, Water-lily Family.

Water-lily root, *Nelumbo nucifera.*

Cruciferæ, Mustard Family.

Radish, *Raphanus sativus* and botanical varieties.

Turnip, *Brassica Rapa.*

Rutabaga, *Brassica campestris* var. *napo-brassica.*

Tuberous-rooted mustard, *Brassica napiformis.*

Horse-radish, *Armoracia rusticana.*

Leguminosæ, Pulse or Pea Family.

Groundnut, *Apios tuberosa.*

Goa bean (tuberous roots), *Psophocarpus tetragonolobus* (see also Group III).

Yam-bean (tuberous roots), *Pachyrhizus erosus* and *P. tuberosus* (see also Group III).

Kudzu, *Pueraria hirsuta.*

Oxalidaceæ, Wood-sorrel Family.

Oka, *Oxalis crenata.*

Euphorbiaceæ, Spurge Family.

Cassava, *Manihot utilissima.*

Tropæolaceæ, Tropæolum Family.

Capucin, *Tropæolum tuberosum.*

Onagraceæ, Evening Primrose Family.

Evening primrose, *Œnothera biennis.*

Umbelliferæ, Parsley Family.

Parsnip, *Pastinaca sativa.*

Carrot, *Daucus Carota.*

Skirret, *Sium Sisarum.*

Tuberous chervil, *Chærophyllum bulbosum.*

Celeriac, *Apium graveolens* var. *rapaceum.*

Aracacha, Peruvian parsnip, *Arracacia xanthorrhiza.*

Convolvulaceæ, Morning-glory Family.

Sweet potato, yam (erroneously), *Ipomœa Batatas.*

Labiatæ, Mint Family.

Chorogi, Japanese or Chinese artichoke, *Stachys Sieboldii.*

Innala, *Plectranthus tuberosus.*

Solanaceæ, Nightshade Family.

Potato, *Solanum tuberosum.*

Martyniaceæ, Martynia Family.

Craniolaria, creole scorzonera, *Craniolaria annua.*

Campanulaceæ, Bluebell Family.

Rampion, *Campanula Rapunculus.*

Compositæ, Composite or Sunflower Family.

Salsify, oyster plant, vegetable oyster, *Tragopogon porrifolius.*

Spanish salsify, *Scolymus hispanicus.*

Black salsify, *Scorzonera hispanica.*

French scorzonera, *Picridium vulgare.*

Girasole (Jerusalem artichoke), *Helianthus tuberosus.*

Chicory, *Cichorium Intybus* (see also Group I).

Gobo, *Arctium Lappa.*

Elecampane, *Inula Helenium.*

Group III. The Fruit Vegetables

in which the fruits or seeds, or both, are eaten.

Gramineæ, Grass Family.

Maize, sweet corn, *Zea Mays* var. *rugosa.*

Chenopodiaceæ, Goosefoot Family.

Quinoa, *Chenopodium Quinoa* (see also Group I).

Ranunculaceæ, Crowfoot Family.

Fennel flower, *Nigella sativa.*

Cruciferæ, Mustard Family.

Rat-tailed radish, *Raphanus sativus* var. *caudatus.*

Leguminosæ, Pulse or Pea Family.

Bean, kidney bean, haricot, *Phaseolus vulgaris.*

Scarlet runner bean, *Phaseolus multiflorus.*

Sieva bean, civet bean, *Phaseolus lunatus.*

Lima bean, *Phaseolus lunatus* var. *macrocarpus.*
Tepary bean, *Phaseolus acutifolius.*
Mung bean, *Phaseolus aureus.*
Urd bean, *Phaseolus Mungo.*
Adzuki bean, *Phaseolus angularis.*
Moth bean, *Phaseolus aconitifolius.*
Metcalfe bean, *Phaseolus Metcalfei.*
Rice bean, *Phaseolus calcaratus.*
Pea, *Pisum sativum.*
Broad bean, *Vicia Faba.*
Peanut, goober (underground fruits), *Arachis hypogæa.*
Lentil, *Lens esculenta.*
Cowpea, *Vigna sinensis.*
Catjang, *Vigna Catjang.*
Asparagus bean, *Vigna sesquipedalis.*
Soybean, *Glycine Soja.*
Chick-pea, garbanzo, *Cicer arietinum.*
Hyacinth bean, *Dolichos Lablab.*
Madras gram, *Dolichos biflorus.*
Chickling vetch, gesse, *Lathyrus sativus.*
Jack-bean, *Canavalia ensiformis*, and probably others.
Ground-pea (of Africa), *Kerstingiella geocarpa* (kandela) and *Voandzeia subterannea* (vandzon).
Goa bean, asparagus pea (edible pods), *Psophocarpus tetragonolobus* (see also Group II).
Pigeon-pea, *Cajanus indicus.*
Yam-bean (edible pods), *Pachyrhizus erosus* and *P. tuberosus* (see also Group II).
Winged pea, *Lotus Tetragonolobus.*
Velvet bean, *Stizolobium* species.
Guar, cluster bean, *Cyamopsis psoraloides.*

Malvaceæ, Mallow Family.

Okra, gumbo, *Hibiscus esculentus.*

Trapaceæ (or Onagraceæ), Trapa Family.

Water caltrop, water chestnut (of Europeans), *Trapa natans.*
Singhara nut, *Trapa bispinosa.*

Umbelliferæ, Parsley Family.

Coriander, *Coriandrum sativum.*

Caraway, *Carum Carvi.*

Dill, *Anethum graveolens.*

Cumin, *Cuminum Cyminum.*

Solanaceæ, Nightshade Family.

Tomato, *Lycopersicon esculentum* and *L. pimpinellifolium.*

Tree tomato, *Cyphomandra betacea.*

Red pepper, chilli, cayenne pepper, *Capsicum annuum* and *C. frutescens.*

Husk tomato, ground cherry, *Physalis pubescens, P. peruviana* and *P. ixocarpa.*

Pepino, *Solanum muricatum.*

Morelle. garden huckleberry, wonderberry, *Solanum nigrum.*

Other solanums yield comestible fruits.

Martyniaceæ, Martynia Family.

Martynia, unicorn plant, *Proboscidea louisiana.*

Cucurbitaceæ, Gourd or Melon Family.

Cucumber, *Cucumis sativus.*

Gherkin, *Cucumis Anguria.*

Mandera cucumber, *Cucumis Sacleuxii.*

Melon, dudaim, *Cucumis Melo* and botanical varieties.

Watermelon, *Citrullus vulgaris.*

Squash, pumpkin, vegetable marrow, *Cucurbita Pepo, C. maxima,* and *C. moschata.*

Siam or Malabar gourd, *Cucurbita ficifolia.*

Wax gourd, white melon, ash pumpkin, *Benincasa hispida.*

Calabash gourd, *Lagenaria leucantha.*

Cassabanana, melocoton, *Sicana odorifera.*

Luffa, *Luffa cylindrica* and *L. acutangula.*

Chayote, christophine, *Sechium edule.*

Balsam apple, *Momordica Balsamina.*

Balsam pear, *Momordica Charantia.*

Snake gourd, *Trichosanthes Anguina.*

Pepino de comer, *Cyclanthera pedata.*

This inventory contains all the leading vegetable-garden plants of the world, and a good number of those of minor importance. It suggests the variety and wealth of the field in plant materials. It would run into many hundreds more if a complete list were attempted. In 1889, Sturtevant (Agric. Sci. iii: 174–8) classified 1,070 species of cultivated food plants, and added that his notes include 4,233 species of edible plants in 1,353 genera and 170 families.* These plants comprise all classes,—grains, fruits, vegetables and others. Undoubtedly these numbers could now be much increased.

In the foregoing lists are 247 entries, of which 114 are leaf vegetables, 59 root vegetables, and 74 fruit vegetables. It displays a fascinating field for labor and study. Here are seeds of unimagined forms, oddities in germination, growths to fix the attention, flowers and fruits representing the vast range of the vegetable kingdom, products in which one may take a personal pride. The number of domesticated forms is sumless, and yet the opportunity for plant-breeding is without end. Who knows the fruits of even the common vegetables? Who can describe accurately even one of the plants, as the botanist would describe it if he had his material properly preserved before him? Where are the herbaria and the museums in which the common things, to say nothing of the uncommon ones, are adequately collected? Plant-growing is so commercialized that we are tempted to give most of our attention to the mechanical and business aspects of the subject, losing our skill as plantsmen. But whatever the develop-

*See also the recent extensive volume issued by the N. Y. Agric. Exper. Station (Geneva), called "Sturtevant's Notes on Edible Plants."

ment of any one of these industries, we must remember that the starting-point is the seed, and that the horticulturist must ever renew his effort to get back to the plant. This effort is not to be conceived as an impersonal task yielding results for commerce and science, but as an ardent affection.

This affection runs not only to the growing of the plants and to the joy of gardening, but also to the appreciation of the good quality that one gets directly from fresh vegetables of merit. It is good to know the plants on which these products grow. As millions of people do not have gardens, so are they unaware of the low quality of much of the commercial produce as compared with things well grown in due season. Most persons, depending on the market, do not know what a superlative watermelon is like. Even such apparently indestructible things as cucumbers have a crispness and delicacy when taken directly from the vine at proper maturity that are lost to the store-window supply. Every vegetable naturally loses something of itself in the process from field to consumer. When to this is added the depreciation by storage, careless exposure and rough handling, one cannot expect to receive the full odor and the characteristic delicacies that belong to the product in nature. We must also remember the long distances over which much of the produce must be transported, and the necessity to pick the produce before it is really fit, to meet the popular desire to have vegetables out of season and when we ought not to want them. There is a time and place for everything, vegetables with the rest. Modern methods of marketing, storing and handling have facilitated transactions, and they have also done very much to safeguard the produce itself and to deliver it to the customer in good condition; but the vegetable well chosen and well grown and fresh from the garden is nevertheless the proper standard of excellence. It is a surpassing satisfaction when the householder may go to her own garden rather than to the store for her lettuce, onions, tomatoes, beets, peas, cabbage, melons, and other things good to see and to eat, and to have them in generous supply.

Yet many vegetable-growers are not directly concerned with the table supply and the general home interest but with the raising of produce for market. Of this range there are two types—market-gardening and truck-growing. The former is the growing of a wide or general range of vegetables by intensive methods near the city, so near that the producer may perhaps drive to the market. The latter (trucking) is the growing of a few specialties on cheaper land by more extensive methods at some distance (often a great distance) from the cities, depending on the long haul by water or rail; of this kind is the growing of large areas in spinach, watermelons, cabbage, kale, potatoes. These distinctions in the business of vegetable-growing were made in the Eleventh Census (Bull. 41, by J. H. Hale; Census of 1890). They are now accepted by American writers.* Yet even in these important commercial practices, now bulking so large in the produce-yield of the country, the relation with the plant is the first consideration.

Having now been introduced to our subject, we may begin at once to grow the plants.

---

*As, for example, R. L. Watts, Vegetable Gardening, copyrighted 1912; L. C. Corbett, Garden Farming, 1913; J. W. Lloyd, Productive Vegetable Growing, 1914; J. G. Boyd, Vegetable Growing, 1917.

THE LABORATORY

Books of practice are now used in colleges and schools as well as directly by growers. The first requisite in the teaching of students in the biological sciences is drill in identification and observation. The student who cannot see what he looks at and accurately describe it, is not ready for lectures or for investigation. It is hoped that vegetable-gardening may be made a means of exact education in natural science, equivalent in its processes with other phases of botany. The student should know the species in the main groups of oleraceous plants. To this end, descriptions of many plants are inserted in the present volume.

The identification and description exercise may well be extended to other species and also to the differing horticultural varieties. All this should be a good preparation for the practical applications, adding to one's proficiency in vegetable-growing as well as opening a world of resources in the objects in nature. To detect and recognize insects and their eggs, plant diseases, the effects of treatments and conditions on the welfare of the plants, requires sharp eyes that are sure of what they see.

The plants themselves, and their many parts, are the primary resource in laboratory work in any branch of horticulture. The growing plants are naturally to be preferred, but they cannot always be had in sufficient quantity and variety, and they soon wilt and lose their significance; a wide range of fairly permanent subjects should be before the student for comparison whatever the season of the year, comprising good herbarium material (not merely leaves), seeds, and accurate pictures of the produce if actual specimens cannot be had. The verification should always extend, however, to the living plants themselves and their products. Whatever the method, the object is to develop the keen and practiced eye, as well as accurate appreciation of record and citation.

The study of the plants does not restrict itself to identification of the kinds and to their taxonomic treatment, although these are the phases specially significant to the beginning student, for he must first know his materials. The physiology and genetics, using these terms in the broadest sense, are subjects of the highest importance; the time must soon come when the accumulated knowledge must be assembled and ably digested.

Be it said at the beginning that the nomenclature of the botanical varieties or races of garden vegetables lies yet in an uncertain state. The search of literature for the oldest tenable trinomial designations has not been made, as it has been made for the names of wild plants. The search will be exceedingly complex, and it will need the services of a trained taxonomist. What classes of literature should be admitted as competent in such inquiries is a subject for discussion before the search itself is undertaken. This field of taxonomy is undeveloped. In the meantime, the writer presents diagnoses of the varieties under the best names he knows, hoping to make a fuller survey of the subject on another occasion.

The varieties under consideration in these technical appendices are mostly the classes or forms presenting such botanical differences that they are capable of preservation and detection on the herbarium sheet. The writer has no sympathy with the practice of giving Latin botanical names to the usual numerous horticultural varieties.

The technical descriptions in this book are all drawn directly from the cultivated plants themselves, and in no case are they copied. This may account for certain discrepancies in comparison with standard botanical characterizations. In this volume we are concerned with the cultigen (the species or the plant of a garden or agricultural ancestry).

It will be noticed that most of the species are credited to Linnæus (Linn. Sp. Pl.). With his Species Plantarum, 1753, begins the modern naming of plants, with the use of the binomial system. This system comprises the genus and the species, the generic name standing first and the specific name second: all onions and their kin are Allium; the species are *A. sativum* (*sativus* is Latin for "planted" or "cultivated"), *A. fistulosum*, *A. Cepa*, and others.

The naming of the species and the botanical forms of plants follows a system characterized by great precision and regularity. It is well for the student to understand the main elements and practices in it, for he is not only enabled to understand but he is trained in accuracy and carefulness of record and reference.

For the most part, English measurements are used in the descriptions. In the minuter weights and sizes, however, metric denominations must be employed. The lowest denomination in avoirdupois weight is one grain, but this denomination is 50 times too heavy to weigh a mustard seed. The grain in apothecaries' or troy weights is still heavier, for there are only 5,760 grains in 1 lb., whereas in avoirdupois weight the pound is divided into 7,000 grains. Therefore, the milligram (mg.) is used for the weighing of seeds. A commercially dry seed of black mustard weighs about 1 mg. (say 1⅓ mg.); so does a small-sized dry turnip seed, while a large turnip seed weighs about 2 mg. There are 1,000 mg. in 1 gram. The pictures of the seeds in this book are mostly enlarged. The flat or rectilinear dimensions are indicated by the figure in parenthesis: (×4) means that the picture is 4 times broader and longer than the seed; (×¼) that the picture is only one-quarter as large as the natural object. Seeds and seedlings are likely to differ between marked garden varieties or races.

THE PLAN OF THE BOOK

The arrangement of the book may now be explained. After the introductory chapter, defining the subject-field, the different vegetables are taken up in groups. They are discussed in groups so that related crops may be considered together, avoiding considerable repetition of advice and contributing to a clearer understanding of the subject. Thus, all melons, cucumbers, squashes, are closely related in cultural requirements, as are the onions, leek and garlic, as well as cabbage, kale, brussels sprouts, cauliflower.

The main principles or considerations are printed *in italic type* at the beginning. Then follows in small type the information that should be available for ready reference, as distances at which plants are to stand, quantity of seed or number of plants to the acre, time of sowing or planting, yields, together with very brief statements of the most important diseases and insects. The condensed paragraphs on the maladies and pests are prepared specially for this edition of the book, all on a uniform pattern, by professors in the New York State College of Agriculture at Cornell University—H. W. Dye for the diseases, C. R. Crosby and M. D. Leonard for the insects.

Following the preliminary matter is the regular reading discussion of the crop. Thereafter is the technical description and record of the plant itself, stated in botanical language for accuracy. The Latin names of the plants, as well as of the insects and the organisms that produce the disease conditions, are always given, for in these days the technical names are a necessary part of our knowledge. These names have much significance and they stand for exact conceptions. Something of the history of our knowledge of the plants is suggested in the synonymy and the records. The records in the text give the names added significance. Students should early learn to think in terms of these names, for their thinking is then straighter. These good names are an index of an educated understanding of the subject.

After the chapters on the oleraceous crops, are parts that discuss the general practices—tillage, fertilizing, marketing, storing, home-gardening, and others.

In his own interest the author should state that these proofs are completed on a sea voyage, without means of reference and verification; but he trusts that serious errors will not arise.

CHAPTER II

PERENNIAL CROPS

Asparagus	**Sea-Kale**
Rhubarb	**Dock and sorrel**
Artichoke	**Udo**
Girasole	

The management of perennial crops differs from that of other vegetable-gardening crops in the fact that they are more or less permanent occupants of the ground, and therefore must be given an area to themselves where they will not interfere with the customary plowing and tilling; in the fact that the chief tillage and care are required early and late in the season; and also because the fertilizing is secured (after the initial preparation of the land) chiefly by surface dressings in spring and autumn. It is advisable, therefore, for cultural reasons, to place these vegetables in a group by themselves, although otherwise they have little in common.

The reader must distinguish between perennial crops and perennial plants. Many perennial plants are treated as annuals in cultivation, as tomato, red pepper, potato, scarlet runner bean, horse-radish, dandelion. On the other hand, some of the perennial crops profit by frequent renewal, as the artichoke. But while the demarcation is indefinite, the gardener readily understands it.

ASPARAGUS

Asparagus is grown for the strong soft young shoots arising in spring; these shoots may be utilized in their natural state (green), or blanched by hilling with earth. A deep, rich, fertile, moist, cool soil, warm exposure, thorough preparation of the land, heavy manuring, thorough tillage in late fall and early spring, are general requisites of asparagus culture. The plants should be allowed to become well established before a crop is cut, and the cutting of the plants should cease in early summer to allow them opportunity to grow and to store up energy for the following year. The tops are mown in late fall, and the land is top-dressed with manure before winter sets in. Asparagus is grown for its young shoots, and the quality is determined by the succulence of these shoots. A good plantation should last ten years and more, at least at the North. Propagated by seeds. Practices in the growing of asparagus vary widely.

In small gardens, asparagus may be set 18 in. apart, in rows as close as 3 ft.; at these distances, about 9600 plants are required to a full acre. In general field culture for green asparagus, the rows are usually farther apart to allow of easier tillage and often 2 ft. in the row. Some growers prefer to plant as wide as 3 by 4 ft., or, 3600 plants. For the growing of blanched asparagus, the rows may be as much as 8 ft. (6 to 8 ft.) and 18 or 20 in. in the row, when about 3600 plants are required to the acre. Seeds are usually sown thickly in rows, and the plants thinned to 3 or 4 in.; 4 or 5 lbs. of seeds are usually sown to the acre. When one year old, the plants are set in permanent quarters, and the following year the first cutting of asparagus may be made. About 2000 (1800 to 3000) dozen bunches (averaging 8 to 12 stalks) is a fair yield to the acre on established plantations. An asparagus bed or field should yield well for 10 to 20 years.

Rust (*Puccinia asparagi*).—Reddish or black pustules are produced on the stems and branches of the plant, killing them prematurely. *Control:* No plants should be permitted to mature during the cutting season and all diseased plants should be cut and burned in the fall. Spraying with bordeaux mixture to which some sticker has been added will aid in control. Certain resistant strains have been developed, especially by the United States Department of Agriculture. These should be used in new plantings.

Beetle (*Crioceris asparagi*).—A gray larva with black head, about $\frac{3}{10}$ in. long, feeding on the young shoots in spring and weakening the plant for the following year. The beetle is about ¼ in. long with prominent orange and black pattern on back, passing the winter in piles of rubbish and under bark; eggs are laid in early spring, on end in a line on young growths of asparagus; they hatch in 3 to 8 days and the young larvæ begin to feed. *Control:* Keep the crop cut clean and starve them out; leave a row or two of asparagus plants on which to poison the larvæ, using arsenate of lead paste, 1 lb. in 20 gals. water; spray the plantation after cutting season is past; let poultry run in the asparagus; clean up rubbish in the fall.

Twelve-spotted asparagus beetle (*Crioceris duodecimpunctata*).—About the size of the common asparagus beetle, reddish orange in color with twelve round black spots on the wing covers. The beetles appear in spring along with the common asparagus beetle and gnaw holes in the tender shoots. The oval eggs are attached by the side, singly, to the leaves. The young grubs enter the berries above; they feed on the seeds, migrating from berry to berry until mature. Pupation takes place in the ground. The insect hibernates as a beetle in dry sheltered places. *Control:* As the larvæ feed inside the berries they cannot be poisoned, but the adults may be destroyed by the same measures as recommended for the common asparagus beetle.

ASPARAGUS MINER (*Agromyza simplex*).—A small maggot that burrows under the epidermis of the asparagus stalk near the ground, sometimes girdling it and causing the stalk to turn yellowish and die prematurely. The parent insect is a small metallic black fly about 1/6 in. long. It appears in New York in May and the female inserts her eggs under the epidermis near the ground. A second brood of flies appears the latter part of July. The insects hibernate as puparia in the old stalks at or below the surface. *Control:* This insect causes little injury in beds being cut, but is sometimes injurious in new beds. No satisfactory method of controlling this pest in commercial plantings is known.

Asparagus is a gross feeder. Land can scarcely be too rich. If the land is originally hard and coarse it should be prepared a year or two in advance by the raising of some thoroughly tilled crop as potatoes and with this crop as much manure as possible should have been used. The asparagus plantation should be made for long use. Therefore it is well to give careful attention to the soil and to the choice of a place that can be permanently set aside for the purpose.

In the home garden, asparagus should be in rows at one side of the plantation, so that it will not interfere with the plowing of the garden area. It usually looks best at the farther side of the garden, where its beautiful herbage makes a background border in summer and fall. The old idea was to have asparagus "beds." The new idea is to plant asparagus in rows as one would plant rhubarb or corn, and to till it with horse tools, if possible, rather than with hoes and finger weeders. For the ordinary family, one row alongside the garden, 75 to 100 feet long, may be expected to furnish a sufficient supply.

As a field crop, it is ordinarily grown in the best and richest soil available. The permanency of the plantation will depend largely on the original quality of the land, the preparation of it, good drainage, the method of planting, and particularly on the subsequent care and fertilizing of the plantation, and in taking care not to cut or harvest it over too long a period. It is the aim to secure large broad crowns. After a dozen years, however, more or less, the size of shoots usually decreases and a new plantation will probably give better results in a good marketable product.

Distances; planting.

Asparagus may be either green or blanched. The difference lies wholly in the treatment. Naturally the shoots are green when they appear above the surface. By hilling over the row with earth, the shoots may be cut through the earth at the side of the ridge before they break out and become green. For such work, the earth should be of a sandy or loamy nature, so that it can be thrown against the row with a banking plow (or a shovel in small plantations); the rows are set as much as 6 feet apart, and often 8 or 10 feet. Green asparagus is better in quality.

The roots of asparagus should be in moist cool earth, with opportunity to forage as far as they will. The roots run horizontally rather than perpendicularly. It is well, therefore, to place the rows not closer than 4 feet. The plants (previously grown from seeds) should be set deep. The custom is to subsoil the land, if it is hard beneath the surface, plowing in a heavy coating of well-rotted manure if necessary. The plants are then set in furrows

6 to 10 inches deep. The crown is covered with loose earth or old compost to the depth of 2 or 3 inches. As the plants grow, the trench is gradually filled. If the trench is filled at first, the young plants may not have sufficient strength to push through the earth. In a commercial plantation, this filling may be performed by the subsequent tillage. Sometimes the furrows are partially filled by running a light harrow over the ground. The plants are usually set in spring, and by the succeeding autumn the furrows should have been filled. The plants should be one-year-old seedlings; two- or three-year-old plants give less satisfactory results.

The distance apart varies greatly, depending on the price and kind of land, the implements to be used in tillage, whether the rows are to be banked, and the personal preference of the grower. In garden plantations the rows may be as close as 3 feet. Usually 4 and 5 feet are allowed between the rows, and a greater distance if the shoots are to be blanched. In the rows, 18 to 24 inches is the usual space, although persons desiring the "hill method" and very large shoots may plant as far as 3 or 3½ feet.

Tillage and care.

Since the crowns of asparagus are so far beneath the surface, it is possible to till the whole area with shallow-working tools late in autumn and early spring. It is essential that this general tillage be given to keep the plantation free of weeds and to maintain the physical texture of the soil. In the growing season, little tillage can be given when the crop is being harvested, it is not practicable to till to any extent; and later in the season, when the tops are allowed to grow, the whole surface is occupied. Some growers disc the plantation just after the last cutting, if the land is hard and weedy; and the cultivator may then be used between the rows before the tops interfere. It is well to dress the plantation heavily in the fall with manure, to which one may add night soil, refuse salt or animal fertilizer, if these are available. It may be well, also, to make another dressing of more quickly available fertilizer early in spring. It is very important that the plantation be given the best of surface tillage for the first year or two, to put it in perfect condition. When the plantation finally comes into full bearing, the asparagus appropriates so much of the plant-food and moisture that there is less annoyance from weeds.

In spring the dressing may be cultivated under, or if it is too coarse for that purpose, the rougher parts may be forked off. After a thorough spring cultivation, it is well again to cover the bed with litter or manure to afford some nourishment, but particularly to conserve the moisture and to produce material for covering the tender shoots in case there is danger of frost. This, however, may be impossible in a large plantation; in such plantations the manure may be applied in spring, at the close of the cutting season, or before winter. Chemical fertilizers are now often used freely in place of some of the manure; but the humus content of the soil must be maintained.

On land to be prepared for asparagus, 20 to 40 tons of manure to the acre are recommended by H. C. Thompson (Farmers' Bull. 829), if the soil is deficient in humus;

if manure is not available, a green-manure crop may be plowed under. In preparation, "for an average asparagus soil 100 to 150 pounds of nitrate of soda, 500 to 1,000 pounds of 16 per cent acid phosphate, and 150 to 300 pounds of muriate of potash to the acre will give good results when applied in connection with manure or leguminous crops." After the plantation is established "a common practice among market gardeners is to apply 20 to 40 tons of manure to the acre broadcast over the bed during the autumn or winter." In addition, Thompson recommends a good complete fertilizer at the rate of 1,000 to 1,500 pounds to the acre at the close of the cutting season.

The energy of the crown and roots is supplied from the foliage that developed in the previous summer. Without a strong growth of top, one cannot expect a good growth of roots and a heavy yield the following year. The tops should be mown late in fall. Some persons allow these tops to lie on the ground as a winter protection. If, however, the plants produce many berries, there will be so many seedling plants as to make trouble; in that case, it is better to burn the tops. It is also well to remove and burn them in order to allow a thorough tillage in autumn. The bed should then be given a dressing as already suggested, both to afford winter protection and to supply plant-food.

The value of asparagus lies in its succulence and tenderness, and these qualities are usually associated with large size of shoot. These attributes are secured by very rich soil and by thorough attention to good tillage, and destruction of beetles and rust.

The crop.

The plants should grow two full years from planting in the field before shoots are cut closely, but a small cutting is often permissible the second year if the plantation is vigorous. It is also easy to injure the bed by cutting it too long a period each season. Whilst the crop is being harvested, however, every stalk should be removed, even though it is too small and poor for eating: the bed should be "cut clean." Only in rare cases should the bed be cut after the 4th of July in the Northern States, and it is usually better to stop before this time. The third season the cutting is for a month or less; subsequently it may run to six or even ten weeks. Thereafter the tops are allowed to grow as they will.

1. A bunch of asparagus.

It is customary to harvest asparagus by severing the shoots 3 or 4 inches beneath the surface by means of a long knife (Fig. 1, adapted from Farmers' Bull. 829). There are special asparagus knives (Fig. 2), but any long butcher-knife will answer the purpose. It is important that this knife be inserted in an oblique direction so as not to injure the new shoots that are rising from the crown. A little experience in the use of the knife will enable one to cut the shoots without injury to the succeeding growths. At the height of the season it may be necessary to cut every day; later two or three times a week may be sufficient. Some of the best growers advise the breaking of the asparagus shoots rather than cutting them. There is then no danger of injuring the crown, and the shoot will not break in the tough and stringy part and therefore the product is sure to be tender and crisp. This is no doubt the better method, but the formal demands of the market make it difficult to sell broken asparagus, notwithstanding its surer quality.

2. Asparagus buncher; also knife or spud for cutting the plants in the field.

Asparagus is sold in bunches 4 or 5 inches in diameter, weighing something over 2 pounds and comprising 12 to 30 stalks. These bunches are tied with soft cord, raffia or tape, although some growers now use rubber bands. Usually the market requires that the butt end of the bunch be cut off square. An average bunch is 7 to 9 inches long. Asparagus "bunchers"—which are forms for holding the bunch and cord, and a knife for cutting the butts—can be had of dealers in gardeners' supplies (Fig. 2). If not marketed at once, the bunches may be stood in a shallow tray of clean water. The shoots should be graded as to size and quality, and they may be washed before bunching.

Seedlings.

One may purchase asparagus plants of dealers. It is usually better, however, to grow one's own plants, particularly if one has a rich piece of land and can give it careful attention. The seed is sown in drills 15 to 18 inches apart (or farther asunder for horse tillage), and it is covered about an inch in depth. Germination is slow.

3. Seeds of asparagus (× 5).

The seeds may be soaked in warm water a day before planting. The plants should be thinned to stand 3 or 4 inches in the row. Give frequent tillage throughout the season. The following spring these plants will be ready for setting in their permanent places. The seeds and the seedlings are seen in Figs. 3 and 4.

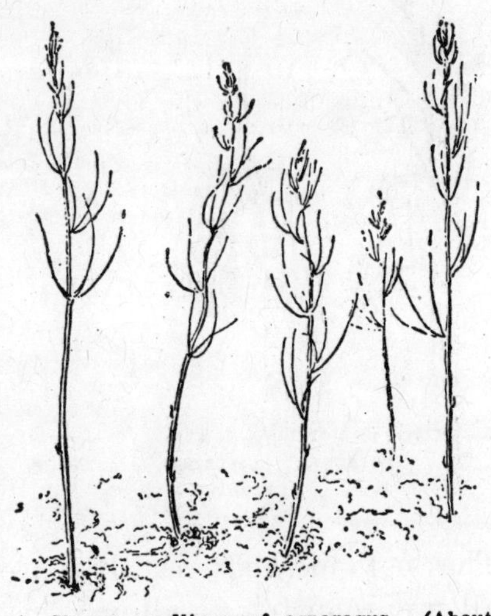
4. Young seedlings of asparagus. (About natural size.)

Seedlings may be expected to vary considerably; it is essential to best results to use only carefully selected seeds. In the selection, the most vigorous and productive plants should be marked and left for seed. Usually only part of the shoots are allowed to remain to each crown, to insure well-developed seed, and often the shoot is topped and only the lower berries saved. The flowers of asparagus are usually imperfect, and one male plant should be left close to every three or four female plants to make sure of pollination. When the berries are fully ripe, the seeds are rubbed or washed out and kept till spring, when they are sown as already explained; or, if very strong plants are required, seeds may be sown under glass and handled to the field in pots.

5. Sterile or staminate flowers of asparagus. (Separate flower × 3.)

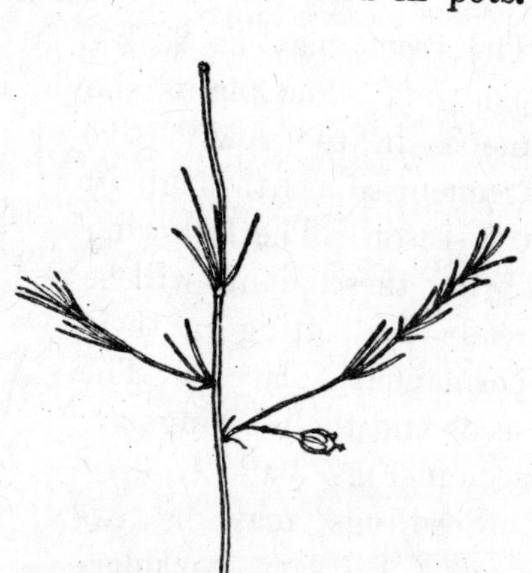

6. Fertile or pistillate flower of asparagus (× ¾).

The pot method readily encourages the discarding of all unpromising seedlings in the transplanting. Male plants are considered to be more productive than the female or seed-bearing plants, but the sexes cannot be certainly distinguished until blossoms appear. Perhaps the inferiority of the fertile plants is due to the lessening of vigor by seed-bearing; when practicable it is well to remove the berries. Eventually the seedling will probably be carried to blossoming period before placing in the plantation; further experience on this point is necessary. The sexes are distinguished in Figs. 5 and 6.

Varieties.

Recognized varieties of asparagus are few, and as the plant is propagated only by seeds (which may not come true to name) the characteristics of the different named kinds are not likely to be clearly marked. The Colossal (Conover's Colossal, Argenteuil of the French) is a standard variety. Palmetto is much grown. Bonvalette Giant (an improved form of Palmetto), Columbian Mammoth, Barr Mammoth, Dreer Eclipse, Giant Reading, Moore Giant, are other good contemporaneous kinds. Improved strains bred by the United States Department of Agriculture are now attracting much attention, known as Washington and Martha Washington.

The Asparagus Plant

Asparagus: a genus of the Liliaceæ or Lily Family, of about 150 species, native in Europe, Asia and Africa, herbaceous or woody, erect or climbing. Aside from the common edible asparagus, the genus contains the "smilax" of florists (not properly a smilax, however) and the so-called "asparagus ferns" of greenhouses. The species are devoid of ordinary green leaves, these organs being represented by small scales or spines and the green stems functioning as foliage. Even the broad leaf-like organs in the florists' smilax are branches, arising from the axils of leaf-scales.

A. officinalis, Linn. var. **altilis,** Linn. Sp. Pl. 313. Garden Asparagus. Perennial much-branching diœcious herb with terete clear green glabrous slightly glaucous stems 4 to 10 ft.

high: root a mass of long fleshy cord-like members spreading from the sides and bottom of a progressive rootstock: shoots arising from the crown in early spring, succulent but subsequently decreasing in diameter, without ordinary foliage, comprising the edible part of the plant; tops dying in autumn: leaves on young shoots triangular-cuspidate, $\frac{3}{16}$ to ⅜ in. long; on the branches represented by very small scarious scales, from the axils of which arise one but usually several short green terete cladodes or cladophylls (commonly regarded as leaves) $\frac{3}{16}$ to ½ in. long and perhaps one elongated branch, the plumose cladodes and stems altogether constituting the foliage: flowers 1 to 4 in the axils of the cladodes, on slender jointed pedicels; male or sterile fls. yellowish green and conspicuous, nearly ¼ in. long, bell-shaped, the perianth 6-toothed about one-third its depth, the stamens 6 and included, pistil present but abortive; female or fertile fls. (on separate plants) less conspicuous, one-half or less the length of the sterile fls., the pistil practically filling the perianth and the 3 stigmas protruding: fr. a globular hanging red 3-celled berry, $\frac{3}{16}$ to ⅜ in. diameter, usually maturing several seeds, the remains of the 6 perianth-lobes appressed on its base; seeds large (⅛ in. or less diam.), rounded at the back and more or less angled or flattened toward the micropyle, black, without prominent surface marks, weighing 15 to 22 mg., retaining germinating power 5 years or more.—Native on coasts and sandy areas, Great Britain, Mediterranean region, to central Asia. The usual native form (var. *maritimus*, Linn.) is a short-branched plant more or less prostrate at the base. The var. *altilis* (Latin: *large, fat, nourishing*) has longer branches and the thick stout stem is erect from the base; known in cultivation and as an escape. It is a plant of ancient cultivation.

RHUBARB OR PIE-PLANT

As a garden vegetable, rhubarb is grown for the large thick acid petioles or leaf-stalks, which are used in spring for sauces and pies. The plant is perfectly hardy; it de-

lights in a deep rich soil. Since its value depends on the succulence and size of the leaf-stalks, every care must be given that will contribute to leaf growth. It is an early spring crop; the land, therefore, should be quick, and the plants should have made a sturdy growth the previous year to have energy to start quickly and vigorously. The top growth is completed by summer. A well-prepared and well-handled rhubarb plantation should last twenty years or more. Propagated by divisions of the root and by seed. It is essentially a northern crop.

Year-old seedlings or divided roots are planted in the field usually 4 or 5 feet in autumn or spring, preferably in spring, requiring about 2,200 plants to the full acre. About 1,500 seeds are contained in an ounce, but 3 or 4 pounds of seed are recommended for the raising of seedlings as rigorous thinning selection must be practised. An acre should yield 3,000 dozen bunches, in full bearing, the bunch usually comprising 3 to 6 stalks, sometimes more if the stalks are small.

RHUBARB CURCULIO (*Lixus concavus*).—A black snout-beetle, ½ in. long, dusted with a yellowish covering which easily rubs off. The insect breeds in dock, sunflower and thistle, but the larvæ are never found in rhubarb. The injury is caused by the punctures which the beetle makes in the petioles from which there exude glistening drops of gum. *Control:* Hand-picking; destroy all wild food-plants in the vicinity of rhubarb.

The effort in the growing of rhubarb is to produce abundantly of large tender leaf-stalks and at the same time to fill the plant with energy for the crop of the succeeding year. The size of the leaf-stalks depends partly on the variety, but particularly on the soil and the tillage. There are only three or four popular varieties, of which the best known are Victoria, Linnæus, and Mammoth Red; but the old-fashioned unimproved rhubarb will often produce a better leaf-stalk when given high cultivation than the best strain of Victoria when grown under neglect. The plant should not be allowed to bloom (the flower-stalks being cut out as soon as they appear), unless it is desired to raise seed.

7. A bunch of rhubarb (petioles).

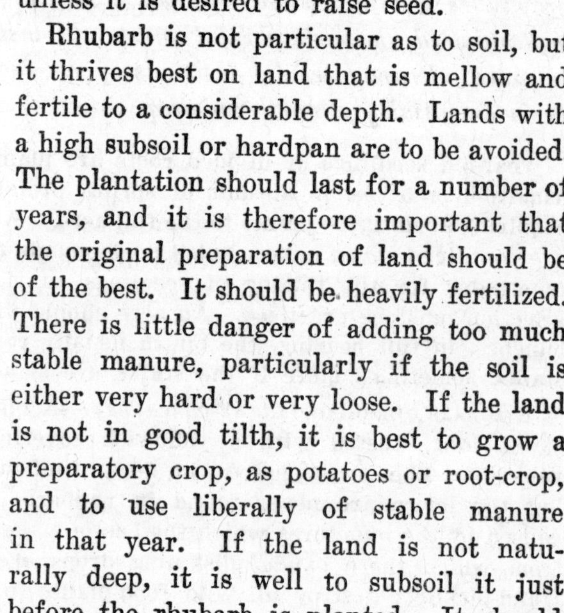

Rhubarb is not particular as to soil, but it thrives best on land that is mellow and fertile to a considerable depth. Lands with a high subsoil or hardpan are to be avoided. The plantation should last for a number of years, and it is therefore important that the original preparation of land should be of the best. It should be heavily fertilized. There is little danger of adding too much stable manure, particularly if the soil is either very hard or very loose. If the land is not in good tilth, it is best to grow a preparatory crop, as potatoes or root-crop, and to use liberally of stable manure in that year. If the land is not naturally deep, it is well to subsoil it just before the rhubarb is planted. It should always be well drained.

The rows should be sufficiently spaced to allow of easy horse tillage,—not less than 4 or 5 feet for the strong-growing varieties. In the row the plants may be placed about 3 to 4 feet apart. Some growers place the rows as far apart as 6 feet, and the plants 3 feet in the row. It is a good plan to leave alleys at intervals in a rhubarb field to allow the entry of wagons. In a single row in the home garden, the plants may be set every 2 feet. A dozen or two good plants should be sufficient for a family.

Good surface tillage, as for corn or potatoes, is all that is demanded. In autumn the bed should be given a heavy dressing of stable manure. This dressing serves the purposes of enriching the soil, preserving the texture of the surface, and affording a winter mulch and protection. Lands heavily mulched do not freeze so deep as those that are left bare, and the plants are likely to start earlier in the spring. This surface mulch may be removed early in the spring and a thorough tillage given the land; or if the land is in good tilth and free from weeds, it may be forked from the crowns and allowed to lie between the rows until the crop is harvested. Some growers hill up the rows in autumn by means of a plow and do not apply a mulch.

The commercial rhubarb season is short. It rarely extends over more than two months. The leaves are pulled, and they separate readily at their insertion if pulled straight and not twisted or yanked. Only the largest and best leaves are harvested. The leaf-blades are at once trimmed off to prevent wilting or softening of the stalk. Other leaves are allowed to remain unless they are very numerous, in which case the larger part of them are pulled to allow the strength to go to the main ones. After the market season of rhubarb is past, the plants are allowed to grow as they will, and tillage is continued. A heavy crop of rhubarb in any year depends to a large extent on the strong leaf-growth of the year before.

To renew rhubarb plantations, the roots are sometimes taken up, more or less divided and reset; but it is usually a better practice to trim the roots where they stand with the plow or the spade, breaking off the strong projecting parts. The purpose is to reduce the overcrowded mass of roots and to start new root growth.

Propagation of rhubarb is by division of roots and by seeds. Ordinarily it is multiplied by means of division. The root may be cut into as many pieces as there are strong eyes, and as much as possible of the root is allowed to remain with each eye. These pieces are planted 3 or 4 inches deep. These pieces of root are usually planted directly in the field, but they may be grown the first year in a nursery. The plants usually grow two years before a cutting is made, and they will not give a full crop until the third year. Rhubarb is readily grown from seeds, but this requires a year's more time and the seedlings are likely to vary. The seeds may be sown early in spring in drills 18 inches apart, or closer if the land is valuable, and the young plants are thinned to about 6 to 8 inches in the row. The plants are set in permanent positions the year following; that is, when they are one year old. In the Northern States rhubarb is usually planted in spring whether from seedlings or root-cuttings, but in milder climates it may be planted in autumn. The seeds (properly fruits) of rhubarb, and seedlings, are seen in Figs. 8 and 9.

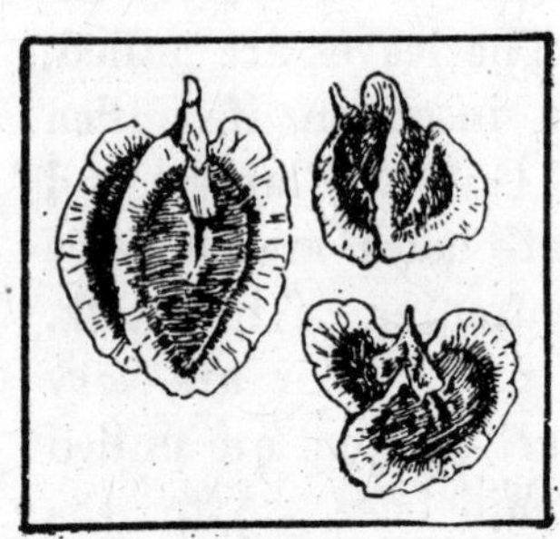

8. Fruits (seeds) of rhubarb (× 1⅝).

By covering the plantation heavily in autumn so that

the ground does not freeze deep, and removing the covering early in spring, it is sometimes possible to hasten the growth and get an earlier first yield. Sometimes barrels or boxes are put over the crown in autumn and banked with leaves or manure for the purpose (Fig. 212). The best results for the early market, however, are obtained by forcing the roots under glass or in a cellar. Strong fresh roots are dug in autumn and set close together on the ground in the forcing-house, the spaces between packed with earth, and the roots allowed to freeze thoroughly before heat is turned on. For the spring crop the roots may remain frozen for some weeks. Marketable stalks should be produced in five or six weeks. Sometimes the roots are forced in the dark in a cellar, having been taken up in autumn and frozen; leaf-blades do not develop, and the stalks have a tender pink semi-blanched appearance, but they are not improved in quality thereby. Forced roots are usually discarded.

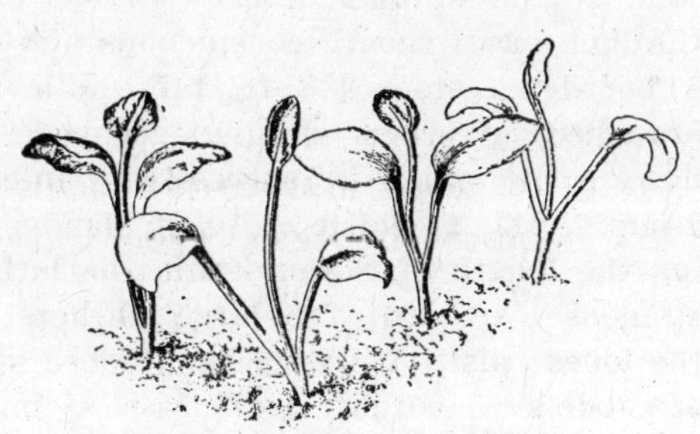

9. Young rhubarb seedlings (× about ½).

The Rhubarb Plant

Rheum, a genus of the Buckwheat Family, Polygonaceæ, strong perennial herbs of about 25 species, in Asia. Some of them are more or less planted as ornamentals, prized for their striking masses of large root-leaves and sometimes for their towering panicles of numerous flowers.

R. Rhaponticum, Linn. Sp. Pl. 371. COMMON RHUBARB.

Stout herb, with large roots (becoming hollow) variously branching from a rhizomatous crown: leaves mostly radical (from the crown at or near the surface of the ground), blade cordate-ovate, concave, the radical ones 12 to 20 in. long and of similar width, margins entire but more or less long-sinuate and usually somewhat undulate, with 3 strong upright and 2 basal ribs issuing from the top of the petiole, glabrous above, lightly pubescent on the nerves beneath; petioles very stout, shorter or longer than the blade, furrowed above and channelled on sides and back, the cross-section concavo-convex, sheathed at the base, the sheaths eventually breaking away; stem lvs. of similar description but successively smaller and the long basal sheath conspicuous and encircling the stem like a boot-leg: stem 4–6 ft. tall, hollow, strict but somewhat branched, glabrous, shining, grooved, the nodes conspicuous: flowers numerous in successive panicles, very small (about 2 mm. long), greenish white, on slender jointed pedicels exceeding the length of the perianth, the latter with 6 obtuse lobes; stamens 9 (8–10), the large anthers equalling or exceeding the lobes; pistil 1, with large 3-lobed stigma: fruits ("seeds" of gardeners) cordate-ovate, ¼ to ½ in. long, strongly 3-angled and winged, brown, glabrous, weighing 14 to 26 mg., tightly inclosing one large 3-sided achene; longevity about 3 years. —Siberia. The above description is drawn from the rhubarb of cultivation, which is commonly referred directly to *R. Rhaponticum*. There is doubt as to the species, however, and the vegetable-garden plant may be a hybrid or mutant race (perhaps represented by *R. hybridum*, Murr.), or even a different species. The species of Rheum are in need of further study. The medicinal rhubarb is from roots imported from Asia, probably from more than one species of Rheum; perhaps the roots of *R. Rhaponticum* are still used to some extent for this purpose. The word "Rhaponticum" means the Pontic rha or rhubarb; Pontus was an ancient region in Asia Minor.

ARTICHOKE

A half-hardy perennial, producing edible heads freely the second year, requiring protection at the North. The plantation should be renewed every two or three years. The strength of the plant is to be conserved not only by good soil and abundant fertilizing, but also by removing extra stalks and not allowing the heads to seed. Propagated by seeds and suckers, preferably the latter when one can select from a good stock.

Plants may stand as far as 3 by 5 feet apart, requiring nearly 3,000 plants to the acre. Suckers are planted at about their natural depth, in spring. Seeds are sown in spring, preferably under glass, at least at the North. Each plant should yield a dozen and more good heads. The product (scales and receptacle) is eaten raw or cooked, usually the latter with sauce or drawn butter.

The artichoke, especially in the South, is often attacked by the artichoke aphis (*Myzus bragii*) and the bean aphis (*Aphis rumicis*). These plant-lice may be controlled by thoroughly spraying the plants several times with "Black Leaf 40" tobacco extract, 1 part in 8 parts of water, in which enough soap has been added to make a suds.

The artichoke is grown for the young unopened burs or flower-heads (Fig. 10), the scales on the outside of the head having thick edible bases and the inside receptacle or "bottom" of the head, after the flowers are removed, being soft and palatable. The leaves and young shoots may also be eaten, when grown and blanched, after the way of celery, but this use of the plant is little known in America.

The artichoke is tender and precarious in the Northern States, although it is grown in favored localities with success by persons who understand the handling of it. While

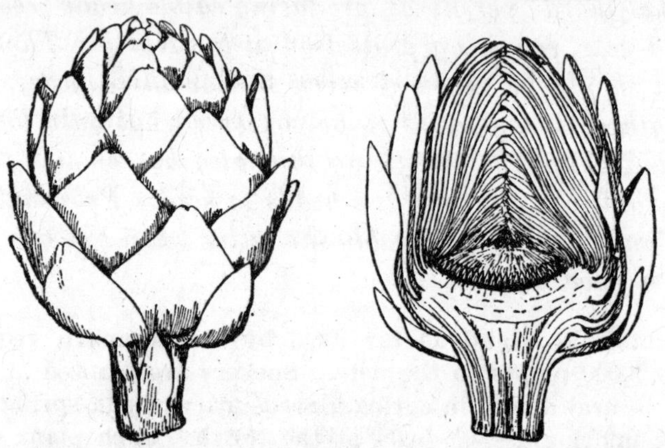

10. Artichokes, showing outside of head and a longitudinal section (× 1/3).

tender, yet too heavy covering of the crown in autumn may smother the plant and kill it. Gardeners sometimes box the plant to protect it from winter winds, but do not fill the box with leaves or manure. It is frequently banked with earth. It should be grown on warm well-drained land and in a protected place. In the Southern States and California the plant thrives and is easy of cultivation. The California product largely supplies the Eastern markets.

11. Fruits (seeds) of artichoke (× 2).

The artichoke is perennial, but the plantation should be renewed frequently. Seeds do not come true to name, and when the grower secures a good strain of hardy and productive plants he should propagate them by means of the suckers that spring from the crown; or he may purchase suckers of reliable dealers.

Seeds give bearing plants the following year, but if they are started early under glass and planted in "quick" soil, a small number of heads may be had the first year. Suckers may give heading plants the first season, but the main cutting may be expected the second year. Removing some of the stalks, if many start, will increase the size of the remaining heads. Figs. 11 and 12 show the fruits (seeds) and the seedlings.

12. Seedlings of artichoke (× about ½).

Distances for planting vary with the grower and the price of land. In rich soil they may be farther apart. Rows may stand 4 or 5 feet, and the plants 2 to 3 feet in the row; 3 by 5 feet, or 3 by 4 feet, are good distances.

The Artichoke Plant

Cynara. About a dozen species of large thistle-like perennial herbs, in the Mediterranean basin, of the Compositæ or Sunflower Family.

C. Scolymus, Linn. Sp. Pl. 827. Plant stout, more or less cottony, forming a clump or stool: stems few to several, erect, 2 to 6 ft., grooved, branching or forking, commonly appearing after the first year: leaves many, mostly radical or basal, dull green and more or less gray-webby above and densely gray-tomentose beneath, divided almost to the winged rachis and the divisions cut and lobed with short spines terminating the long narrow lobes, the radical ones 2 to 3 ft. long and a foot or more broad, arching at maturity; stem lvs. similar but successively smaller, decurrent: heads large (3 to 4 in. diam.), terminal, globular, erect, often subtended by bracts, producing a brush of numerous purple tubular florets; involucre scales imbricated, ovate to ovate-lanceolate, entire, obtuse or emarginate; receptacle thick and fleshy, bearing many bristles: fruit (seeds) oblong-ovate, ¼ to ⅜ in. long, somewhat flattened, smooth, striate and spotted, weighing about 40 to 70 mg., retaining vitality 5 to 7 years.—Southern Europe and northern Africa; tending to run wild in parts of California. It is a plant of relatively modern cultivation. Sometimes called "globe artichoke" to distinguish it from the girasole. Very closely related to the cardoon (*C. Cardunculus*, Linn.), also of S. Europe and extensively naturalized in S. America. The cardoon is a taller and stouter plant; a form with thick leaf-stalks is cultivated for food after the manner of celery. Some botanists consider the artichoke and cardoon to be forms of one species.

GIRASOLE

Hardy plant grown for its underground tubers, which may be used as a vegetable, as are potatoes, or for stock feed. Requires no special treatment, and will persist indefinitely, and spread, if left to itself. Propagated by planting the tubers.

As grown in this country, the girasole is seldom tilled. The tubers are planted whole, 1 to 2 feet apart, and the plants allowed to shift for themselves beyond an occasional destruction of big weeds. Better results are to be expected when the tubers are planted in rows far enough apart for horse tillage, and 12 to 16 inches in the row. The plant requires the entire season in which to make its tubers, and the product is not dug till the tops begin to die. Tubers left in the ground

are not injured by frost. Planted in autumn or spring. Under regular cultivation, crops have been reported at the rate of 9 to 20 tons, and even more, to the acre.

The girasole is one of the tuber-bearing native sunflowers, long cultivated by the Indians and often highly recommended for more general cultivation because of its heavy yields and its ability to grow on indifferent land and with little care. It readily responds, however, to good land and treatment. There are improved strains, and undoubtedly it could be readily modified by systematic selection. The plant tends to become a weed, and farmers often turn hogs into a field infested with it, as they root for the tubers. The plant can be eradicated by thorough tillage, by means of which the tops do not have an opportunity to grow. If the field is plowed in the fall, many of the roots will be exposed and they may be picked out. In fact, this is one of the best means of harvesting the crop.

The girasole provides a very palatable food. It is strange that it has not met with better favor. The weedy character of the plant and the fact that potatoes have been abundant are probably reasons for its neglect. Its real service, however, is not in competition with the potato, but as another food plant of very distinct attributes. The tubers, produced underground, vary greatly in size and shape; Fig. 13 shows a common form.

13. Girasole, an underground tuber (× ½).

The girasole is commonly known as Jerusalem artichoke, but as it is not an artichoke and has no relation to Jerusalem, the name should be dropped. In fact, "Jerusalem" in this case is supposed to be a corruption of girasole, an Italian name. The French name, topinambour, is too formidable to become popular in English.

The Girasole Plant

Helianthus, the sunflowers, comprises about 70 species, as now recognized, natives of the western hemisphere. The common garden sunflower, *H. annuus*, yields edible seeds and its herbage provides more or less fodder. Several species produce underground tubers, one of which has long been known as a food-plant. Helianthus is one of the Compositæ.

H. tuberosus, Linn. Sp. Pl. 905. Girasole. Topinambour. Perennial, producing tubers on the ends and branches of underground stems or rootstocks, as well as midway on the rootstocks: stem erect, 5 to 10 ft. tall, striate, hirsute: lvs. opposite or the upper ones alternate, petioled, long-ovate to ovate-oblong, upper ones narrower, acuminate, serrate-dentate, rough above, more or less thin-pubescent beneath, with a pair of strong lateral ribs or nerves from the base, narrowed either abruptly or gradually into a somewhat winged petiole: heads few or many terminating the branches, 2 or 3 in. across, with conspicuous light yellow veined pointed rays; involucre of two or more series of lanceolate pointed ciliate scales, the outer ones spreading; receptacle with scales subtending the achenes; ray florets neutral (sexless), the disc florets perfect and yellow, pappus of small deciduous scales: fruit (seed) oblong, pubescent, nearly or quite ¼ in. long, usually only a few (sometimes none) developed in each head.—Canada and U. S. It is doubtful whether Linnæus meant to designate this plant in his description of *H. tuberosus;* his references do not certify to it, and he writes "habitat in Brasília," although the Brazil of his day was apparently a broad geographical term and not necessarily the country now known by that name.

SEA-KALE

A perfectly hardy perennial grown for its excellent young leaves and shoots, which are blanched as they appear in spring by banking with earth or covering with inverted pots or other tight receptacle. The soil should be deep and rich and rather moist. After cutting, the subsequent treatment is for the purpose of putting energy into the plant for the next year. Propagated by seeds, division, and root-cuttings.

Planted at least 3 feet apart either way, and preferably somewhat farther if sufficient land is available. At 3 x 3 feet, about 4,800 plants are required for an acre. A good crop may be expected the second or third year from cuttings or seeds. The plant should give good results for about 10 years.

There appear to be no important diseases or insects on sea-kale in this country.

Sea-kale is little known in this country, although it is deserving of popularity. It is particularly prized in England, where the culture has been highly developed. After the plants are well established, the young shoots are blanched by covering the crown to the depth of a foot or more with loose fine earth in early spring. Sometimes the shoots are allowed to grow upward

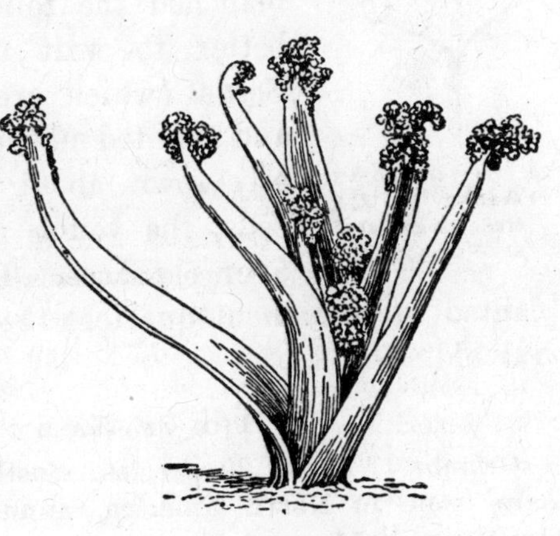

14. Shoots of sea-kale.

into a dark receptacle, as into a box inverted over the crown. Sea-kale may be forced after the manner of rhubarb. Fig. 14 shows the characteristic young growth at the edible stage.

After the early spring shoots are removed, the plant is allowed to grow as it will for the remainder of the season for, as in asparagus and rhubarb, the vigor of the young shoots of any season depend, to a large extent, on the vigor and energy of the plant in the preceding year. The soil should be deep and rich, and rather moist. An autumn top dressing is beneficial.

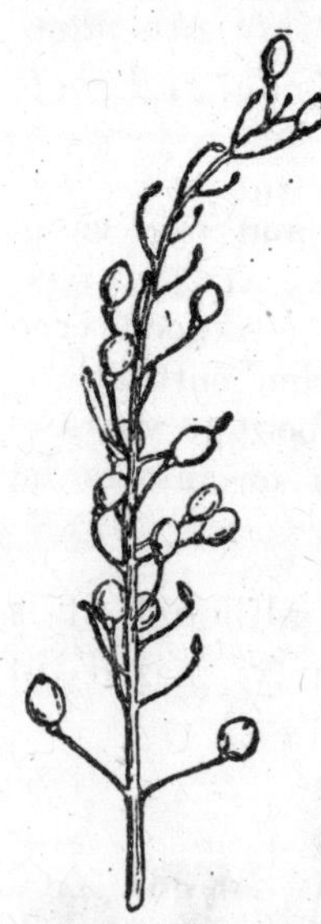

15. Seed-pods (known in the trade as seeds) of sea-kale (× 1/3).

Propagation is simple either by root-cuttings or seeds. Vigorous roots are cut into pieces 4 or 5 inches long and planted directly in the field in spring. If the land is strong, some of the shoots may be blanched the following spring, but it is better to wait till the second spring. Seeds (which are really 1-seeded fruits and planted unshelled, Fig. 15) are usually sown about 1 inch deep, in seed-beds, the young plants being thinned to 5 or 6 inches. The seedlings are transplanted to permanent quarters the next spring, when one year old.

The Sea-Kale Plant

Crambe. About 20 species, mostly native in Europe and Asia, none in North America, annual and perennial herbs; Cruciferæ or Mustard Family.

C. maritima, Linn. Sp. Pl. 671. SEA-KALE. Fleshy perennial, glabrous, glaucous-blue, with thick cord-like deep roots: stem erect, to 3 ft. high, much branching above, more or less grooved, many-striate when dried: leaves thick, petioled, variable in shape but mostly oblong-ovate in outline, variously lobed and notched; lower leaves long-stalked and cabbage-like and often 2 ft. or more long, with strong midrib and many prominent side ribs, nearly entire in outline or shallowly wide-lobed along the sides, the margins coarsely and irregularly toothed or notched, undulate; stem leaves smaller and usually more deeply lobed, variously notched, the upper ones short-petioled: flowers about ½ in. across, white and showy, in terminal broad corymbed racemes, on long stout upright pedicels that elongate in fruit; sepals oblong, hyaline-margined, obtuse, about half the length of the 4 obtuse veined petals which have an oval limb and clawed base; stamens 6, all anther-bearing, 2 shorter, the longer ones with 1 to 3 notches or branches at or above the middle; pistil 1, oblong-columnar, with a large globose stigma, comprising two joints, of which the lower one is short and barren and the upper one somewhat broadened at the middle and fertile: fruit (the "seed" of seedsmen) a globular or globular-oblong pod ¼ or less to ⅜ in. diam., borne on an apparent very short stalk above the receptacle but which is the abortive lower point of the 2-jointed silique, smooth, the walls thick and spongy; seed single, large and filling the cavity, suspended on a white stalk that arises from or near the bottom of the cavity and extends alongside the seed to the top; pod and contents weigh about 100 mg. for a fair full-grown specimen; full germinating power probably is retained for only a year or two.—Sea-coasts and cliffs, western Europe; introduced to cultivation probably within 200 to 300 years, at least in England. The above description is drawn from the cultivated plant, which differs considerably in appearance from the wild plant observed by the writer on sea-cliffs, the latter being more squat, with lower stature and lopping simpler branches, the leaves smaller, more crinkly and less cabbage-like.

DOCK AND SORREL

Perennial deep-rooted hardy herbs grown for the radical leaves appearing in spring, prized for greens. The plants require no special care, but the better the soil the more abundant will be the supply of foliage. They may be placed at one side of the garden and remain undisturbed for a few or several years, until they begin to run out. Propagated by seeds and division.

Some of the weedy docks are gathered in spring for "greens." The roots of some of them provide old family remedies. The sorrels are known for their acid leaves. They are members of the large genus Rumex (Polygonaceæ, Buckwheat family). The sorrels are diœcious plants (sexes separated on different plants), while the docks are larger and have perfect flowers or the plants may be monœcious. (sexes separated in different flowers on the same plant). A few species are cultivated for the edible foliage, but they are little known in this country, and technical descriptions are not necessary. Some of them are desirable additions to the garden because they yield a pleasant food in very early spring, and, once planted, remain for years.

The spinach dock or herb patience (*Rumex Patientia*), native in Eurasia and somewhat run wild in North America, is a very stout herb with a deep taproot and flower stalk reaching 5 to 6 feet high. Seeds may be sown in spring in a row where they are to stand, and leaves may be taken the following spring. The broad crisp leaves appear early in April, when there is nothing green to be had in the open garden, and they can be cut continuously for a month or more.

The garden sorrel is a developed form of *Rumex Acetosa,* native in Europe and scattered in this country. The common variety is Large Belleville. It has thinner, lighter green and longer-stalked leaves than the spinach dock, with spear-like lobes at the base, and the plant is not so tall and stout. The leaves are very sour, and will probably not prove to be so generally agreeable as those of the spinach dock; but they are a week or ten days later, and afford a succession. It is grown the same as the spinach dock, but some leaves may be harvested the first year from seed. The male plants are usually preferred, as they do not reduce themselves by seed-bearing.

Both the spinach dock and sorrel may be grown about 12 inches apart in the row. Sometimes they are propagated by suckers that arise near the crown. The seed stalks should be kept down, and only part of the leaves should be cut at any one time if the energy of the plant is to be conserved to the utmost.

Other species of Rumex are sometimes cultivated, as the French sorrel, *R. scutatus,* by the French, and the dentate dock, *R. dentatus,* by the Chinese.

UDO

The udo is a Japanese plant introduced into this country nearly twenty years ago, and now considerably known as an early spring vegetable. The plant is of the Ginseng or Aralia family (*Aralia cordata,* Thunb.), a strong hardy perennial; it sends up strong shoots in spring, and if these shoots are allowed to grow through a box of light sand, much after the way of growing witloof, they make a delicate blanched vegetable, eaten after being boiled, or prepared for salads. An unpleasant flavor is removed by boiling ten minutes in salt water and then changing the water; or if wanted for salad by being cut into thin slices and placed in ice-water for an hour.

Udo is readily propagated by seeds. At three years, shoots may be taken, and thereafter for a number of years if the plants are given room and not allowed to run to seed. Plants should stand about 4 feet from each other. The tops spread widely, like the wild spikenard (*Aralia racemosa*) and reach 5 or 6 feet in height. Udo is a promising vegetable.

CHAPTER III

SPINACH AND OTHER GREENS

Spinach	**Mustard**
Orach	**Purslane**
Chard or leaf-beet	**Dandelion**

Potherb crops, or greens, are grown for their leaves: therefore they must make quick growth in order to be crisp and tender; the ground must have good surface tilth and much available plant-food; the application of soluble nitrogenous substances is usually important, particularly when the growth is nearing completion. Most potherb crops demand a cool season; and nearly all of them are partial-season crops, and are therefore treated as succession- or companion-crops.

To the plants discussed in this chapter, several others might be added. New Zealand spinach is not a spinach, but a member of the Fig Marigold family (Aizoaceæ); it is *Tetragonia expansa* of the botanists. It is annual; it endures hot weather and therefore may be substituted for spinach in summer, being sown at intervals. Kale (see Chapter IV) is really a potherb plant; and it would not be great violence to include cabbage in this group. Several docks and sorrels are grown as potherbs, but as these are perennial they are discussed in Chapter II. The potherbs are among the oldest of the vegetable-garden plants, and the number used first and last is legion. The need for green food is common to all peoples. They are cheap foods to grow, in comparison with seed-foods, as they usually require only a part of the season in which to grow. Most of the potherbs are of very simple culture.

SPINACH

Spinach is essentially a spring and autumn crop. It delights in cool moist weather. It quickly runs to seed in summer. It is grown mostly in drills. It is usually a succession-crop. Propagated by seeds, which germinate quickly. It is a true annual, but may be carried over winter by starting it in autumn, as it is very hardy. The crop requires a moist soil, well supplied with quickly available fertility.

Seed is sown about 1 in. deep from late August to November, according to locality, or at the earliest moment in spring, in rows about 1 ft. (8 to 14 in.) apart, and thinned to about 6 in., making a stand of about 87,000 plants to the acre. Sometimes it is sown broadcast on clean land, and not thinned. From 10 to 15 lbs. of seed are required to sow an acre in drills, and nearly or quite that much if broadcasted. In a continuous growing season, the plants should be ready to harvest in 6 to 8 weeks. The yield of a good crop should be 200 to 250 barrels; the number of "heads" can be estimated from the distances planted.

Blight or mosaic.—Great losses of spinach are incurred because of this disease. The symptoms are similar to those of other hosts affected by mosaic, and may be recognized by the mottling and malformation of the foliage, the dwarfing, and finally the premature dying of the plant. Insects are now known to carry the virus from diseased to healthy plants, as well as to act as virus-bearers during the part of the year when spinach is not grown. Neither the soil nor the seed are considered as carriers or hibernating places for the contagium. *Control:* The only recommendations possible are the elimination of the aphis. Experimental breeding for blight-resistant spinach is being conducted and may later prove effective in reducing the losses now caused by the disease. (See McClintock, T. A. and L. B. Smith, True nature of spinach blight and relation of insects to transmission. Jour. Agr. Research 14: 1-60. 1918.)

SPINACH APHIS (*Myzus persicæ*).—A pale yellowish green plant-louse that infests the underside of the leaves, often ruining the crop. It also transmits the mosaic disease or blight of spinach. *Control:* Spray with "Black Leaf 40" tobacco extract, 1 pint in 100 gals. water, in which 5 or 6 lbs. soap have been dissolved, taking care to hit the underside of the leaves.

BEET LEAF-MINER (*Pegomyia hyoscyami*).—See under beet, page 164.

Spinach, or spinage, is the standard plant for spring and fall greens. For home use it may be had in summer by making successional sowings in rather cool and moist ground; but as a commercial crop, it is not grown in warm weather. Formerly spinach was brought to early maturity in the North under glass on a rather large scale, but of late years it is grown in such quantities about Norfolk and other parts of the middle country and the South that it is seldom grown in frames in the North except for home use. From southern fields it comes both as a winter and an early spring crop. Fig. 16 is a good spinach plant.

The winter and early spring spinach is usually grown from seeds sown in the field in September, or later than this in the Central and Southern States. The land should be rich; also well drained, that the plants may not "heave" by frost. It is customary to plow the land into low ridges or beds 6 to 9 feet wide, to secure perfect surface drainage. Lengthwise in these beds the spinach is sown in rows about 12 inches apart, the distance depending on the means employed for tillage; in some cases, 18 inches is left between the rows, and in other cases only 8 inches. The distance between the plants, after thinning, is usually 4 or 5 inches. The plants should become thoroughly established before winter, having made

16. Spinach at good edible stage (× 1/3).

a spread of leaves of three or four inches at least. The crop is usually left uncovered in the North, even as far north as New York State; although if material is at hand, it may be covered lightly with straw or litter to prevent heaving and thawing. On the first opening of spring the spinach resumes growth. In fact, in mild seasons it may grow throughout most of the winter. It should be ready for use in April and May, and be off the ground early in June, even in the Northern States, leaving the land for other crops. In the South it is marketed from late November to March and early April.

Since spinach is prized for its crisp tender leaves, it is a crop that profits by an application of soluble nitrogenous fertilizers. It is customary, in some parts of the country, to sprinkle the ground early in the spring with a weak solution of nitrate of soda or sulfate of ammonia, using 50 to 75 pounds of the fertilizer to the acre at each of two or three successive applications. These applications may be made at intervals of ten days to two weeks. The applications are often applied by means of a street sprinkler or similar arrangement. Other growers apply dry fertilizer, broadcast, in liberal applications, as much as 1,000 to 1,500 pounds or more to the acre, depending on soil and season. Sometimes the beds are top-dressed with manure in the fall, and the leachings from the manure start the plants quickly in spring. Hen-manure is sometimes used.

17. Fruits of the smooth-seeded spinach (× about 3).

18. Prickly-seeded (-fruited) spinach (× 2).

For home use, and sometimes for market, plants are started in spring in a warm position, the seed usually being sown where the plants are to remain. It is more easy to secure a good stand by this spring sowing, but the plants do not mature so early. Spinach is sometimes started under glass and transplanted to the open; and it is frequently grown to edible maturity in frames. Sometimes beds of fall-grown spinach are covered with sash in February or March to hasten the plants. There is always more or less loss of fall-grown plants in the Northern States.

Two general classes of spinach are familiar to gardeners, the smooth-seeded and the prickly-seeded. The latter tends to fall into disfavor because of the trouble of sowing it, owing to the very sharp spines on the fruit (or "seed"); it has been preferred for autumn sowing because very hardy, but smooth-seeded kinds are coming to be popular for this purpose. The savoy-leaved spinachs (smooth-seeded) are valued for the large and wrinkled leaves. Strains or varieties of spinach have been developed that run tardily to seed; they are known as the long-standing kinds; they are specially useful for spring planting. The figures (17 and 18) show the two kinds of seeds, and Fig. 19 the seedlings.

19. Seedlings of spinach (× about ½).

In its undeveloped state, both types of spinach bear relatively narrow halberd-shaped or spear-shaped leaves, having strong spreading lobes at the base. The modern purpose in the selection of stock is toward "round-leaved" types, those in which the leaves are broader and lack the basal lobes. Even in varieties developed with this purpose, lobed leaves usually appear freely, even on the same plant with the prevailing round leaves; but the lobing is mostly less marked and the leaves are broader than in the older types.

Spinach is mostly diœcious—the sexes separated in flowers on different plants. After flowering, the staminate or male plant usually ceases to grow and dies, while the pistillate or female plant continues to grow to ripen its crop of seed. This may account for some of the "poor plants" in seeding spinach rows.

THE SPINACH PLANT

Spinacia. A genus of four or less species, annual herbs, of southwestern Asia, member of the Chenopodiaceæ or Goosefoot Family, and therefore closely related to the beet.

S. oleracea, Linn. Sp. Pl. 1027. (*S. spinosa*, Moench, Meth. 318. 1794.) PRICKLY-SEEDED SPINACH. Annual, diœcious; plant smooth and glabrous throughout, tap-rooted, producing abundant crown-leaves in the cool season when young, in warm weather soon sending out an erect simple or branched leafy stem (and sometimes supplementary stems) 6 in. to 2 ft. tall: leaves all petioled, various in shape and size, the margins entire, acute or obtuse at the apex; radical leaves in the presumably more primitive races narrowly oblong to ovate-oblong, in the more developed races ovate to round-ovate and sometimes several inches long, the petiole shorter or longer than the blade, base of blade obtuse and semi- or unequally cordate or truncate or with downward-extending or outward-extending pointed narrow lobes, sometimes with extra lobes below and above as if the leaf were inclined to be compound; stem leaves smaller, alternate, oblong to broad-ovate, becoming lanceolate in the inflorescence, very various in size and lobing or in absence of lobing, the petioles usually conspicuously long: flowers apetalous, small and practically uncolored (green), the staminate mostly forming leafless spikes or panicles of sessile or stalked glomerules, the pistillate flowers several to many and sessile in the axils of leaves or of leafy bracts; staminate perianth with 4 obtuse hyaline-margined divisions (divided to base) and 4 exserted stamens opposite them, the pistil rudimentary; pistillate perianth 2-notched and close-pressed about the single 5-styled pistil, the styles exserted, the perianth bearing 2 to 4 spines on its exterior: fruit a small brown achene inclosed within the persisting enlarged closed and indurated spiny perianth, the entire structure constituting the "seed" of gardeners, ½ to ¼ in. in spread; this seed (fruit) weighs 10 to 20 mg. and has a germinating vitality of about 5 years.

Var. **inermis**, Peterm. Pflzschluss. 377. 1846. (*S. inermis*, Moench. Meth. 318. 1794. *S. glabra*, Mill. Gard. Dict. Spinacia No. 2. 1768. *S. oleracea* var. *glabra*, Guerke, Richt.-Guerke, Pl. Eur. ii, 138. 1897.) ROUND-SEEDED SPINACH. Fruits "smooth," *i.e.*, without spines: plant supposed usually to make closer tufts of larger root leaves.—Whatever may have been the distinctions in foliage, size and habit between the two races of spinach in earlier times, in cultivation at present the characters appear to be largely merged except in the smoothness or spininess of the fruit; and even in these fruit characters the difference may not be great, for in some strains the spines are very short, and marked rudiments of spines also may be observed frequently on round-seeded kinds. In defining the two kinds, Philipp Miller in 1768 characterized *S. oleracea* as "spinach with arrow-pointed leaves and prickly seeds," and *S. glabra* as "spinach with oblong oval leaves and smooth seeds." He did not speak of "round leaves." Spinach is a plant of relatively recent domestication, and it is not greatly modified.

OTHER GREENS

Many kinds of plants aside from spinach are used as greens or potherbs. Some of the common weeds are much prized for this purpose in the rural districts, particularly the common white pigweed or lamb's quarter, pusley or purslane, dandelion and dock. Shepherd's purse is a favorite food plant in China, where it is cultivated. The amaranths supply vast numbers of people in other parts of the world with green food. Chicory tops, in the form of witloof and otherwise, are much eaten. Many plants are adaptable to such uses; we shall probably learn to prize them as time goes on.

Orach, a luxuriant annual of the goosefoot (pigweed) tribe, is grown for the large succulent root-leaves. It is essentially a cool-season plant, the seed being sown early in spring and the foliage used before midsummer. The plant sends up a strong flower-stalk, and thereafter it is of no use as a potherb; to avoid the flowering habit, seeds should be planted very early, and successional sowing may be made. There are green-leaved (white orach) and red-leaved forms. As young plants they make handsome pot specimens, particularly the red-leaved kinds.

Orach is *Atriplex hortensis,* Linn., of Asia, with triangular-ovate long-stalked leaves which have sinuate or irregularly dentate margins, and usually a halberd-lobed or truncate-lobed base; upper stem leaves oblong to lanceolate. The smooth and glabrous erect graceful flowering stems rise 3 to 5 feet; the fruits ("seeds") are large, flat, winged, disc-like, circular to ovate. Var. *rubra,* DC., is the red-leaved orach.

Chard, or *leaf-beet,* is one of the best of potherb plants, particularly for summer, as it withstands heat. It ordinarily requires nearly a full season in which to mature, although it will give a supply of edible foliage from early summer until autumn. The chard has very broad and thick leaf-blades and midribs, which are usually white or tinted rather than green (Fig. 20). Sometimes these are blanched by tying up the bunch of foliage. Seeds are sown early in spring, as are ordinary beet seeds, and the plants are thinned as used until finally they stand 6 to 12 inches in the row. The rows should stand 18 to 24 inches, as the plants produce very large tops. Small plants of the common beet, as explained on page 164, are often used for greens, but they are inferior to the developed forms known as chard.

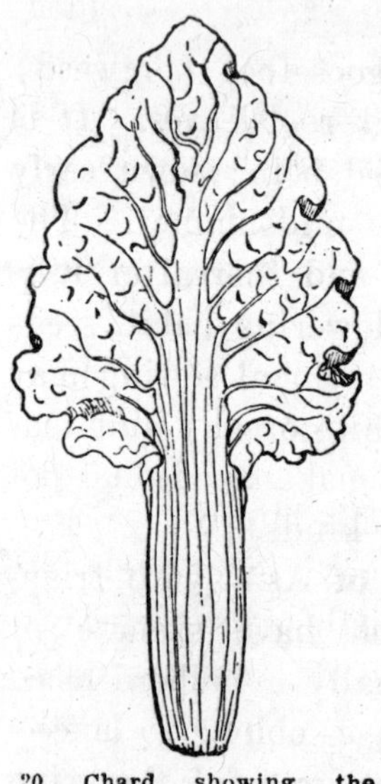

20. Chard, showing the wide edible petiole and midrib (× ⅛).

From mature plants the leaves are taken as wanted, care being exercised not to strip the crown at any gathering. The plant should continue to produce throughout the season, and crowns remaining over winter often grow in spring, although the second year they run quickly to seed. Fall-sown plants, if well established, often pass the winter in safety. Sometimes they are carried over in coldframes, for early spring crop; and the plants may be started under glass late in winter, and transplanted, for the same purpose.

Chard is a beet, ***Beta vulgaris*** var. ***Cicla,*** for which see page 170. The beet leaf-miner sometimes attacks it.

Mustard is much used for greens in home gardens, and it is also grown to a large extent in parts of the South, where the climate is too hot for many other potherb crops. Some of the improved varieties of curled-leaved mustard are amongst the best of all potherb plants. In many other countries the mustards supply immense quantities of food, being eaten with rice and other basic materials.

The seeds are sown very early in spring, and the tender bunch of foliage is ready for use in May or June. In fact, even in the Northern States, on sandy warm land the seeds may be sown in autumn and the plants will be ready for use in early spring, although the seeds may not germinate in the fall. In midsummer the plants run to seed. Care should be exercised not to let the plants seed themselves too freely, as they are likely to escape into unoccupied areas and become weedy.

The kinds of mustard are many, representing several species of Brassica (Cruciferæ or Mustard family). Some of them yield oil from their seeds—used extensively as food and in the arts. They are so little appreciated as potherb vegetables in this country, however, that technical descriptions of them are unnecessary at this point, and the more so as the botanical status of some of them is yet unsettled (see pages 96 to 98). In the South, the Southern Giant Curled mustard (*Brassica japonica*) is much used, largely taking the place of both spinach and lettuce. The Ostrich Plume is of this race. The Broad-Leaf (*Brassica rugosa*) is a most robust plant, and gives a large amount of excellent herbage quickly. The young leaves of white and black mustard (*Brassica alba* and *B. nigra*) are sometimes employed as potherbs.

Purslane, or "pusley," has been much improved by the arts of the plant-breeder, although the wild purslane is prized as a potherb. The ordinary pusley of the field is a weak-stemmed plant trailing on the ground (Fig. 21), whereas the Improved, or French purslane, grows more or less erect, and has very thick and succulent stems and large leaves (Fig. 22). It is easily grown in any good quick garden land from seeds sown in early spring where the plants are to stand. It matures quickly, and, unlike many other kinds of potherb plants, it is not injured by warm weather. However, the crop is usually harvested

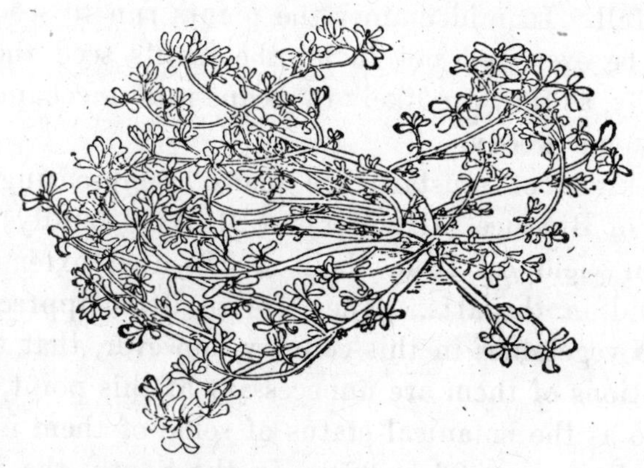

21. Common purslane or "pusley."

before midsummer, as greens are not in demand at that time. Sometimes it is started under glass and transplanted. Although the seeds are small, they germinate strongly. There seems to be little danger of the cultivated purslane self-sowing and becoming a weed. If kept moist and vigorous, the plant may be cut more than once. The plant rises one foot or more and spreads widely or lops with age.

The cultivated purslane is *Portulaca oleracea,* Linn. var. *sativa,* DC. It is probably a result of domestication, although the point needs further investigation. There are erect forms of wild purslane, as *P. oleracea* var. *erecta*, Edgew. & Hook. f., in India, but the leaves are narrow. De Candolle speaks of the var. *sativa* as if native in India and South America, and also as cultivated in Europe. The contrast in habit and stature between the wild purslane and the cultivated kind affords a marked example of the supposed effects of domestication. The winter purslane is a different plant of the same family (*Montia perfoliata*, Howell), native of western America to Mexico. It is sometimes grown for autumn and winter use, being sown in summer, or treated as a winter annual for early spring.

Dandelion.—The dandelion is a great favorite for spring greens, being cut from meadows and yards for the purpose. It seems not to be generally realized, however, that the plant has been greatly improved in size and vigor as a potherb, and that it is much grown abroad and also to a considerable extent as a market crop in this country. Some of the varieties with large leaves and others with cut or frilled leaves are great improvements on the wild plant, and the foliage is often handsome for garnishing as well as useful for food. Some of

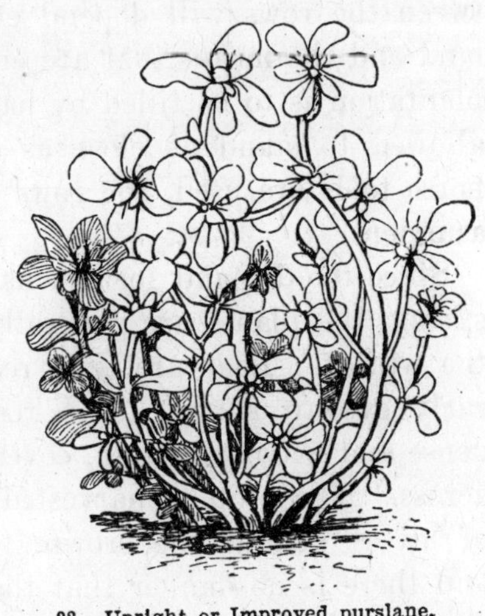

22. Upright or Improved purslane.

the forms resemble curled endive. Others are "heading" dandelions, the leaves forming a firm tuft or core.

In cultivation the dandelion is treated as annual, although the plant is perennial or biennial. The seed is sown in early spring and the crop is harvested in autumn, or plants are allowed to remain in the ground until the following spring. Although dandelion will grow anywhere, it must have deep rich soil and good tillage if it is to make large and succulent foliage. Occasionally the seed is sown in seed-beds or in frames, and the plants are transplanted to the field; but usually the seeds are sown where the plants are to stand. The young plants are thinned until they stand one foot apart in the row. The distance between the rows will depend entirely on the value of the land and the means that are employed for tilling. If the plantation is to be tilled by hand tools, the plants may be allowed to stand as close as one foot each way; but if horse tools are used, the rows should be two or more feet asunder.

Since the demand for greens is usually greatest in early spring, the plants are generally allowed to stand through the winter. They are then ready for use as soon as the early growth starts. The rosette of foliage should be dense and wide-spreading, covering a space 12 to 20 inches across. The crop is harvested by cutting off the rosette of leaves just at the crown. The land is then plowed, and there is no danger that the plant will become a pest. The small and inferior plants not fit for sale should also be cut to prevent them from going to seed and becoming a nuisance. Sometimes a light crop is harvested the first year, the leaves being mown off so as not to injure the crown; even in this case, the main crop may be had the following spring.

The roots of the garden-grown dandelion are sometimes taken up in autumn and removed to the hotbed or forcing-house, and greens may be had in cold weather. They may be forced in this way in a dark place to provide blanched leaves. Even in the field the leaves may be tied up to blanch the inner part of the crown, much as endive is treated.

The cultivated plant is a developed form, or series of forms, of the common dandelion, *Taraxacum officinale*, Weber. The dandelion itself is with us a naturalized weed from Europe.

CHAPTER IV

COLE CROPS

Brussels sprouts	**Cauliflower and broccoli**
Kale, borecole and collards	**Kohlrabi**
Brussels sprouts	**Pe-tsai**

All cole crops are hardy and demand a cool season and rich soil, and abundance of moisture at the root. Propagated by seeds, which germinate rapidly; except the kales and kohlrabi, all are seed-bed crops, and even kales are often started in beds. Each plant requires considerable space to develop well. Cole crops are grown for the vegetative above-ground parts rather than for fruits or roots.

The cole crops constitute a natural group from the fact that they are closely related. They are all brassicas, and most of them are races of the same species. They are frost-hardy, and thrive particularly well in the cool of the year, although most of them are grown in summer.

CABBAGE

Cabbage is grown for the dense rosette or "head" of leaves. Cool soil which is deep and has power to hold much moisture, continuous growth from start to finish, frequent and thorough surface tillage, extra care in the selection of seed, avoiding the root-maggot, club-root, and rot by means of rotation and special treatments, destroying the cabbage-worm as soon as it appears,—these are essentials in cabbage growing.

One ounce of cabbage seed contains over 8,000 seeds, but not more than one-third or one-half of these seeds may be expected to make good plants. Early varieties are set 18 x 24 inches, or 24 x 24 inches (about 12,000 plants to the acre); late varieties 2 x 3 feet (about 7,000 plants). Four to six ounces of seed are usually required for an acre. The yield can be estimated from the number of plants to the acre, as a plant produces only one head. The heads of early varieties, which are set close, weigh from 2 to 4 lbs., of the late varieties 5 to 6 lbs. Due allowances must always be made for uneven crop, insect depredations, and other losses. For early cabbage, 8,000 heads to the acre are considered a good crop. In field culture, the yields run 10 to 20 tons to the acre, with 15 tons as perhaps a fair average in the hands of good men; 25 tons, and even more, are sometimes secured.

CLUBROOT (*Plasmodiophora brassicæ*).—The most striking symptom of clubroot in the field is a flagging of the leaves of affected plants on sunny days. Such plants regain their normal appearance overnight, but soon wilt again. The roots of diseased plants show characteristic malformations or swellings which frequently attain large size. Cauliflower, turnips, radishes, shepherd's-purse, wild mustard and other related plants are affected. *Control:* A clean seed-bed is essential. Slaked lime at the rate of about 75 bushels to the acre applied every few years is advisable if the disease has appeared in a field, since an alkaline reaction is unfavorable to the development of the organism. This should be applied the fall previous to planting. Diseased plants should be destroyed by burning. A long rotation, during which cruciferous weeds and cultivated crucifers are not permitted to grow, is important.

BLACK-ROT (*Bacterium campestre*).—The yellowing of affected leaves followed by a blackening of the veins is the first indication of the disease. Affected leaves may later fall off. Leaf petioles, leaf-scars, and stems of affected plants show blackened dots, where the sap tubes are discolored. Practically all cultivated crucifers and many cruciferous weeds are susceptible. *Control:* Seed disinfection is necessary. Formaldehyde solution made by adding 1 ounce of formaldehyde (40 per cent) to 2 gallons of water, or 2 teaspoonfuls to one pint of water, may be used. Seed should be immersed for 15 to 20 minutes in the solution, the formaldehyde washed off in clear water, and the seed spread out in a thin layer to dry. Mercuric chloride solution (1 to 1000) made by adding 1 ounce of mercuric chloride to 7½ gallons of water, or 1 tablet to 1 pint of water, is also satisfactory. Seed should be soaked for 15 minutes, the mercuric chloride washed off in clear water and the seed spread out in a thin layer to dry. It is desirable not to place the seed in direct sunlight and to stir them at intervals during drying. A clean seed-bed, care in the destruction of diseased material, and crop rotation are necessary.

BLACK-LEG (*Phoma lingam*).—The disease develops on leaves, stems and roots but characteristically attacks the stems and taproot below the surface of the ground. In advanced stages the dead areas are covered with tiny black fruiting bodies. *Control:* Same as for black-rot.

YELLOWS (*Fusarium conglutinans*).—Affected plants are stunted, the leaves turning a pale yellow. Usually the symptoms appear earlier and are more severe on one side of the plant, so that there is a warping and curling of stems and leaves. There is a darkening of the vascular bundles of the stem and the lower leaves of diseased plants drop early. *Control:* Seed disinfection (as recommended for black-rot and black-leg) is important to prevent the introduction of the fungus, and a disease-free seed-bed is essential. Planting must be into disease-free soil. The Volga and Houser are considered resistant. The Wisconsin Hollander is a resistant selection for the winter crop.

Green cabbage worm (*Pontia rapæ*).—A velvety green caterpillar about 1 in. long that eats holes in the leaves and often burrows into the forming head. *Control:* The U. S. Department of Agriculture recommends spraying the plants with the following formula: water, 50 gals.; soap, 4 lbs.; arsenate of lead (paste), 4 lbs., or powder, 2 lbs. In small quantities: water, 1 gal.; soap, 1 inch cube; arsenate of lead (paste), 1 oz., or powder, ½ oz. Since the cabbage head grows from inside the plant there is no danger from poisoning. If the outer leaves are removed before cooking, spraying is safe to within three weeks of harvest. If spraying is begun early in the season there will be little damage from late broods of worms.

The poison may be applied in the form of a dust, using 1 part powdered arsenate of lead in 4 parts air-slaked lime. In gardens the dust may be shaken on the plants by means of a cheesecloth bag. Apply thinly while the dew is on the leaves.

When only a few plants are grown, hand-picking is often the cheapest and easiest way to destroy the worms. In the home garden, pyrethrum, hellebore and hot water (130° F.) are convenient and useful remedies.

Cabbage looper (*Autographa brassicæ*).—A green looping caterpillar, marked with longitudinal white stripes, about 1¼ in. long when full-grown. The caterpillars eat out holes in the leaves and often bore into the forming head. *Control:* The caterpillars are difficult to poison as they dislike foliage coated with an insecticide, and as they crawl about freely can easily avoid the poison. The best results have been obtained by spraying with paris green, 1 lb. in 80 gals. of water to which the resin-lime mixture has been added. Some growers lightly dust with pure paris green with satisfactory results.

Diamond-back moth (*Plutella maculipennis*).—Small pale green caterpillars, about ⅜ in. long when full-grown, that eat holes in the leaves from beneath. The injured part dies, turns brown and drops out, leaving the leaf riddled with holes. *Control:* Spray with 2 lbs. paris green and 6 lbs. soap in 100 gals. water or with arsenate of lead (paste), 8 lbs. in 100 gals. of water.

CROSS-STRIPED CABBAGE WORM (*Evergestis rimosalis*).—A bluish gray caterpillar marked with distinct transverse black stripes, about ½ in. long when full-grown, that eats holes in the foliage somewhat like the green cabbage worm, and often attacks the tender central leaves and forming head; restricted to the Southern States. *Control:* Same as for the green cabbage worm.

CABBAGE WEBWORM (*Hellula undalis*).—Dull grayish yellow caterpillars marked above with five conspicuous brownish purple longitudinal stripes, about ⅝ in. long when full-grown; restricted to the Southern States. They feed on the underside of the leaves, bore into the leaf-stems and developing head, and usually cover their feeding grounds with a silken web. *Control:* Where this pest is likely to be troublesome, keep the plants well sprayed with paris green, 1 lb. in 50 gals. water, or with arsenate of lead, 4 lbs. (paste) in 100 gals. water, to kill the young larvæ. After the webs are formed, it is impossible to poison the caterpillars. In severe cases collect and destroy cabbage stumps left in the field after the harvest.

GARDEN WEBWORM (*Loxostege similalis*).—A dull green caterpillar about 1 in. long, marked on the back and on each side with a pale line, and with numerous small shining black spots on the back; restricted to the Southern States and to the Mississippi Valley. The young caterpillars skeletonize the leaves on the underside, covering them with webs; the older larvæ devour the entire leaf. *Control:* Same as for the preceding.

PURPLE-BACKED CABBAGE WORM (*Evergestis straminalis*).—A bristly purplish brown to dark greenish caterpillar, ¾ in. long, marked on each side with a yellowish stripe, that feeds on the leaves, webbing them together, and sometimes burrows into the stem and crown; restricted as a pest to the maritime provinces of Canada. *Control:* Spray with arsenate of lead (paste), 2 lbs. in 50 gals. of water.

ZEBRA CATERPILLAR (*Mamestra picta*).—A brightly colored caterpillar, 2 in. long when full-grown, black, with two bright yellow stripes on each side of the body. It often attracts attention, but rarely causes serious damage. *Control:* Spraying with an arsenical as for the green cabbage worm.

CABBAGE APHIS (*Aphis brassicæ*).—A mealy grayish-green plant-louse that often occurs in dense masses on the underside of the leaves and on the tender leaves on the heart of the plant; most abundant and destructive in seasons of drought. *Control:* Thorough spraying with so-called whale-oil or fish-oil soap, 10 lbs. in 100 gals. water, or with "Black Leaf 40" tobacco extract, 3/4 pint in 100 gals. water with 4 or 5 lbs. soap added. Use long leads of hose equipped with short extension rods and direct the spray by hand. For effective work, a pressure of 150 to 175 lbs. should be maintained and enough of the spray applied to wet the lice thoroughly.

TURNIP APHIS (*Aphis pseudobrassicæ*).—A plant-louse closely related to the cabbage aphis and often confused with it. *Control:* Same as for the cabbage aphis.

SPINACH APHIS (*Myzus persicæ*).—See under Spinach.

CABBAGE ROOT-MAGGOT (*Phorbia brassicæ*).—Small whitish maggots about 1/3 in. long which taper toward the head. They first attack the tender rootlets and then burrow in the main root, causing the plants to wilt and die. In the North they are most destructive to early cabbage in the field and late cabbage in the seed-bed. *Control:* Seed-beds are best protected from maggot attack by screening them with cheesecloth covers. The bed is surrounded by 6 or 8 in. boards placed on edge and the cheesecloth is stretched over the top, being supported by galvanized wires running over short posts. Early cabbage in the field may be protected by placing tarred paper discs around the plants when they are set out. Recent experiments in Canada indicate that corrosive sublimate, 1 part in 1,000 parts water, has a repellent effect on the young maggots. Two or three applications are required to keep the plants free from injury.

HARLEQUIN CABBAGE BUG (*Murgantia histrionica*).—A stink-bug ⅜ in. long, mottled red, black or yellow-orange, that in both the adult and immature stages attacks the plants, puncturing the leaves and stems, sucking out the juices and apparently poisoning the tissues. *Control:* Practice clean farming; destroy all cabbage stumps and other refuse after the crop is harvested; reduce hibernating shelter to a minimum; leave a few piles of rubbish in the field in the fall as traps. After the bugs have collected in these piles they should be burned. In the spring plant trap crops of kale, mustard or rape that will come up before the main crop, and when the bugs collect on these plants, kill them by spraying with clear kerosene.

Cabbage is a major oleraceous crop. It is used in one form or another in every household. It is both early and late. It practically covers the year. It is adapted to a wide range of country. It is useful for stock feed. It is grown by the home gardener, market-gardener, trucker, general farmer. A good cabbage head (Fig. 23) is a comely and handsome object, with flowing lines, excellent colorings, and attractive modelling.

The cabbage crop produces an enormous gross tonnage. Aside from the harvested heads, the leaves, stumps, roots and discards make great bulk and weight. Land must have good sustaining power to produce this herbage; and as the major part of the weight is water, the moisture-content must be unfailing. Make the land rich, prepare good depth to hold moisture, and keep the cultivator moving. Use every means to save the soil-moisture. If the nearly mature heads cease growing and are then started into growth again by means of tillage or rains, they are likely to crack.

Cabbage thrives on a great variety of soils. "Good corn land," if thoroughly prepared, should yield heavily in cabbages. Liberal fertilizing is usually essential to good results. Intensive growers often apply 1,000 to 2,000 pounds to the acre of chemical fertilizer to the early crop,

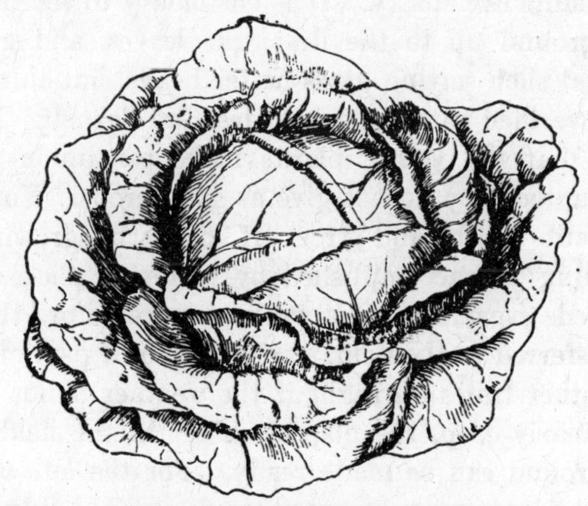

23. Cabbage of the oblate type.

with a liberal supply of nitrogen to hasten growth; for the late crop, with a longer season, less amounts may be supplied, although a heavy yield demands good feeding. Stable manure is much used for cabbages, sometimes as much as 40 loads to the acre. Late cabbage often follows an early crop of something else, as of peas or strawberries; early cabbage is often followed by late crops, as of turnips or fall-set strawberries.

Propagation; tillage.

For the early crop, the plants are raised under glass. For the main-season or late crop they may be started in seed-beds in the open. Seeds for late cabbages are sometimes planted directly in the field where the crop is to stand, but this is unwise for the young plants cannot receive proper care and the bugs get them. See that the young plants are stocky. It is customary to set the plants in the ground up to the first true leaves, and gardeners think that such setting gives better heads, but this opinion was not verified in three years' tests at Cornell. It is important that the young plants make continuous growth, for if stunted they do not give as good crops. Young cabbage plants withstand frost if properly grown. This "hardening" is accomplished by removing the sash from the hotbeds every day for a week or more before the plants are transferred to the field, sometimes for a part of the day and at other times all night if the weather is not too cool. For the early crop, the plants are set in the field as soon as the ground can be made ready. For the late or winter crop, the plants may be set in midsummer, July in New York. In small areas, transplanting is by hand, but in the larger areas it is performed by machines. Seeds and seedlings are seen (Figs. 24, 25).

24. Seeds of cabbage (× 5).

For general field crop, the early setting is raised under cheesecloth at the North, to protect from insects and other dangers. The last transplanting in the field in central New York for main field crop is seldom later than July 1. If plants are grown under protection so that the loss is small, four acres may be set from one pound of seed. Intervals between the rows under general farm conditions are commonly 3 feet and in the row 22 to 24 inches.

The plants need tilling very often if they are to grow rapidly. It is well to go through them the first time with a hand cultivator, as the plants are so small that a horse cultivator will cover some and damage others. But when the plants are well started, the horse and cultivator are employed. As the plants are but two feet apart for early crop, and the cultivator needs careful handling, let a boy lead the horse. Nearly all the work is performed with the horse, except a very little near the plants. Although they are very strong and rapid growers, few plants are more sensitive to neglect than the cabbage, or more favorably affected by extra good care. For very intensive work, the small early cabbages are sometimes set as close as 15 by 24 inches; in this case, hand tools are mostly used.

25. Seedlings of cabbage (× ½).

Harvesting; storing.

To harvest, the head is bent over and the stalk severed at the base of the head by means of a large sharp butcher-knife. The stumps are usually left standing until the field is cleaned for winter or for another crop. The trimmings are sometimes used for stock food. Soft springy heads are not mature enough for market, although they are sometimes shipped to meet an advance price; they do not keep long in good condition. A good cabbage head should feel firm and hard when pressed by the fingers; it should be free of decayed spots, cracks and blemishes.

For market-garden and truck-growing purposes, cabbages are usually shipped and sold by barrel or by crate; but the general late farm crop, used for kraut and for cattle feed, is handled by bulk in wagon, motor-truck and car.

26. Cabbages buried in a trench

Cabbages are extensively stored for winter use and sale. The first requisite to success is to store only such kinds as will keep, exercising as much choice in this respect as in the storing of apples. The early cabbages are naturally not of this kind. The flat or drumhead types usually do not keep well. The Danish Ballhead types are solid and long keepers. A successfully stored cabbage should be plump, not shrivelled, free from disease, full of natural moisture. Cabbages are stored either in the ground (buried) or in buildings. They should be free of rot when put in storage, and without bruises or injury from rough handling. Keep water from the middle of the head. The heads should be kept as cool as possible, without actually hard freezing. Be sure that they do not dry out.

A method of burying by a successful cabbage-grower is as follows: Dig a trench about four feet wide and at least one foot deep. Pull up the cabbage without shaking the dirt from the roots and retaining all the leaves. Place the heads in the trench with the roots up, close together, and wrap the leaves closely around them. Throw a few inches of straw over them and then cover with earth,—not more

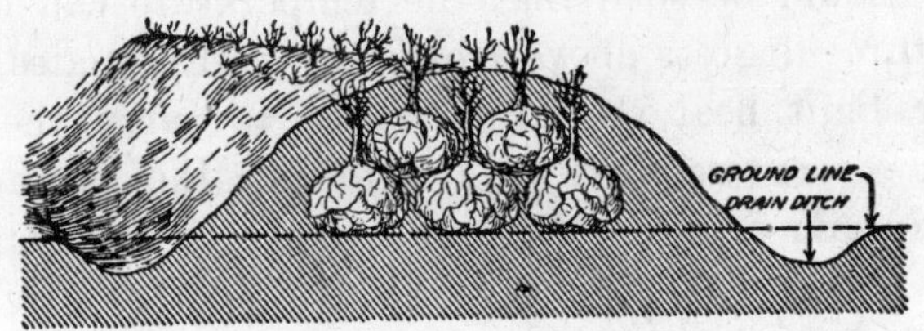

27. Cabbages buried on the surface.

than three or four inches at first. Two dangers must be guarded against: If too warm they will surely rot; or if they freeze too hard they will be spoiled when the frost comes out in the spring. After the weather becomes cold, freezing somewhat, put on more earth. A foot will do no harm in a cold climate. The entire lot may be lost by too hard freezing. If possible, dispose of the entire crop in the fall, even if obliged to sell at a low rate. The accompanying pictures (Figs. 26, 27) show methods of burying cabbages. The former is "cabbage in a trench for home use," from R. W. De Baun, N. J. Extension Bull., Vol. 1, No. 12 (1917), and the latter a "method of storing cabbage on a small scale at the North," by L. C. Corbett, Farmers' Bull. 433 (1915). Sometimes cabbages are stored temporarily, for a month or so, by inverting them when dry on the sward of a pasture or mown meadow and covering with straw.

For storing cabbages in a large way, special buildings are constructed. Sometimes the cabbages are piled in bins, but better results may be expected when the heads are laid on shelves, one layer deep or perhaps two or three layers. The outer loose leaves and all the stumps should be removed. The building should not freeze, and the ventilation should be such that the temperature can be kept two or three degrees above frost. In cold climates, provision for light heat should be made to carry the house through severe weather. All water on the cabbages, as from drip and condensation and leakage, is to be avoided.

Varieties; seed-growing.

Varieties of cabbage are many. The Wakefield types are prized for the early crop. For autumn and early winter use, and for kraut, the Drumheads and Flat Dutch type are popular. For winter storage the Danish Ballhead is extensively grown, from imported seed. Copenhagen, Empire Early, All-head, Enkhuizen are popular kinds for general field culture. The red cabbages are grown chiefly for pickling. The savoy cabbages, characterized by puckered or blistered leaves, are prized by amateurs for the delicate flavor; in this country they are grown mostly as an autumn crop.

Success with cabbage depends largely on the quality of the seed. It is better to purchase seed from reliable seedsmen and specialists than to attempt to grow it. Stored cabbages, with stump and roots intact, are planted in furrows in spring, the head being cut deep (usually crosswise) to allow the flower-shoots to come through. The stumps themselves, with head removed, often throw up flowering tops. The flowers mature rapidly, and seed is ripe in early summer.

KALE OR BORECOLE; COLLARDS

As compared with cabbage, kale requires less exacting care, is hardier, and the seed is usually sown where the plants are to mature. Kale is grown for its large leaves. It is raised mostly as a spring crop, seeds being sown the previous autumn; or as an autumn crop, seeds being sown in spring.

Plants usually are thinned to stand a foot or two in the row if very large plants are desired for the yield of individual leaves; or if the whole plant is to be gathered at once, the distances may be as close as 6 to 12 inches. The rows in gardens may be 2 feet apart; in large plantations they may be somewhat farther to allow of horse tillage. The yield to the acre in commercial plantations is 200 to 300 barrels, with 250 to 300 barrels perhaps an average fair crop.

Kale is affected by the insects attacking cabbage, particularly by aphis, and often by harlequin cabbage bug.

Kale may be likened to a cabbage plant that produces no head. In fact, it is a form of the cabbage species that is very near the original type. Greens from kale are prized in the market only very late or early in the season when many other kinds cannot be had in quantity. Small tender plants are best for eating, but leaves are often taken at intervals from older plants. This crop is much prized in England; the cool mild climate is well adapted to it.

In the North, kale is ordinarily sown in the spring, the seeds being placed where the plants are to stand. The rows may be far enough apart to allow of horse cultivation, and the plants may eventually stand, after the thinning process, from ten to twenty-four inches apart, allowing each plant an opportunity to develop to its best. The plants are not used until late fall or even winter. Often they are allowed to stand in the field all winter and the hardiest kinds are not injured by freezing, not even in the Northern States, if they are well matured, although a light mulch on the ground is beneficial. The older leaves and leaf-stalks are usually improved by being frozen. The tenderest leaves are picked from the plants at intervals, or the whole plant may be harvested at once.

For early spring use the seed ordinarily is sown in late summer or early autumn in the South and Middle South, and the plants stand out of doors in winter and are ready for use very early in the spring. In the northernmost States, however, these young plants are likely to perish unless protected under frames; therefore fall-sown kale is relatively little known in the colder parts of the country. It is grown on a very extensive scale about Norfolk, Virginia, and elsewhere

28. Scotch kale, showing a plant of large size.

South, and is shipped to the northern markets from New Year's until the opening of spring. In the Norfolk region, August is a favorite month for sowing.

The so-called Scotch (Fig. 28) and Siberian kales are chiefly grown in this country. Other forms, much taller and producing heavy yield of herbage, are grown for cattle in some countries.

Collards.—In the Southern States a kale-like plant known as collards is much grown, particularly in those regions so warm that good cabbages cannot be raised. The plants are grown as are cabbage plants, the seed being sown very early in spring, usually in a seed-bed under protection, in order that the plants may get a good growth before hot weather sets in; or they may be sown in mid summer for the fall growth in places farther north, where seasons are shorter. The leaves are ready for eating in the fall, or in very mild climates the plants may be left till spring. True collards are large plants, and 3 x 4 feet is not too great distance for them to stand. Sometimes young cabbage plants are raised for greens and are known as collards.

BRUSSELS SPROUTS

The culture demanded by brussels sprouts is essentially that required by kale, except that the plants are always grown as a fall crop and they are usually started in seed-beds. The crop requires a longer season than cabbage. The plant is grown for the small heads along the main stalk.

Plants stand 18 to 30 in. asunder in the row, and the rows are usually 3 ft. apart; dwarf varieties may stand closer. A good plant should yield 1 qt. of sprouts or heads. In the

Long Island sprouts region, plants are commonly spaced 30 x 36 in.; seed is sown June 1 to 15; 2,000 qts. to the acre is a fair average yield, but 3,000 qts. or even more are sometimes obtained.

The diseases and insects are those that prey on the cabbage.

Brussels sprouts is closely allied to kale, but along the straight strong stem little buds or miniature cabbages are borne, and these are the edible parts (Figs. 29, 30). A good "sprout," as one of the buds is called, averages one to two inches in diameter. When the sprouts are small and tender, they constitute one of the best and most delicately flavored vegetables of the cabbage tribe. The sprouts are gathered as they mature, from the bottom of the plant upward, and are sold by the quart. The adjacent leaf is cut off as soon as the sprout attains considerable size.

29. Plant of brussels sprouts before harvesting.

In the North the seeds ordinarily are sown rather late that the plants may not mature too early, for the sprouts are most prized in late autumn and winter. A large part of the growth is made in the cool weather of fall. If seeds are sown in June, the plants may be set in the field after the manner of cabbages in late July or August. In the Middle States the plants may be left out of doors in winter as the light freezing does not injure the sprouts. In the northernmost States, however, plants are usually dug late in the fall and planted out in pits, something after the method employed with celery and leeks.

30. A single sprout, nearly natural size

A good crop of brussels sprouts is dependent very largely on the strain of seed, as the plants tend to run down when careful selection in seed-raising is not practiced. A strong plant of the ordinary varieties makes a stalk 2 to 3 feet high, producing sprouts from near the base to the large canopy of leaves at the top. There are dwarf varieties, however, that grow 16 to 18 inches high that are in favor in short-season climates.

CAULIFLOWER; BROCCOLI

Cauliflower is grown for its white tender heads formed of the shortened and thickened flower-parts. From cabbage, the culture differs chiefly as follows: The plant is more particular as to climate, requiring a relatively cool moist season; it is mostly less hardy; it demands a constant supply of soil-moisture; care must be exercised that the heads do not sunburn; it is vitally important that the very best strain of seed is used. It is a crop of special localities.

A good distance for main-crop cauliflower is 2 by 3 ft., requiring upwards of 7,000 plants to the acre. The early smaller kinds may be 16 to 24 inches in the row. An ounce of seed for the production of 1,000 plants is a standard recommendation. An acre should yield 5,000 good heads.

The diseases and insects are those of cabbage.

Cauliflower is difficult to grow to perfection in the hotter and dryer parts of the country. Its requirements are similar to those of the cabbage except that it is injured by hot suns and dry weather, and it therefore needs a cool and moist atmosphere. Along the seaboard of the Northeastern States, near the Great Lakes, and in the Puget Sound region, cauliflower is grown with success, as it is also in special locations in many parts of the country. Wherever irrigation can be practiced, it may also be grown successfully. In the American climate the effort is usually made to secure the crop early or late and thereby to avoid growing it in the heat of midsummer. When thus grown, its range of adaptability is much extended. Under this system, the early crop is usually off in June or July. This crop is secured by growing the early varieties, as the Snowball and Paris, and by starting the plants under glass. The late crop is matured in autumn from seeds sown in summer in seed-beds. For this crop some of the later and larger-growing varieties may be used. In the southernmost parts of the United States cauliflower is grown as a winter crop from autumn-sown seeds.

Every effort should be made to conserve the moisture by deep preparation of the land in the first place and by frequent surface tillage thereafter. Low but well-drained bottom lands are usually chosen in order that the plants may have a constant supply of moisture. On Long Island, however, where the cauliflower is very largely grown, this precaution is unnecessary, since the atmosphere is moist from proximity to the ocean and the water-table is not deep; in other coast regions the same may be true. In small areas, mulching is sometimes advised to hold the moisture. In home gardens, of course, the plants may be watered. Land for cauliflower should be in a high state of fertility.

Some of the practices in the growing of cauliflower on opposite sides of the continent may be compared. In Rhode Island a large grower plants seeds about the middle of May, 1 ounce to 300 feet of drill, ¼ inch deep, the plants about 15 to the foot and not thinned; trans-

31. Head of cauliflower, trimmed for the market. See also Fig. 234.

plants to field by July 1 for largest crop; rows 3½ feet apart, plants in the row 16 inches; applies 1,500 to 3,000 pounds 4–8–4 fertilizer (no manure), all put on with wheelbarrow side-dresser in strip 12 inches wide on either side of row; expects 75 per cent good heads when set on time (by July 1) but far less for later plantings; early cauliflower, marketed in July and August, expects smaller percentage perfect heads. In eastern Washington, a grower sows seed beginning of March in hotbed for early crop, transplants to coldframes and sets in field late in April; seed for late crop sown in the open May 10 to 14, plants set in field June 20 to July 1; rows 30 inches apart, plants 18 inches in row.

The head of cauliflower is usually protected from the sun and whitened by tying the outer leaves over it. Plenty of room for ventilation should be allowed under the leaf-canopy, otherwise moisture may collect and the head may decay. The heads are harvested by cutting off, as are cabbages; the leaves are then trimmed to form a border or cup, as in Fig. 31. The crop is harvested in barrels or crates. Heads should be wrapped and handled with much care. They cannot be stored any great length of time. A good head has a regular "curd" or substance, without breaks, uneven growths, or "buttons."

Probably no other vegetable so quickly runs down from poor seed as the cauliflower. It is therefore exceedingly important that the choicest strain of seed be secured if the best results are to be attained. The best cauliflower seed is expensive, running as high as five to eight dollars an ounce; but cheap seed gives a smaller percentage of heading plants and the heads are usually irregular and broken. The cauliflower has a tendency to "button" or to throw up irregular growths from the head. This is due to poor seed, dry soil and too great heat, and also to allowing the plants to become checked and then starting them into growth by renewed tillage. The cauliflower seed of the market is grown in the Old World, the best of it coming from Denmark; but the Puget Sound country is attracting attention as a region for the growing of cauliflower seed. Good seeds may be grown under glass.

There is a family of long-season and late-maturing cauliflowers, relatively little grown in this country, known under the general name of broccoli. This plant requires the entire season in which to mature, and in Europe it is often allowed to stand over winter and to make its heads in spring. The heads are usually smaller than those of cauliflower.

KOHLRABI

The treatment required by kohlräbi is that demanded by flat turnips. It is usually not transplanted. The plant is grown for the tuberous stem, which must not be allowed to become tough; rapid growth is essential.

The plants usually stand, after thinning, 6 to 10 in. apart, the rows being 18 in, to allow of the use of the wheel hoe or farther apart if horse tillage is to be employed. An ounce of seed should yield about 1,500 plants; if grown as a field crop for stock, 4 to 5 lbs. of seed are usually allowed to the acre, and the crop may be 500 or 1,000 bushels.

The diseases and insects of cabbage may attack kohlrabi.

32. Kohlrabi (× 1/3).

Kohlrabi produces a turnip-like tuber just above the ground. It is grown mostly as a stock food and is relatively little known in North America outside of Canada. However, it is a very excellent garden vegetable, of delicate flavor, if used before the tubers become large and stringy, when they are yet globular or oblate; as the plants mature the tuberous part becomes elongated. They should be used when two to three inches in diameter; it is essential that they should have grown quickly and continuously, otherwise they are hard and bitter. Successive sowings may be made at intervals of two or three weeks to continue the table supply. Do not hill up the earth about the tuber. White Vienna is the leading garden variety. Kohlrabi is cooked the same as turnips. The plant and a leaf are shown in Figs. 32 and 33.

33. Leaf of kohlrabi, showing its long petiole and characteristic blade.

PE-TSAI

Grown as a potherb for its great tuft of leaves and the solid heads, and also as salad for the blanched and tender cores. It requires rich quick soil, abundance of water, cool season. In warm weather and on poor dry land it runs quickly to seed. Germination and growth are rapid. Good strains of seed are important.

As yet, no standard practices have been developed in North America for the rearing of this crop. Its culture is to be likened to that of kale. Plants may stand eventually 10 to 18 in. apart in the row, the plants being thinned for greens. Worms and aphis are to be expected, as for cabbage.

Under the name of Chinese cabbage and celery cabbage, this plant is now attracting much attention, although a full report was made on it, after repeated trials, more than twenty-five years ago by the Cornell Experiment Station. It has long been more or less known in Europe, and in China it is an ancient vegetable of major importance. To foreigners in China it is known as Shantung cabbage, from the province where it is extensively grown. To the

34. Pe-tsai as commonly grown for salad and greens (× 1/6).

Chinese it is known as pe-ts'ai, peh-ts'ai, po-ts'ai, the first word or element meaning "white," and ts'ai a green-vegetable or leaf-vegetable.

It is unfortunate that the name "cabbage" has become associated with this plant, for it represents a different species (if, in fact, not a different genus) from the cabbages, and it has none of the characteristic strong odors of them, nor is it so heavy for the digestion. It is a sweet delightful vegetable when properly grown, and as we learn how to raise and utilize it we may expect it to come into general use. Well-grown and neatly blanched pe-tsai is superior to lettuce as a salad. Undoubtedly we shall need to give special attention to seed-selection for American conditions.

35. Pe-tsai as grown in China, the long-headed form (× 1/10).

With us pe-tsai seems to be known mostly as a mass of loose foliage (Fig. 34), often developing a core of white tender leaves, not unlike cos lettuce in appearance. The growers of Shantung produce solid heavy heads (Figs. 35, 36, 37), sometimes weighing 5 to 7 pounds. The seed is usually sown by them in August, often following millet. Land is well prepared, and bean-cake or other fertilizer is applied in the row. Seed is sown in rows; as the plants attain considerable leafage, they are thinned, the young plants being used as a potherb. If weather is dry, the plants are watered. The remaining plants are left to form heads. If the rains of autumn are too heavy, the water is drained away. Too much wet makes a soft and yellow plant. By the approach of winter the

36. Longitudinal section of Fig. 35.

heads are formed. The plants are pulled, the outer loose leaves removed, and stored in an outside cellar for winter use and sale.

As known in this country, the crop is started very early in spring for use in warmer weather, or in August or September for producing dense heads.

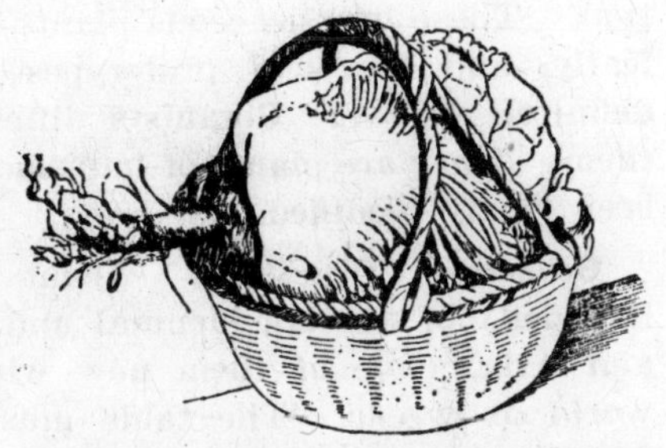

37. Pe-tsai as grown in China, the round form.

Botanically, pe-tsai is *Brassica pekinensis;* see the description of it on page 96.

THE COLE PLANTS AND THEIR KIN

The true coles (*i.e.*, generically kale plants, German *kohl*) are the thick-leaved blue-green plants of the kale-cabbage-cauliflower group, grown for their leaves or above-ground parts. To this group belongs also the kohlrabi (word the German form of *cole-rape* or *caulo-rapa*), with a thickened stem; and properly also the rutabaga is a cole plant, although not commonly so regarded in this country. The rutabaga (the word is of Swedish origin), known familiarly in N. America as "baga," is the Swedish turnip or "swede" of the English, and the *kohlrübe* (cole-turnip) of the Germans; and it is also called turnip-rooted cabbage, recognizing thereby the cabbage appearance of the foliage and flowers. It should be said that in America the word cabbage is restricted to plants that produce heads (the word is associated with the Latin *caput*, a head) but elsewhere it has a wider application in the cole crops.

The cole plants are of the genus Brassica; to this genus belong also the turnips; also the mustards, although certain of them are separated by some botanists in the genus Sinapis. Therefore it is best to consider all these plants together, in this account mentioning only the kinds of common cultivation. These brassicaceous plants are difficult to define botanically, and the wild prototypes of some of them are not definitely known. Botanists differ in their interpretation of them. They are plants of immemorial domestication, and have been vastly modified.

Brassica. *Cruciferæ.* About 40 species (if Sinapis is included) of annual, biennial and perennial herbs, of Europe and Asia, some of them now widely spread throughout the world as weeds. The table mustard of commerce is made from the seeds of some of the species. The botanical characters of separation between the species lie to an important degree in the size, shape and position of the pods, and in the shape and length of the beak or top end of the pod beyond the valves or detaching sides. The seeds in these plants are globular in general form, without conspicuous surface markings; they are essentially black in the cole plants and turnips, but may be brown or lighter colored in the mustards; they weigh 1 to 5 mg. and the vitality is about 4 or 5 years. Of the cole plants, including rutabagas, the average seeds are approximately 3 or 4 mg. in weight; of turnips about 2 mg.

A. Plant glaucous-blue or blue-green (varying to red), the foliage usually thick and more or less fleshy, the mature leaves glabrous; larger leaves on the flowering stems usually clasping: flowers large (1/3 to 1 in. long), whitish yellow, cream-yellow or ochroleucous, the petals long-clawed, sepals mostly firmly erect and not spreading.—COLES.

B. Flowers large (mostly exceeding ½ in.) and very light colored (sometimes almost white), the inflorescence elongated at anthesis (4 to 10 in. long).

Leaves large, mostly thick: stem not thickened.

1. *B. oleracea.*

Leaves relatively small, thinner: stem tuberous.

2. *B. caulorapa.*

BB. Flowers smaller (not exceeding ½ in.), yellower, the part of the inflorescence in bloom at any time rarely exceeding 2 or 3 in. and usually shorter than this.

3. *B. campestris.*

AA. Plant green or essentially so, slightly or not glaucous, the foliage thin and often sparsely setose-hairy on the ribs; stem leaves various: flowers small (3/8 in. or less long), bright yellow or sulfur-yellow, the petals less prominently clawed, sepals separating or spreading.

B. Stem leaves clasping or the petiole with a broadly expanded base.—TURNIP. 4. *B. Rapa.*

BB. Stem leaves petioled or sessile.—MUSTARDS.

C. Pod glabrous (not hairy).

D. The ripe pods long, spreading away from the stem.

E. Leaves more or less lobed or notched, but not deeply cut.

Radical and lower blades tapering to winged midrib. 5. *B. pekinensis.*

Radical and lower stem leaves distinctly petioled. 6. *B. rugosa.*

EE. Leaves deeply cut. 7. *B. japonica.*

DD. The ripe pods short, closely appressed to the stem. 8. *B. nigra.*

CC. Pod hairy. 9. *B. alba.*

1. **B. oleracea,** Linn. Sp. Pl. 667. Glaucous perennial with woody and often branching stem 2 to 5 ft. tall, native on the sea-cliffs and shores of western Europe: lvs. thick, large, 1/2 to 2 ft. long, obovate or oblong in general outline, often with several small lobes along the petiole, the margins irregularly lobed or sinuate, often obscurely dentate, and usually more or less undulate and crisped: flowers large (3/4 to 1 in. long) in an elongated panicle, whitish yellow: pod 3 to 4 in. long, 1/4 in. across at maturity, with a conical beak 1/4 to 3/4 in. long, the valves with a strong central rib.—As *B. oleracea* itself is not cultivated, and apparently not eaten in the wild state, a full description is not necessary here. In the wild it gives little suggestion of the cabbages, brussels sprouts and cauliflowers, although it is much like some of the kales. Under domestication this species has produced a multitude of forms, some of the main races of which may be described.

Var. **ramosa,** Alef. Landw. Fl. 234. 1866. Tree Cabbage or Tree Kale. Thousand-headed Kale. Stem erect, 3 to 6 ft. or even more, woody at the base, more or less branched above, the leaves scattered rather than in a terminal clump or rosette. —Grown mostly in Europe, and chiefly for cattle forage.

Var. **acephala.** DC. Syst. Nat. ii, 583. 1821. KALE, COLLARD. Plant very short to tall, the stem simple or only sparingly branched: leaves various, aggregated toward the top of the stem, oblong to oval to roundish in outline and lobed toward the base, in some forms much crisped and curled, the midrib and petiole usually thick and stout.

Var. **gemmifera,** DC. Syst. Nat. ii, 583. 1821. BRUSSELS SPROUTS. Stem erect, 1½ to 3 ft. tall, bearing large edible buds 1 in. or so in diam. in the axils: leaves short and broad, short-oblong to nearly circular, usually with one or two large rounded lobes near the base but sometimes unlobed, the margins of the main leaves not notched or dentate, petiole not winged.

Var. **capitata,** Linn. Sp. Pl. 667. CABBAGE. Plant low and squat, with a very short stem, producing one large compact terminal head 4 to 12 in. in diam.: leaves large, spreading, oblong-obovate to nearly circular, the main ones mostly unlobed and the blade tapering into a short margined petiole, margins undulate and more or less obscurely toothed.—A race or subvariety is Var. *sabauda,* Linn. (Var. *bullata,* DC.), the Savoy cabbages, with blistered or bullate leaves (the word *sabauda* means *Savoyan*). Recently there has come into cultivation a "green-glazed" cabbage, with bright green shining foliage.

Var. **botrytis,** Linn. Sp. Pl. 667. BROCCOLI. CAULIFLOWER. Plant of the stature of Var. *capitata,* but bearing long-oblong or elliptic mostly undivided upright or incurving leaves with margins entire or minutely denticulate, and the flower-clusters (malformed stems and flowers) rather than the leaves condensed into a head.—Sometimes broccoli is separated as Subvar. *cymosa,* Duchesne, and cauliflower as Subvar. *cauliflora,* DC. (The word *botrytis* means "a bunch of grapes"; here it refers to the forms in the broccoli or cauliflower head.)

2. **B. caulorapa,** Pasq. Cat. Ort. Bot. Nap. 17. 1867. (*B. oleracea* var. *gongylodes,* Linn. Sp. Pl. 667. *B. oleracea* var. *caulorapa,* DC.) KOHLRABI. Plant low and erect, 1 to 2 ft. tall over all, the stem thickened just above the ground

and turnip-like, foliage arising from the tuber: leaves small and thinnish, the blades 4 to 8 in. long, oval or round-oval to oblong, the margins prominently toothed or notched, the base more or less irregularly lobed or shaped, the petiole slender and thin and often bearing a few detached small leaf-lobes, the base expanded and clasping.—Probably an offshoot of the composite species *B. oleracea*, but marked in its stem and foliage characters; grown for the stem tuber (the ante-Linnean name *gongylodes* means "roundish").

3. **B. campestris,** Linn. var. **Napobrassica,** DC. Syst. Nat. ii, 589. 1821. (*B. oleracea* var. *Napobrassica*, Linn. Sp. Pl. 667. *B. Napobrassica*, Mill. Gard. Dict. ed. 8 no. 2. 1768. *B. Napus* var. *Napobrassica*, Reichb, in Moessl. Handb. Gewachsk. ed. 3, ii, 1220. 1833.) RUTABAGA. SWEDISH TURNIP. Plant in flower or fruit 2 to 3 ft. high, branched, erect but sometimes falling with the weight of seed: root a fusiform or oblong (rarely globular) tuber with a long neck: radical lvs. long-stalked, 12 to 24 in. long over all, the blade oblong in outline, strongly pinnate-lobed, the terminal lobe broad and obtuse, the others successively smaller downward and semi-opposite or scattered, some of the smaller parts entirely separate and remote on the petiole, the margins variously and irregularly dentate or notched, the mature leaves mostly wholly glabrous but sometimes bearing scattered setæ on the ribs, the small leaves immediately succeeding the seed-leaves more or less sparsely hairy; upper stem leaves becoming oblong to lance-oblong, strongly sessile-auriculate, notched, dentate, or nearly entire: flowers light yellow, in elongating clusters: pod about 2 in. long exclusive of the conical beak, which is about ⅜ in. long.—Sometimes the white-fleshed and yellow-fleshed rutabagas are separated, in which case the former may take the name Subvar. *communis*, DC. and the latter Subvar. *Rutabaga*, DC.; the botanical origin of these races is not cleared up.

Brassica campestris itself is a weed in and near cultivated areas, not producing an enlarged root. Rape is often considered to be of the same species, *B campestris* var. *Napus*, Babington (*B. Napus, Linn.*).

4. **B. Rapa,** Linn. Sp. Pl. 666. (*B. campestris* var. *Rapa*, Hartm. Handb. Skand. Fl. ed. 6, 110. 1854.) TURNIP. Plant green, slightly or not at all glaucous, the foliage usually roughish to the hand: root tuber flattened or globular, sometimes oblong white- or yellow-fleshed, the top part often purple, the neck short: root leaves not thick, mostly long-pinnatifid, the lobes in several irregular uneven pairs and successively smaller downward, but sometimes tapering gradually from the broad blade to a narrowly winged petiole and without large lobes; leaves usually sparsely setose-hairy on the ribs beneath, at least in the young expanding foliage; upper stem leaves obovate to oblong to lanceolate in outline, the margins of the larger ones irregular and notched, often narrowed toward the base, clasping: flowers small (¼ to ⅜ in. long), bright yellow, the clusters short in anthesis: pods about 1½ in. long exclusive of the slender conical beak.—Nativity undetermined. (Rapum is a Latin word for turnip.)

5. **B. pekinensis,** Rupr. Fl. Ingr. i, 96. 1860. (*Sinapis pekinensis*, Lour. Fl. Cochin, 400. 1790. *B. Pe-tsai*, Bailey, Bull. 67 Cornell Exp. Sta. 190. 1894.) PE-TSAI. An erect green soft-foliaged annual of quick growth: radical leaves many, large, veiny and crinkled, 12 to 20 in. long, oblong or broadly obovate in outline, the top broad and rounded, tapering below and vanishing to the lower end of the very broad whitened midrib, the upper margins wavy, the lower margins jagged-notched; stem leaves multiform, sometimes broad and clasping, sometimes merely sessile, sometimes petioled, in shape various, the margins notched or crinkled or in the upper leaves entire: flowers light yellow, about ⅜ in. long, the cluster short in anthesis: pod stout, 1 to 2 in. long exclusive of the short cone-shaped blunt beak.—Probably native in China. See page 88.

6. **B. rugosa,** Bailey, Bull. 67 Cornell Exp. Sta. 191. 1894; Prain, Bull. 4, Dept. Land Rec. and Agr., Bengal, 11. 1898.

(*Sinapis rugosa,* Roxb. Fl. Ind. iii, 122. 1832.) BROAD-LEAF MUSTARD. Plant green, producing abundance of foliage, annual: radical leaves large and quick-growing, more or less hairy when young, usually blistered or bullate, 1 ft. or more long and three-fourths as broad, obovate or oval, angled or notched, separately cut or lobed below on the narrowing sides, the petiole broad or stout; lower stem leaves of similar shape, large (blade 4 to 5 in. long and nearly as broad) notched and angled, distinctly stalked; upper stem leaves oblong to lanceolate, nearly or quite entire, usually sessile or tapering to base, sometimes clasping: flowers about ¼ in. long, bright yellow, the clusters short in anthesis: pod 1 to 2 in. long, exclusive of the rather slender acute beak.—Probably native in China; usually cultivated as "Chinese mustard."

7. **B. japonica**, Sieb. acc. to Miq. Prol. Fl. Jap. 74. 1865-6. (*Sinapis japonica,* Thunb. Fl. Jap. 262. 1784. *B. nigra* var. *japonica,* Schulz, in Pflanzenr. iv, 105, p. 79. 1919.) CURLED MUSTARD. Very like *B. rugosa,* and perhaps a form of it, distinguished by the frizzled and cut foliage.—An old garden favorite. Now grown under several forms, as Giant Southern Curled, Fordhook Fancy, Ostrich Plume, California Peppergrass, the last one with finely cut leaves.

8. **B. nigra**, Koch, in Roehl. Deutschl. Fl. ed. 3, iv, 713. 1833. (*Sinapis nigra,* Linn. Sp. Pl. 668.) BLACK MUSTARD. Tall branching annual, 3 to 10 ft. high, with slightly glaucous glabrous or sparsely hairy stem which is often reddish: leaves oval to oblong, obtuse or short-acute, notched and variously lobed, slender-petioled: flowers light yellow, about ¼ in. long, terminating slender racemes: pod short (½ to 1 in. long over all), 4-sided, stout, with a short conical beak, becoming closely appressed to the rachis of the raceme: seeds small, brown or brown-black, weighing about 1 mg.—Europe; now widely spread as a weed. Employed as a source of table mustard, manufactured from the seeds; sometimes mentioned as grown for the early radical leaves, for greens, but there are better kinds.

9. **B. alba**, Rabenh. Fl. Lusit. i, 184. 1839. (*Sinapis alba,* Linn. Sp. Pl. 668.) WHITE MUSTARD. Erect more or less hairy

annual 2–3 ft.: lvs. oblong, all petioled, obovate or oval in outline, deeply pinnately lobed, the margins bluntly notched: flowers about ⅜ in. long, light yellow, terminating elongating racemes: pods squarrose (at about right angles with the rachis), hairy, with a flat beak longer than the body: seeds few, large, yellowish or light brown, weighing 3 to 4 mg.—Europe; sometimes run wild. Cult. sometimes for greens, particularly under the name of White London mustard.

CHAPTER V

SALAD CROPS

Lettuce
Endive and chicory
Cress
Corn-salad
Parsley
Chervil
Celery

As a general statement, it may be said that salad plants require cool moist soil, and a quick continuous growth if the best results are attained. They are often benefited by a special application of quickly available fertilizers during growth, particularly of nitrogen in those species desired chiefly for a rapid growth of leaves. Most of them do not require occupation of the ground the entire year.

The plants included in this chapter are a somewhat miscellaneous company, and it is difficult to state principles that apply to all of them. They are closely connected with the potherb crops. Celery and lettuce have little in common, but the above grouping seems to be as satisfactory as any. Some of the plants are used both as salads and potherbs, as endive; but they are placed in the group to which their most common use assigns them. A salad is eaten uncooked; a potherb or "greens" is boiled. Horse-radish is properly a salad plant, or a relish plant.

On the necessity of giving extra care to the rearing of salad plants, Waugh writes (Bull. 54, Vt. Exp. Sta.): "Doubtless all vegetables ought to be fresh; but with salad plants the demand is imperative. A good salad cannot be made from wilted or stale plants. For this reason the best salads are practically prohibited to people who do not have their own gardens. The plants should be freshly picked within half an hour of meal time. Up to this time they should have been rapidly and vigorously grown. A rich spot of ground, plenty of water, clean and thorough culture with favorable weather, must combine for best results. Dry, tough, wilted, weed-choked plants are not worth gathering. Yet most of the true salad plants reach edible maturity so quickly that any reasonable attention should secure good returns. Here again it is not time and money that are required for success, but a little thoughtful promptness of action."

In these days, when we begin to know something of the value and office of vitamines, contained in the herbage of plants, we should have a new appreciation of the importance of salads and potherbs to the welfare of mankind. It is the result of long and tried experience that many races of men have come to place great reliance on green food.

LETTUCE

Lettuce is a hardy, cool-season, short-season succession- or companion-crop, requiring mellow moist soil, quickly available fertilizers and continuous growth from start to finish. In this country it is grown in the open ground throughout the season, and it is also extensively forced under glass. It is very easy of cultivation in rich and well-prepared land.

Lettuce is commonly grown in rows 8 to 14 in. apart, and thinned eventually, as the young plants are taken out, to 8 to 12

in. in the row; grown as a field crop tilled by horse, the rows may be spaced as far as 18 in. For early use start in forcing-house, frame or kitchen. Sow in succession till warm weather. In late summer or September, sowing may be made for the autumn crop. In the South it may stand out over winter and resume growth in spring. Calculate on 1,000 plants for each ounce of seed. Most of the forcing varieties, started under glass are good for early use, as Tennisball, Boston Market, Simpson. For summer use, plant varieties that withstand heat, as Deacon, Hanson, Summer Cabbage, Cos. A good commercial acre should yield upwards of 30,000 heads.

RHIZOCTONIA, OR BOTTOM-ROT (*Rhizoctonia solani*).—Plants in any stage of development may be affected. Rusty slightly sunken areas on the leaf-stalk where it comes in contact with the ground and the total rotting of the leaf-blade are indicative of this disease. The entire head may later rot and remain as a blackened erect stump. Frequently the disease causes a damping-off of seedlings. *Control:* Soil sterilization in the greenhouse will prevent the development of bottom-rot. Thorough drainage and frequent cultivation to dry out the surface soil will reduce somewhat the development of the disease in the field. The more erect types of lettuce are apparently less affected.

DROP, OR SCLEROTINIA ROT (*Sclerotinia libertiana*).—Affected plants become water-soaked and collapse with a soft rot in a few hours after showing evidence of this disease. White felts of mycelium with black fungous bodies imbedded in them develop on the under surface of the leaf. This is a serious disease of field and greenhouse lettuce, the seriousness being increased by the fact that the organism will attack almost any host. *Control:* Thorough soil sterilization, when practicable, will control the disease. Prompt removal of affected plants and drenching the soil with copper sulfate solution has met with considerable success. All refuse should be removed and destroyed.

GRAY MOLD, OR BOTRYTIS ROT (*Botrytis cinerea*).—Usually but one leaf or one side of a plant is first attacked. The disease may spread until the destruction of the entire plant results. A characteristic gray fungous growth from which the name of the disease is derived occurs on the rotted tissues. The parasite is most destructive in greenhouses. *Control:* Care in ventilating and watering will do much toward preventing the development of this rot. The prompt removal of all débris is desirable.

ANTHRACNOSE (*Marssonina panattoniana*).—Leaf lesions appear first as somewhat circular water-soaked spots which later become brown. In the later stages, the affected tissues die and drop out, giving the leaf a shot-holed appearance. On the midrib the brownish spots are sunken and elongated. *Control:* Prompt removal of affected plants, together with rotation of crops in the field, is desirable. Sanitation in the greenhouse is important. Slightly higher temperature than is usual, together with careful ventilation, will check the disease.

MILDEW (*Bremia lactucæ*).—Yellow areas are evident on the upper side of affected leaves and a white mildew is present on the under surface of such spots. Mildew is primarily a greenhouse disease, although it may occur in the field in cool weather. *Control:* Care in ventilating and watering will prevent the development of this disease. A slight increase in temperature may tend to check its development.

TIP-BURN.—A blackening of the leaf margins, frequently evident only on the inner leaves, is characteristic of tip-burn. Apparently this disease is not due to a casual organism but to unfavorable environmental conditions. *Control:* Careful watering and ventilating in the greenhouse will aid in preventing this trouble. There is some indication that an excess of nitrate and excessive applications of fertilizers in midsummer may increase the development of tip-burn.

CABBAGE LOOPER (*Autographa brassicæ*) and CELERY LOOPER (*Autographa falcigera*).—Both of these common looping caterpillars sometimes attack lettuce. As an arsenical cannot be used, hand-picking is the only available measure.

PLANT-LICE (several species).—Lettuce both in the greenhouse and in the field is liable to infestation. Tobacco dust is employed. In the greenhouse fumigate with nicotine.

Cutworms and slugs sometimes attack lettuce. See pages 430, 437.

Lettuce is the standard salad plant. It is good in itself, and both market and kitchen practices are well understood. It needs no explanation. The culture is also simple. It does not occupy the land the entire year. It is a

38. Common head lettuce, seen from above.

succession-crop or companion-crop. It is grown in the North spring to autumn and the South autumn to spring. It readily adapts itself to forcing in glasshouses. It grows well in hotbeds and frames. It may be made to stand much frost. For all these reasons, it is a year-round crop.

Lettuce is commonly grown as a seed-bed crop. The early crop is usually started in the house or in hotbeds and transplanted to the field; or some of it may mature directly in the hotbed or frame. In some cases, particularly for the midseason and later crops, the seed may be sown where the plants are to stand. In large-area lettuce-farming in the Northern States, principally on reclaimed muck land, seed is sown directly in the field and the plants (if Big Boston) are thinned to stand 10 to 14 inches in the row, the rows 14 inches or more apart. Two or three pounds of seed are required to the acre. In good weather and on well-prepared land, the crop is ready to harvest in six to eight weeks. Sowings are made every week or so till the beginning of August.

39. Cos or Romaine lettuce (× 1/6).—Lactuca sativa var. longifolia.

Lettuce may be followed by cabbages, early cauliflower, celery or various other succession-crops. Sometimes lettuce is transplanted between the plants of early cabbages or cauliflowers, since it will mature before the other plants need all the space. If one's soil is moist, and particularly if the exposure is somewhat cool, the ordinary spring lettuce may be grown with success throughout the summer. Successional sowings may be made as often as once in ten days to three weeks. The earliest spring lettuce taken from the open is usually started in frames or forcing-houses, or sometimes in boxes in the house. It transplants easily.

The crop may be grown in autumn from seeds sown late in August or in September. In such case it is best to sow in a seed-bed, because the moisture conditions can be controlled better, and a field is usually too dry at that time of the year to give quick germination. It is essential that lettuce make a quick and succulent growth to be at its best. For the late spring and summer crops the seed is usually sown rather thickly and the thinnings are used on the table. The plants that are to attain the largest size should stand as much as a foot apart.

Letttuce usually does best in soil that is loose and warm, or one that the gardeners call "quick." Heavy lands, and particularly those with much clay, are ill-adapted to the crop. To secure a quick growth, it is sometimes advisable to apply nitrate of soda soon after the plants are set. The nitrate is usually sprinkled broadcast on the surface and raked or cultivated in. An application at rate of 200–300 pounds to the acre may be made with good results. The surface should be kept well tilled to conserve the

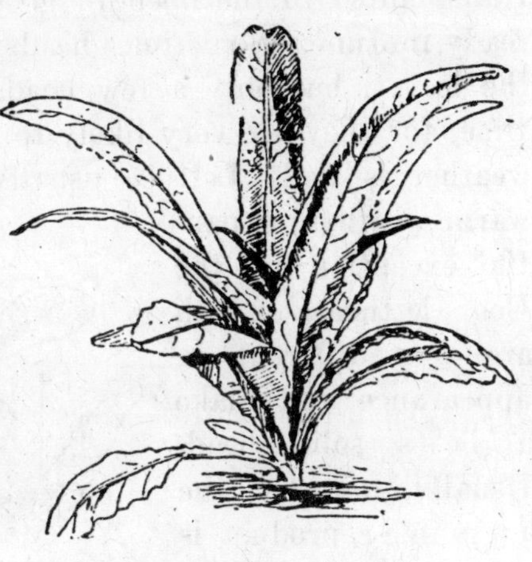

40. Asparagus lettuce (× 1/5). Lactuca sativa var. angustana.

moisture and to promote all those activities that result in rapid growth.

Although the lettuce product is usually spoken of as a "head," there are many kinds of leaf-clusters, and some of the kinds are known technically as "head lettuce" or "cabbage lettuce." The Boston Head lettuce is one variety, as Simpson and Grand Rapids are others. "Head lettuce" is grown the same as other kinds, special care being exercised to get good seed. Started indoors in April and transplanted to the open in good warm soil, the crop is ready in June. Sometimes heads are blanched by tying up the leaves, but only a few heads should be treated at a time, for they are very likely to decay, particularly if the weather is wet. Lettuce usually does not head well in warm weather; a partial exception are the Cos lettuces, which are very different in appearance and make a less solid head. Usually, however, the summer product is "leaf lettuce" or "bunching lettuce," the product of many non-heading varieties. Figs. 38, 39, 40 are widely different forms of lettuce; Figs. 41 and 42 show the seeds and seedlings.

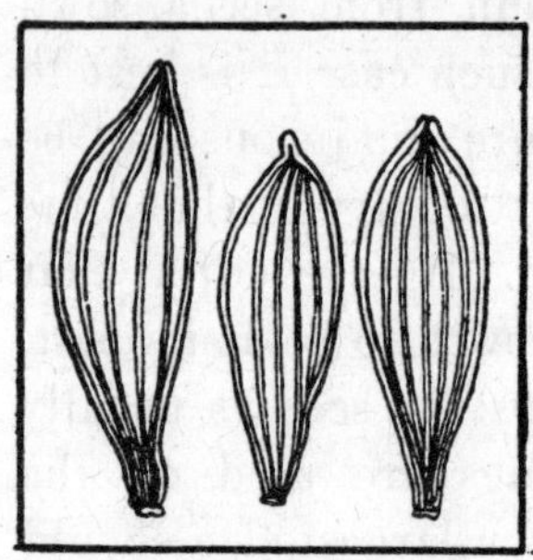

41. Seeds (properly fruits) of lettuce (× 6).

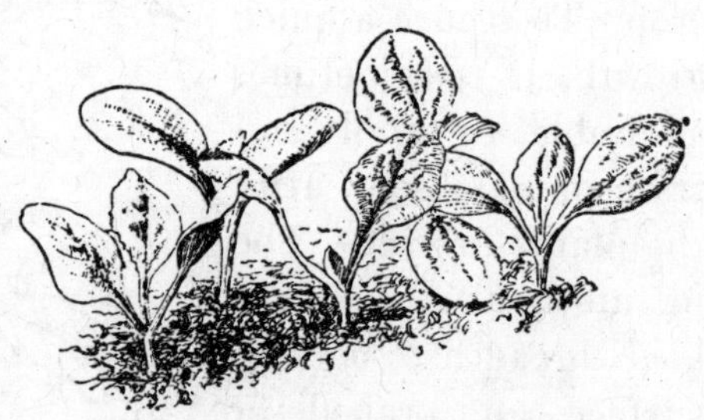

42. Seedlings of lettuce (× ½).

The Cos lettuces, or Romaines, produce rather loose heads, but the midribs are usually very broad and in the interior of the head are likely to be somewhat blanched. Gardeners sometimes tie up the heads at the top to further the blanching, but the plants must be watched carefully to avoid rot. Romaine is grown the same as other lettuce, but it is likely to stand longer in the field before running to seed. Sown late, it makes very acceptable autumn salad.

For market, the lettuce plant is cut just above ground, the outer leaves are removed and the heads or bunches are shipped in ventilated crates or barrels. The plants should not be cut for this purpose in the middle of the day, for they soon wilt.

Seed of lettuce is grown extensively in California. Yields vary with the variety and the handling; one pound of marketable seed may be had from 30 to 60 plants.

The Lettuce Plant

Lactuca. ***Compositæ.*** **Genus of weedy herbs, perhaps 100 species in many parts of the world, some of them native in the United States and Canada and others introduced weeds; annuals, biennials, perennials.**

L. sativa, Linn. Sp. Pl. 795. ***L. Scariola,*** **Linn. var.** ***sativa,*** **Clarke, Compos. Ind. 263. 1876. Garden Lettuce. Annual erect smooth herb with milky juice, producing a rosette or cluster of radical leaves; stem 3–4 ft. high, leafy, branching above, the many slender branches bearing numerous clasping-conduplicate cordate mostly acute bracts: radical leaves (used in salad) various, 5 to 10 in. long, thin, spreading, roundish to oblong to obovate to lingulate, obtuse and usually very blunt, margins plane or undulate, entire or sinuate-dentate, often somewhat lobed or erose toward the narrowing base, the petiole very short or none, the blade with many prominent ribs aris-**

ing from the broad midrib; stem leaves similar in shape to the root leaves of the particular variety, alternate, clasping-auriculate, mostly finely apiculate-serrate, passing into bracts toward the inflorescence: flower-heads erect, on short or long pedicels, about 12- to 16-flowered, opening in morning and closing about midday, florets all perfect and each with a yellow 5-toothed ray; receptacle naked; involucre cylindrical, becoming conical in fruit, scales lanceolate to ovate, all appressed, the outer ones successively shorter; ovary lenticular, bearing many white pappus bristles at its constricted summit; style-branches short: achene ("seed" of gardeners) white or black, lenticular-oblong, broadest toward the top, strongly several-nerved, bearing a long slender beak on which the pappus is carried; when the beak drops or is removed in threshing, the remaining "seed" is 1/8 to 3/16 in. long and weighs 1 to 1½ mg., retaining its vitality about 5 years.—Unknown in a native state and considered to be a modification of *Lactuca Scariola*, Linn., an Old World weed now also widely spread in this country. In lettuce fields "rogues" now and then occur strongly suggestive of *L. Scariola.* Lettuce has been cultivated so long that its history is inexact. Var. **capitata**, Linn. Sp. Pl. 795 (*L. capitata*, DC., Prodr. vii, pt. i, 138. 1838). HEAD LETTUCE, has radical leaves forming a more or less dense ball. Var. **crispa**, Linn. Sp. Pl. 795 (*L. crispa*, DC. l.c.). CURLED LETTUCE, has the leaves cut and fringed or crisped.

Var. **longifolia**, Lam. Dict. iii, 403. 1789. (*L. romana*, Garsault, Trait. Pl. et Anim. Usage Med. ii. 196, t. 315. 1767.) COS LETTUCE. ROMAINE LETTUCE. Plant forming an upright columnar or loaf-shaped loose head, the radical leaves obovate to oblong, rounded or obtuse, 8 to 12 in. long and 4 to 6 in. broad, the midrib usually very wide; stem leaves long, mostly oblong or obovate, obtuse.

Var. **angustana**, Irish, Cyclo. Amer. Hort. 867. 1900; Bailey, Gent. Herb. 1:49. 1920, with botanical diagnosis. (*L. angustana*, Hort.). ASPARAGUS LETTUCE. Plant not forming a compact head: radical and lower stems narrow- or oblong-lanceolate, long-attenuate, entire or irregularly sinuate-dentate, plane, 8 to 12 in. long and 1½ to 3 in. wide; upper leaves lanceolate-attenuate, amplexicaul.—Grown for its thick edible stem; quickly runs to seed. It is little known in N. America.

ENDIVE AND CHICORY

Endive affords a good supplement to lettuce, since it is essentially a summer and fall crop and thrives at a season when lettuce is somewhat difficult to grow to perfection. The culture is not unlike that of lettuce, except that the plant requires a longer time in which to mature. It is more popular as an autumn and winter crop, seeds being sown in summer. The plant is used both as salad and greens.

To obtain large heads or tops, plants should stand 12 to 16 in. apart each way, but they are often grown as close as 8 or 10 in. They may be grown in rows 18 to 20 in. apart for easier tillage, but the plants should not be crowded if they are not eaten when young and small. One ounce of seed should supply a row 100 to 150 ft. long. Two months or less should produce edible tops.

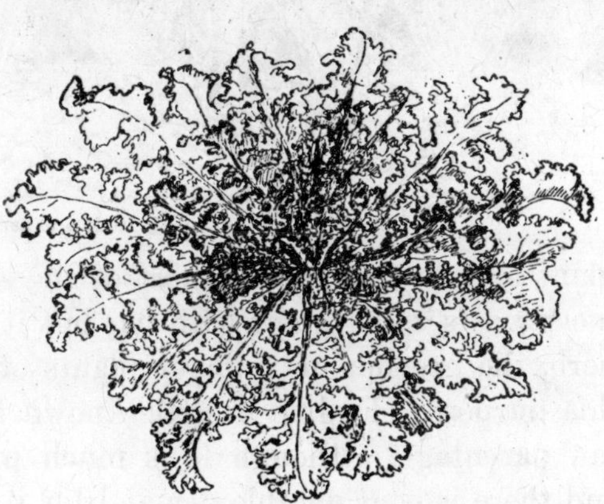

43. Young plant of endive (× 1/5).

Diseases and insects appear not to be troublesome to endive.

In respect to soil, tillage, distance apart and other treatment, the care of endive differs little from that of lettuce. Seeds may be started under glass and transplanted to the open, although such plants are thought to run more quickly to seed; good tops may be had in late spring and early summer. Or seeds may be sown in June for plants to be used in August and September. Seeds may be sown in summer for the autumn and winter crop, and this is the better adaptability of the plant.

The top may be harvested entire (Fig. 43) or only certain leaves taken at intervals; by the latter method the

44. Endive blanching under paper covers.

plant may be kept going most of the season. It is known mostly as a salad plant with us; but it is an excellent potherb, the greener or younger plants often being taken for this purpose. Endive is little known to people of American parentage, although it is much prized by foreigners, and there is considerable demand for it in the larger cities. It deserves to be better understood.

The green rank leaves are likely to be bitter and tough. It is customary to blanch the interior leaves of the crown or head by gathering all the leaves into a bunch and tying them near the top. This tying is performed two or three weeks before the plant is desired for use. In very hot and wet weather the heads are sometimes blanched in ten days; but under ordinary conditions it requires nearly or quite twice that length of time. If heavy rains and cloudy weather follow the tying, the crowns must be examined frequently to see that they are not decaying. After the interior leaves are well blanched, they must be used quickly or decay will set in; they should be dry when tied. The later plants, taken up in autumn, are sometimes blanched by being set in cellars or pits or coldframes; or if the heads are packed securely in well-ventilated barrels, they may blanch in transportation.

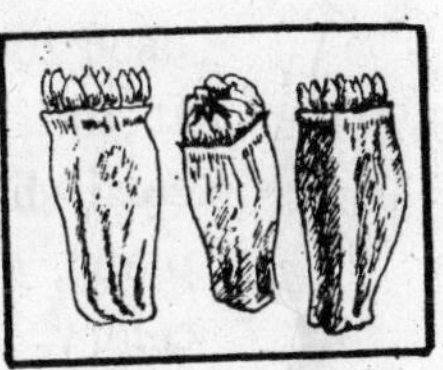

45. Seeds (fruits) of endive (about 4½).

On the blanching of endive, S. N. Green comments as follows (Mo. Bull. Ohio Exp. Sta. No. 32), with particular reference to treatment of the crop grown under glass (Fig. 44): "The blanching of the greenhouse grown crop is the most difficult part of the culture of endive under glass, and as yet no perfectly satisfactory method has been found. In the field, especially in the cool fall months, blanching is not difficult. Any sort of a covering that will

46. Seedlings of endive (× ½).

exclude the light seems to answer the purpose. Planks, mats or straw are commonly used. In other cases, each individual plant is tied up, the outer leaves being drawn towards the center and a rubber or string being used to keep them in place. Loss is apt to occur if the leaves are too closely compacted, and the cord should be somewhat below the center of the plant, allowing the blanching to proceed as with celery. In the greenhouse, where the soil is damp, the air moist and the ventilation slow when blanching by any method, loss by rot is sure to occur unless much precaution is taken. Careful ventilation and temperature regulation are necessary. In a general way, the lower the temperature the slower the blanching process, from 2 to 3 weeks or more being necessary. We have found for our conditions that a paper-covered frame gives satisfactory results. This excludes the light, allows fair circulation of air and there is little loss from rot." The method may provide a suggestion for other than glasshouse conditions.

47. Witloof (× 3/8).

48. Seeds (fruits) of chicory (× 4).

The achnese or "seeds," and the young plants coming from them, are seen in Figs. 45 and 46; it is interesting to compare them with chicory (in Figs. 48 and 49).

Chicory

Chicory is very closely related to endive, but the leaves (for salads and greens) are mostly desired in winter or spring from roots that have been grown for the purpose and taken up on the approach of cold weather. The effort is to grow strong roots and to have them in prime condition at the end of the growing season. The culture is simple, as for carrots or parsnips.

While the culture of chicory (or succory, an old name) is easy, the grower must know for what purpose he is to rear the plant. The purposes may be four: (1) to obtain the green leaves to be used as potherbs; (2) to produce barbe-de-capucin ("friar's beard") and witloof, which are the colorless leaves arising from stored roots; (3) to secure the young green roots themselves, of certain varieties, for cooking and eating, a use very little known with us; (4) to raise roots to dry for the making of a substitute for coffee. The last category does not come within the scope of this book. Only the first two uses may be considered here.

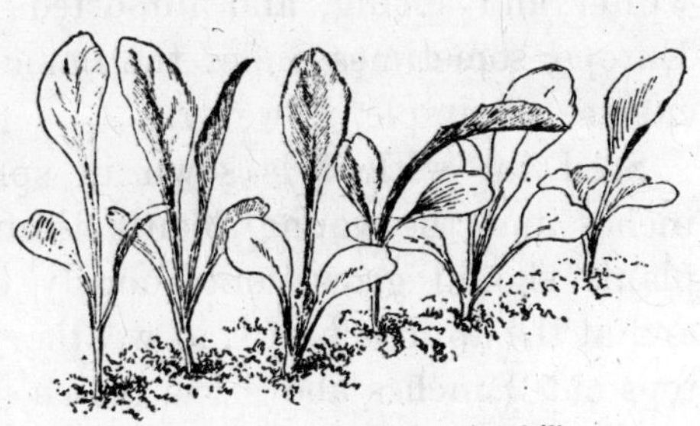

49. Chicory seedlings (× 2/3).

The roots are grown as are parsnips or carrots, and harvestings of leaves may be made throughout the growing season. One may also leave the roots in the ground over winter and gather the crown of leaves in the spring, or one may take them to the cellar or greenhouse and secure the leaves in winter. It is usually preferable to grow a new lot of plants each year.

For the production of blanched leaves, the strong roots are usually taken up in autumn. The roots are buried in a sloping direction in sand in pit or cellar, the crown projecting an inch or so above the earth. The place should be kept dark. In a month or less, the small leaves are produced.

Witloof and barbe (barbe-de-capucin) are different forms of white forced chicory tops. Witloof ("white leaf") is a compacter head than barbe, being raised from a strain or variety of roots grown for the purpose; the looser and more leafy tuft or head of barbe may be produced from ordinary chicory roots. The culture and handling of the two products are essentially the same. Witloof is a delicate slightly bitter salad, much prized in winter and spring, and imported into this country from Europe, sometimes under the name "French endive." Its culture is simple, however.

Seed for witloof is sown in spring in rows about 18 inches and the young plants thinned to 6 inches. The plants should grow continuously throughout the season; and at the approach of cold weather the roots are lifted, the tops cut 2 inches above the crown, and the roots stored in a cellar, so that they will remain dormant till needed. When the forcing is begun, the roots are trimmed on the bottom so that they are 8 or 9 inches long; the roots are set upright in sand or soil in boxes or beds, being very close together; about 8 inches of clean sand are placed over the crowns; the tops soon begin to push through if a temperature of 55° to 60° is maintained and sufficient moisture is provided; in two weeks the cone of leaves should be ready for the table. A good head or cone is about 6 inches long (Fig. 47).

The Endive and Chicory Plants

Cichorium. *Compositæ.* Annual, biennial and perennial

herbs, of probably a half dozen species, in Europe and Africa, of which two are cultivated.

C. Endivia, Linn. Sp. Pl. 813. ENDIVE. Annual, perhaps also biennial, usually with a strong taproot, forming a cluster or rosette of brittle edible foliage, juice milky; stem 2 to 3 ft. tall, very leafy, loosely long-hairy (particularly on line beneath the leaves), branching, the branches soft and often more or less fasciated: leaves oblong, obovate-oblong or ovate-oblong in outline, narrowed to the base to a short winged petiole, 8 to 12 in. long and 3 to 5 in. broad, sometimes sparsely hairy on the midrib beneath, in cultivated forms deeply sinuate many-lobed and crisped, the lobes sometimes 1 in. broad and in other forms multifid and almost thread-like; stem leaves similar but successively smaller, alternate, passing into lanceolate broad-based clasping bracts: flower-heads axillary and others terminating short or long branches, about 12- to 16-flowered, florets perfect and purple-rayed; receptacle naked; involucre short-cylindric, scales in about two rows of which the inner are lanceolate-subacute and erect and hyaline-margined and the outer ones leafy and broad, spreading or recurved and ciliate-margined, the head often subtended by two short-spreading obtuse ciliate bracts; ovary obconic, bearing at its top a rim of pappus-scales like a scalloped edging inside which arises the hairy corolla-tube; style-branches purple, long and curving backward or coiled: achene ("seed") oblong but enlarging toward the top, 3 to 4 mm. (about ⅛ in.) long, angled and ribbed, glabrous, carrying the scalloped pappus-crown which may be broken or wanting in the commercial seed and which is one-sixth to one-eighth the total length of the achene and crown, the achene weighing 1½ to 2 mg. and retaining its vitality 8 to 10 years.—Probably Asian but by some botanists supposed to be a culture-form of *C. pumilum*, Jacq. (*C. divaricatum*, Schousb.) of the Mediterranean region and by others of *C. Intybus*, the chicory.

C. Intybus, Linn. Sp. Pl. 813. CHICORY. Perennial with hard long taproot which is much thickened in some of the cultivated races, taller, stiffer and more virgate than *C. En-*

divia, the stem and branches much less leafy, stem loosely hairy below but glabrous or nearly so toward the top, the hard elongated branches practically leafless: leaves various, the lower ones mostly oblong-oblanceolate, 6 to 15 in. long and 2 to 3 in. broad, obtuse or very short-acute, narrowed to a clasping bracts: flower-head sessile in clusters and also ter-above and beneath, the margins nearly or quite entire to sinuate-dentate or runcinate or jagged, the upper leaves of similar character but smaller and passing into ovate-lanceolate clasping bracts: flower-head sessile in clusters and also terminal on peduncles, much like those of *C. Endivia* but larger (½ to 2 in. across in full bloom), blue and sometimes pink or white, the involucre scales narrower and the outer ones much less foliaceous, the achenes more ribbed and with a shorter pappus-border.—Europe, and now extensively run wild, making one of the finest displays of blue in the forenoon when the blossoms are open; the flowers do not expand a second time. Some of the frizzled cultivated forms known as chicories may belong more closely to *C. Endivia*.

The Latin name of chicory is *intubus* (or *intybus*); the word *endivia* is probably derived from it. Endive and chicory are both of relatively recent domestication.

CRESS

Cresses are grown for their piquant leaves, which are used in salads and garnishings. Two kinds are in common cultivation, members of the Cruciferæ or Mustard Family. They may be considered separately, as they differ in cultivation. Other plants known as cress need not be discussed here, as they are little grown for food in North America.

Garden Cress

The garden cress is a short-season annual, a cool-weather plant, grown for its root leaves. Usually the leaves are not desired in summer. Seeds may be sown as soon as the ground is fit in spring, for the plant is hardy or half-hardy; they germinate very rapidly. A rather cool and rich soil is to be chosen, for the value of the foliage will depend, to a large extent, on the vigor of its growth. Late in the season and in warm weather the plant runs quickly to seed. For autumn use, the seeds may be sown late in summer and in early fall. It is easily grown in pots or boxes in the house in winter. Cress is sown in rows a foot apart, and thinned as it grows.

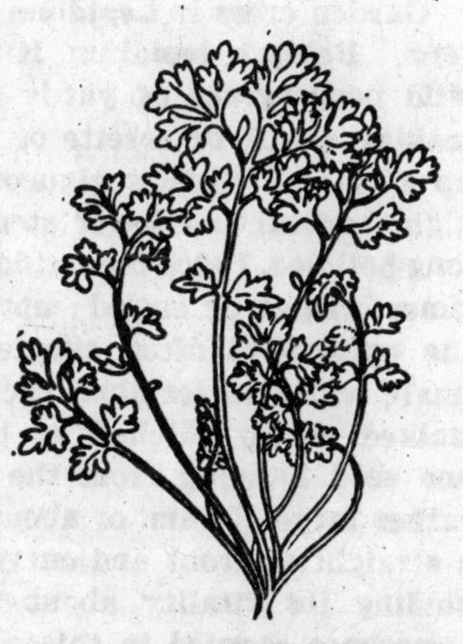

50. Young growth of garden cress (× ⅓).

Leaves fit for use may be had in six to eight weeks from the sowing of the seed, under ordinary conditions. If leaves are removed carefully, the plant continues to bear.

51. Seeds of garden cress (× 6).

52. Seedlings of cress (× ½).

Sowings should be frequent, to provide succession. There are a number of varieties, some of them with beautifully curled foliage. The garden cress is less popular in

America than abroad. Figs. 50 to 52 show the garden cress.

Garden cress is Lepidium sativum, Linn. Sp. Pl. 644. *Cruciferæ.* Being a lepidium, it is therefore closely related to the wild peppergrass of yards and waste places. Plant annual, making a tuft or rosette of leaves used in salad, soon sending up a smooth slightly glaucous erect branching stem 1 to 2 ft. high: radical and lower stem leaves oval or oblong in outline, long-petioled, twice pinnatifid into narrow lobed divisions, sometimes crisped or curled; upper leaves once pinnate or ternate, the uppermost often simple, long-linear and entire: flowers small, white, in terminal racemes: pod a flattened broadly oval stalked silicle notched at the top, about 6 mm. long, with one seed hanging from the top in each of the 2 cells: seed rather large (2 mm. or about $\frac{1}{12}$ in. long), smooth, brown, with a straightish front and curved back, weighing about 2 mg. and holding its vitality about 5 years.—Native in Europe, and sometimes escaped in this country.

The winter and spring cress, of the cruciferous genus Barbarea, is rarely grown. Upland cress grown by the writer many years ago, from American commercial seed, was Barbarea; recently he has planted seed under this name, and it is *Lepidium sativum.* The spring cress of cultivation is probably mostly *Barbarea verna,* Aschers. (*B. præcox,* R. Br.). It is usually biennial, the young plants becoming established from seeds dropped in summer, and sending up the flower-stalks early the following spring. In cultivation, it is treated as an annual or as a winter perennial. The seeds may be sown late in the season and the young plants are ready for use the next spring; or seeds may be sown in earliest spring. The plant is perfectly hardy.

Water-cress

Water-cress is a prostrate perennial, rooting at the joints, with small roundish leaves, thriving in very moist places and in running water. It is readily propagated by seeds, which may be scattered along cool brooks, and by bits of the stems planted in the earth. In order that it may reach its best development, the water should be pure, cool, and clean. When once established in a permanent place, it will persist indefinitely, taking care of itself. When a natural brook is not to be had, it may be grown in a moist shady place in the garden where it may be watered frequently. Sometimes it is grown in the pit of an abandoned hotbed, into which water may be run with a hose. If the ground is kept moist, or even wet, the plant will thrive and it will not be necessary to have it covered with water. The plant is best grown, however, by being colonized along brooksides and about springs. If the colonies are picked or harvested very closely, the plants will suffer.

WATER-CRESS SOWBUG (*Mancasellus brachyurus*).—A grayish aquatic species of sowbug more or less shrimp-like in form that attacks the submerged portions of the plants, cutting the roots and stems. *Control:* There is no practicable method of controlling sowbugs in natural streams and ponds. Some growers, however, have been able to overcome the difficulty by growing the plants in broad shallow beds sloping towards the center, where a trough ten inches square lined with boards extends the whole length of the bed. When the sowbugs become abundant, the water is shut off for twelve to twenty-four hours, allowing the beds to drain. Water is retained in the trough, in which the sowbugs soon accumulate in great numbers. They may be destroyed by the addition of a liberal quantity of copper sulfate solution. Less injury will result if water is drained off soon after the cress has been gathered.

For our purpose we may use for water-cress the botanical name *Roripa Nasturtium-aquaticum*, Hayek. (*Sisymbrium Nasturtium-aquaticum*, Linn.). Thereby we come upon the most complex situation in nomenclature in any of the common garden vegetables. It is well to state the case briefly in outline, that the student may comprehend the nature of these tangles. The question is involved with the botany of the plant and also with the application of current rules of nomenclature. The primary problem is whether the water-cress should be associated with other plants in a more or less composite genus, or whether it should be separated wholly or largely by itself. In some respects it is unlike the plants with which it has been associated, by Linnæus himself in Sisymbrium and by subsequent authors in Nasturtium, Radicula and Roripa. If it is separated, the question of the generic name to be adopted is not simple. In the necessary dismemberment of the Linnæan genus Sisymbrium, it would seem that the water-cress should go into another genus inasmuch as it apparently does not typify the genus Sisymbrium as Linnæus intended it. The plant happens to be the first species described by Linnæus under Sisymbrium, however, and for this reason certain authors hold it in that genus as Linnæus has it, *S. Nasturtium-aquaticum.* In this disposition, Sisymbrium may be regarded as a monotypic genus, the water-cress being the only species. This arbitrary resolution of the case is not commonly followed. If another genus is desired for it, recourse may be had to Cardaminum of Moench, 1794, or to Baeumerta, Gærtner, Meyer & Scherbius, 1800, both names being proposed exclusively for the water-cress. Radicula of Hill, 1756, Roripa of Scopoli, 1730, and Nasturtium of Robert Brown, 1812, are proposed for multiple segregates from Sisymbrium and in them the water-cress has found lodgement. The plant is commonly known in the trade as *Nasturtium officinale*, but this name cannot hold under any interpretation, as in present practice the Linnæan specific name, *Nasturtium-aquaticum*, must be used with the generic name. The report of the International Botanical Congress of Brussels, 1910, recommends the retention of Nasturtium for the water-cress, rather than the older generic names Cardaminum and Baeumerta, on the assumption that the changes would be fewer or at least that the situation would be better understood; but under Nasturtium the proper combination of names apparently has not been made, any more than it has under Cardaminum; in either case, therefore, a new name must result if the plant is removed singly from Sisymbrium. If the plant is to be associated with others in a genus, the clearest destination seems to be in Roripa (for reasons not necessary here to explain), and the plant is so disposed of in this book. The synonymy may be displayed as follows: **Roripa Nasturtium-aquaticum**, Hayek, Sched. Fl. Stir. Exsicc., 3.14 lief. (Dec., 1905) 22. *Sisymbrium Nasturtium-aquaticum*, Linn. Sp. Pl. 657. *Cardaminum Nasturtium*, Moench, Meth. 262. 1794. *Baeumerta Nasturtium*, Gærtn. Mey. & Scherb. Fl. Wett. ii, 467. 1800. *Nasturtium officinale*, R. Br. in Ait. Hort. Kew. ed. 2, iv, 110. 1812. *Nasturtium aquaticum*, Wahl. Svensk. Bot. t. 624. 1823–5. *Cardamine Nasturtium*, O. Kuntze, Rev. Gen. i, 22. 1891. *Roripa Nasturtium*, Beck. Fl. N. Œst. ii, 463. 1892. *Radicula Nasturtium-aquaticum*, Britt. & Rendle, Brit. Seed Pl., 3. 1907. *Baeumerta Nasturtium-aquaticum*, Hayek, Fl. Steierm, i, 498. 1909. Perennial, creeping or floating, smooth, emitting long white roots at the modes: leaves odd-pinnately compound, of 1 to 4 lateral pairs; terminal lobe oblong to orbiculate, entire, undulate or obscurely toothed; lateral leaflets usually much smaller: flowers white, in very short terminal racemes that elongate in fruit, small: fruit a curved linear long-pedicelled pod: seeds small (about 1 mm. across), brown, oblong-orbicular, tuberculate, weighing less than 1 mg., and holding vitality about 5 years.—Europe; widely naturalized in this country in ditches, rills and pools.

CORN-SALAD

Corn-salad or fetticus is used both as salad or potherb, chiefly the former, the thick bunch or rosette of root leaves being employed for the purpose. It is a hardy cool-season plant, of easy culture except in hot weather.

It may be grown as a mid-spring crop from seed sown the same season; as a fall crop from seeds sown in late summer or early autumn; as a very early spring crop from plants allowed to stand over winter.

For the mid-spring crop, corn-salad should be sown as soon as the land can be fitted. It quickly runs to seed in hot and dry weather. Plants should stand about 6 inches apart in the row. An ounce of seed should yield 2,000 to 3,000 plants. The plant matures in six to eight weeks, giving a bunch of leaves somewhat like small-leaved spinach.

For the late or main supply the seeds may be sown, at the North, in the latter part of August or early part of September. It will provide edible herbage late in the season, and in a mild climate or open winter it will survive and yield acceptable crop in early spring; or it may be protected over winter by leaves or straw, much as spinach is handled; it may be grown and carried over in frames.

Corn-salad is the cultivated form of **Valerianella Locusta**, Betcke, Anim. Bot. Valer. 10. 1826. *Valerianaceæ.* It is commonly known in horticultural literature as *V. olitoria*, Poll. Hist. Pl. Palat. i, 30. 1776. (*Valeriana Locusta* var. *olitoria*, Linn. Sp. Pl. 33.) It is a small glabrous annual, native in Europe, where it grows among the corn (grain), whence the name, "corn-salad"; it is run wild to some extent in North America: plant making a tuft or mat of oblanceolate or oblong obtuse root leaves 2 to 3 in. long, which are entire or toothed; stem leaves similar, successively smaller, opposite, sessile, some of them narrowed to the base: stem 1 ft. or less high, at length much branched, bearing very small light blue 5-lobed flowers in dense heads terminal on forking branches: fruit ("seed") nearly orbicular but with a short 2-pointed beak, somewhat flattened sidewise, $\frac{1}{16}$ in. long, light brown, furrowed up the middle, where 1 lenticular seed is

borne, the fruit weighing ½ to 1 mg.; vitality about 5 years. Figs. 53 and 54 show the seeds (properly fruits) and seedlings.

PARSLEY

In this country, parsley is the most popular of the garnishing herbs. The leaves are used also for salads and for flavoring. The plant is biennial, but the foliage is gathered the first year, and the plants are then destroyed unless seed is wanted.

53. Seeds (fruits) of corn-salad (× 7).

The seed is slow to germinate, and it is best to sow in a seed-bed unless the ground is in excellent tilth and is moist to the top. Some growers soak the seeds before sowing, in tepid water. Thin or transplant to 8 to 12 inches apart each way. Make successive sowings. It usually requires three months from sowing to bring good foliage for gathering. The strongest established plants may be covered with sash, and leaves may then be gathered all winter. The plants will stand considerable frost. It is a good plan to lift a few roots in late fall and set them in pots or boxes in the house; from these a winter supply may be secured.

54. Seedlings of corn salad or fetticus (× about 2/3).

For market the leaves are tied in small attractive bunches. The various forms of curled parsley are most popular, although the plain-leaved is as good. Parsley fruits and seedlings are shown in the figures (Figs. 55, 56).

55. Seeds (fruits) of parsley (× 4).

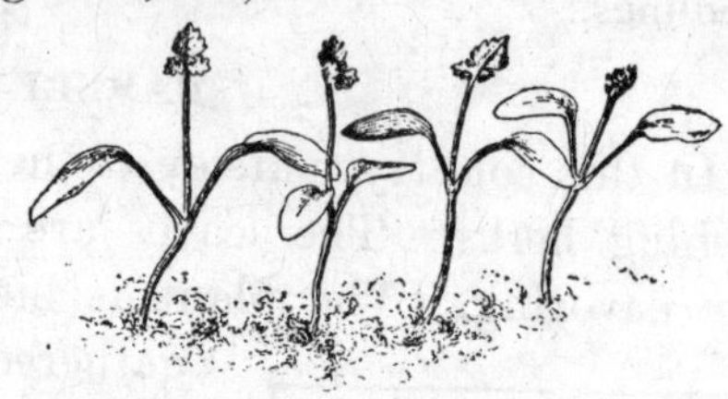

56. Parsley seedlings (× ½).

Parsley is one of the Umbelliferæ, **Petroselinum hortense**, Hoffm. Gen. Umb. 163. 1814, known also as *Apium Petroselinum*, Linn. Sp. Pl. 264, and *Petroselinum sativum*, Hoffm. Gen. Umb. 177. It is native in the Mediterranean region, but is sometimes escaped in this country: glabrous biennial or short-lived perennial, making many radical leaves which are prized in cookery and for garnishing: stem 18 to 30 in. high, much branched: leaves ternately decompound, the ultimate leaflets wedge-ovate, deeply cut and petioled: flowers small, greenish yellow, in compound umbels: fruit ("seed") one of the two separated carpels, oblong-convex with one style curving backward from the top like a little hook (often broken in commercial seeds), ribbed on each edge, 3-ribbed on the back, about ⅛ in. long and weighing 1 to 2 mg.; vitality 3 years. Var. **crispum** (*P. sativum* var. *crispum*, DC., Prodr. iv, 102. 1830) has leaves cut, curled and crisped. In the Moss-curled parsley the leaves are very finely divided and somewhat bunched. Var. **radicosum** (*P. sativum* var. *radicosum*, Alef. Landw. Fl. 153. 1866) is the turnip-rooted parsley, grown for the thick parsnip-like tapering root.

SALAD CHERVIL

The salad chervil is an annual plant much like parsley, popular in Europe, but little known in this country. It is used for garnishing and seasoning, for which the curled-leafed variety is the most prized (Figs. 57 and 58).

The plant is of easy culture, giving a cutting of leaves in six to eight weeks from the seed. It does not thrive in hot dry summers, and therefore should be grown as a spring or fall crop, unless the particular location is cool, as in partial shade or with a northward exposure. It is hardy, and where winters are not severe can be carried over the cold season by light coldframes or even by protection of brush. The plant reaches

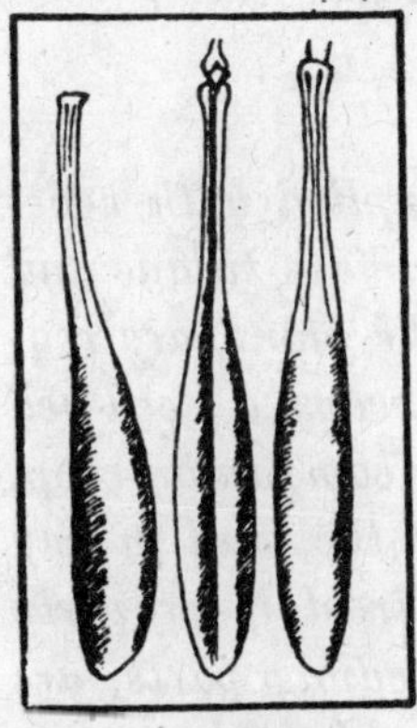
57. The long fruits ("seeds") of salad chervil (× 5).

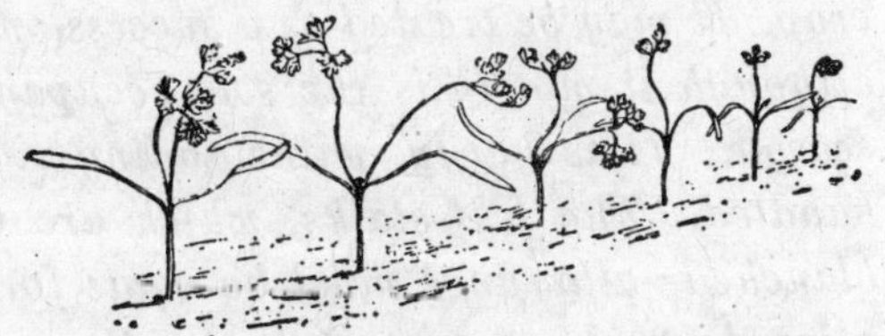
58. The slender seedlings of chervil (× 2/3).

a height of nearly two feet when mature, but the young foliage is most desired. The plants should stand 8 to 12 inches apart. For turnip-rooted chervil, which is another plant, see Chapter VII.

Salad chervil is **Anthriscus Cerefolium**, Hoffm., Gen. Umb. 41. 1814. *Umbelliferæ.* It is known in literature also as *Scandix Cerefolium*, Linn. Sp. Pl. 257; *Chærophyllum sativum*, Lam. Encyl. i, 684. 1783; *Cerefolium Cerefolium*, Britton, Ill. Fl. ed. 2, ii, 629. 1913. It is a fine-leaved soft annual of quick growth, native in Europe and sometimes run wild in North America: stem erect, branching, 1 to 2 ft., glabrous: radical and main stem leaves ternately decompound, the ultimate leaflets small (½ to ¾ in. long); ovate to orbicular and deeply cut: flowers white, minute, in compound umbels: fruit ("seed" of commerce) representing 1 of the 2 carpels broken apart, linear, with the very slender often whitish beak about 3/8 in. long (beak sometimes broken off in commercial samples), black, smooth, grooved on inner face, weighing about 1 to 1½ mg.; vitality 1 to 3 years. (The word *Cerefolium* is an old substantive signifying "pleasant leaf.")

CELERY

Cool, very rich and moist land well supplied with vegetable matter, deep preparation, the best surface tillage and the most careful attention to all care of the plant, are requisites of good celery culture. It is always a seed-bed crop. It may be treated as a succession- or companion-crop, although it mostly is the sole occupant of the land in any season. It is hardy, withstanding light frost if properly handled. The leaf-stalks, which are the edible parts, are blanched; allowance must be made for the blanching operation by wide spacing between the rows. The crop must be stored from freezing if kept in winter.

Celery is planted 6 to 12 inches apart in the row. The rows vary from 2 to even 6 feet. Sometimes the rows are double, the two being 6 in. apart. In the self-blanching system, the plants are set 6 to 8 or 10 inches each way; at 7 x 8 in., about 112,000 plants are required to the full acre. There is usually much loss in seeds and young plants, and it is therefore advisable to sow the seed very thick. One ounce of seed to 200 feet of row in the seed-bed is a liberal allowance. Some gardeners estimate 2,000 good plants from each ounce of seed, but this allows for an unusual amount of loss. An ounce should give 5,000 to 10,000 good plants, after allowing for several times that amount in loss. One pound of celery seed should give enough strong plants to set four to five acres. In single-row planting 6 in. apart, and the rows 5 ft., as in earth-banking, more than 17,000 plants stand on a full acre. The yield from an acre should be 400 to 600 dozen bunches of 3 or more stalks each, if the distance of planting is 3 ft. by 6 in.

LATE BLIGHT OF CELERY (*Septoria petroselini*).—Irregular brownish spots containing small black fruiting bodies are produced on leaves and leaf-stalks. Frequently the spots unite, causing the entire leaf to become dry and papery. BACTERIAL BLIGHT.—Lesions differ from those of late blight in that the spots are smaller, more regular in outline, darker brown in color and have no black fruiting bodies scattered over them. CERCOSPORA BLIGHT (*Cercospora apii*).—Characteristic ashen gray spots develop. Later the entire leaf may become somewhat yellowed and ashen gray and velvety. *Control:* All of the above blights are controlled by thorough spraying with bordeaux mixture 5-5-50. If the disease is present in the seed-bed one or more applications there is advisable. Field spraying with two nozzles to the row, the two being placed some distance apart and at such an angle that the two sprays overlap at the top of the row and thoroughly cover the sides, is advisable. Applications should begin about three weeks after transplanting and should continue at about weekly intervals, depending somewhat on weather conditions, till blanching time.

STORAGE ROT (*Sclerotinia libertiana*).—Frequently plants of celery in storage become water-soaked in appearance, and on this softened tissue white felts of mycelium containing hard black fungous bodies develop. *Control:* The introduction of wounded plants or those showing beginning of decay, is to be avoided. The maintenance of proper storage conditions is imperative.

CARROT RUST-FLY (*Psila rosæ*).—A slender straw-colored maggot, $\frac{3}{10}$ in. long when mature, that eats off and destroys the fibrous roots of young celery plants. A second brood appears in late summer and bores in the taproot. No practicable method of control is known.

BLACK SWALLOW-TAIL BUTTERFLY (*Papilio polyxenes*).—A beautiful green caterpillar about 2 in. long, each segment with a black band near the front margin enclosing six yellow spots; feeds on the leaves of celery and is most destructive to young plants. *Control:* Hand-picking is the most dependable remedy.

CELERY LOOPER (*Autographa falcigera*).—A looping caterpillar about 1¼ in. long, pale translucent green with a dark median line bordered on each side with three light lines. It sometimes feeds on the leaves of celery. *Control:* Hand-picking is the only available measure, as arsenicals cannot be used on celery.

TARNISHED PLANT-BUG (*Lygus pratensis*).—A small inconspicuous brownish bug, about ¼ in. long. The adults often attack celery plants that are blanching, puncture the tender stalks, producing large brown wilted spots and a blackening of the tissues at the joints. No satisfactory method of control is known.

THE NEGRO BUG (*Thyreocoris pulicarius*).—A short, broad, shining black strongly convex stink-bug about $\frac{1}{10}$ in. long, that often attacks celery, puncturing the stalks and stunting or killing them. The injury to celery is mostly done by the adults which have bred on various weeds such as beggarticks, tick-seed, etc. *Control:* Destroy all weeds in the vicinity of celery on which the bugs may breed. Spraying is not effective since many of the bugs burrow in the soil where they cannot be reached.

PARSNIP WEBWORM (*Depresaria heracliana*).—See under Parsnip.

SPINACH APHIS (*Myzus persicæ*).—See under Spinach.

Celery is practically a universal table supply in North America, prized for its crisp aromatic leaf-stalks, as well as for the decorative character of the finer parts of the foliage. The seed is sometimes used in cookery for flavoring, particularly in the preparation of soups. The whitened leaf-stalks are usually eaten raw, but they are also cooked in different ways.

Celery is commonly grown on bottom lands because it then receives a sufficient and constant supply of moisture. Usually, also, such lands are very fertile. Celery of excellent quality can be grown on uplands; but ordinarily more care is required in securing deep tillage and in conserving moisture, and more expense is entailed in adding fertilizers. Successful commercial celery growing on high lands is usually possible only when much stable manure is added and when irrigation is practiced; the overhead method of irrigating is well adapted to the crop. Under those conditions, however, the celery grown on high lands may be fully as good as that raised in reclaimed marshes. Level black-soil marsh or bottom lands, in which the water-table does not fall below 2 or 3 feet in summer, are usually chosen for commercial celery growing. In all celery growing, every effort must be made to conserve the moisture. Furrow irrigation may be employed where rainfall is deficient.

For home use celery can be grown in any well-tilled and rich garden soil. Home gardeners are often specially successful with it in city and village lots. Under such circumstances, particular attention can be given to trenching or other deep preparation of the land and to consistent care from first to last. Well-rotted stable manure may be used freely.

Field management.

Celery is grown as a short-season crop; that is, it may not occupy the land the whole growing season. The main crop is sometimes planted as a succession, early cabbages or other spring crops having been grown on the land. In the case of lowland fields, however, the celery crop is commonly the only one grown, since the land is usually too wet in the spring to allow of any early planting. In some celery-growing regions, two or three crops of celery are raised on the land at the same time, the later or main crop being planted between the rows of the early crop. The main or late crop, which is used for winter consumption, may be planted in the field as late as the middle or last of July in the Northern States. The early crop may be set in the field as soon as the weather is settled in spring, but there is relatively small demand for very early celery. The young plants should not be subjected to hard frosts.

Commercial fertilizers are used to supplement liberal supplies of stable manure. When the manure cannot be obtained, such fertilizers may be used to supplement the humus supplied by good rotation or change of land. Compounds rich in nitrogen are usually advised. In fact, nitrate of soda alone is used, in several applications, as much as 150 or 200 pounds each time. The rich bottom lands, however, may not require such supplements.

"Celery luxuriates in a soil rich in vegetable matter," writes Voorhees (Fertilizers, rev. ed. 295). "A heavy application of the basic mixture (page 383)—a ton to the acre, used at time of setting the plants—may be followed with advantage by frequent and reasonably heavy top-dressings of nitrate of soda, 100 pounds to the acre or more, and well worked into the soil."

Ordinarily, frequent level tillage is practiced until the plants are ready for the hilling or other blanching process. Some growers, however, prefer to mulch the land heavily enough to retain moisture and keep down weeds. Stable manure a few inches deep is one of the best covers, but straw and other materials are also employed. The manure should be kept from direct contact with the plants.

59. Celery "seeds" (× 10).

Celery is always a transplanted crop. The seeds are small (Fig. 59) and slow to germinate; and the seedlings are delicate (Fig. 60). It is only in a well-prepared seed-bed that satisfactory results can be expected. This seed-bed should have perfect surface tilth and retain moisture to the top. Preferably, it should be protected from hot and dry winds. Some persons prefer to have the bed partially shaded; but if the shading is too dense, the plants are likely to be soft and tender when taken to the field, and they are killed by sun-scald. It is advisable, whenever possible, to have the seed-bed in such place that it can be watered every evening if necessary; but care must be exercised that the watering is not so heavy that it packs and puddles the earth. Sometimes the bed is covered with boards, brush or straw, to maintain the moisture until germination has taken place. This may be advisable, but if the covering is left on too long, the plants make a very weak and spindling growth and are worthless. If covering is used, it is well to remove it gradually as the plants germinate.

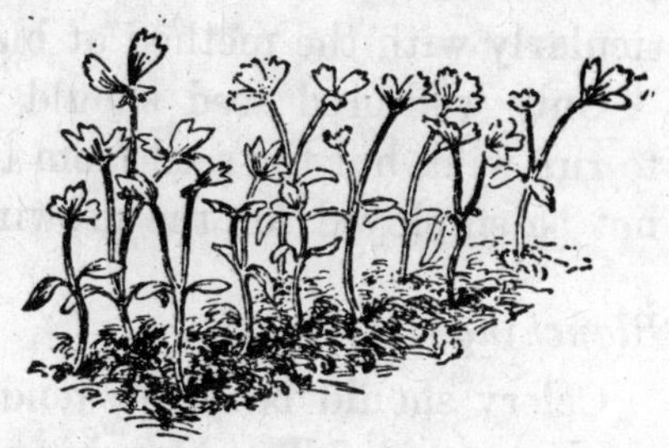
60. Celery seedlings (× 2/3).

The perfect seed-bed, however, is one that does not need a cover, but which holds the moisture of itself.

The early crop is commonly started under glass at the North, transplanted to the open in six weeks to two months. Plants for late crop are started in seed-bed in the open.

To secure stocky plants, they should be transplanted once or twice in the seed-bed, or they may be thinned until they finally stand at 2 or 3 inches apart. The labor of transplanting is so great that growers of large areas prefer to secure stocky plants by the thinning process and then by shearing off the remaining plants when they become too tall. The plants may be cut back a third their growth by shears or sickle, or on large beds with a scythe or mowing-machine. Transplanting is preferable whenever it can be managed.

The plants should be 4 or 5 inches high and stocky and dark green when they are planted in the field. Plants are usually set 6 to 12 inches in the rows, and the distance between the rows varies with the price of land and particularly with the method of blanching.

Only well-bred seed should be sown. The plant tends to run wild, but the seed from this depreciated stock should not be employed for the growing of a crop.

Blanching.

Celery should be crisp, tender and well blanched to be used as salad. The blanching is accomplished by excluding the light. There are four common methods of blanching celery in vogue at present: by the use of boards or paper; banking up with earth; close planting; blanching in pits or storage. It may be said that green unblanched celery may be used for cooking, and in some countries the plant is not blanched to the extent to which it is known in North America.

Blanching by means of boards is employed for the early or summer celery, because protection from frost must be supplied to the celery that remains in the field after the first of October, and the boards usually do not afford sufficient protection; and the early self-blanching varieties are likely to decay or at least not to stand so long if banked with earth. Boards one foot wide and one inch thick and about 12 or 14 feet long are used. If the boards are much longer than this, they are awkward to handle. These boards are set on edge close against the crown of the plant, one on either side of the row, and the tops are tipped together until they are only two or three inches apart or until they rest against the plants. The boards are held in this position by cleats nailed across the top, or by wire hooks. The first "boarding" is made when the celery is only tall enough to show a few of its leaves above the boards. The plants shoot up for light, making slender soft stalks. The foliage fills the space between the boards and excludes the light from above. In ten to twenty days in warm "growing" weather, the celery may be blanched by this method. In any means of blanching in summer one must see that the plants do not rot at the heart, as they are likely to do if they are too wet at the core. The board method of blanching celery is one of the most economical and is now extensively used in the large celery fields. Growers usually find that it pays to obtain a good quality of lumber and to use it year after year. Some commercial growers think it best to have the lumber dressed on both sides. In the boarding system the rows may be put only far enough apart to allow of good horse tillage, say from 2 to 3 feet.

Paper is sometimes employed rather than boards. Rolls of building paper are sawn across to make strips one foot wide. The strip is then unrolled against the row and held in place by means of stakes. Good paper well taken care of should last for two or three crops.

A different use of paper is to wrap and tie each plant in stiff strong manila or similar stock. Of course this is adapted only to small areas. Large tiles are sometimes set on the plants for the same purpose.

Blanching by earth usually gives a somewhat better quality of celery; but this method is expensive and it cannot be employed so well in midsummer, since the plants are more likely to rot at the heart. Usually two or three "handlings" or bankings are given. When the plants have spread so much as to make a crown or head a foot or eighteen inches across, the celery is "handled" by gathering the leaves in the hand and holding them whilst earth is shoveled against the plant so as to cover it two-thirds or more of its height. In ten days or two weeks the "handling" is repeated. In late years the banking of celery, particularly in large areas, is performed by

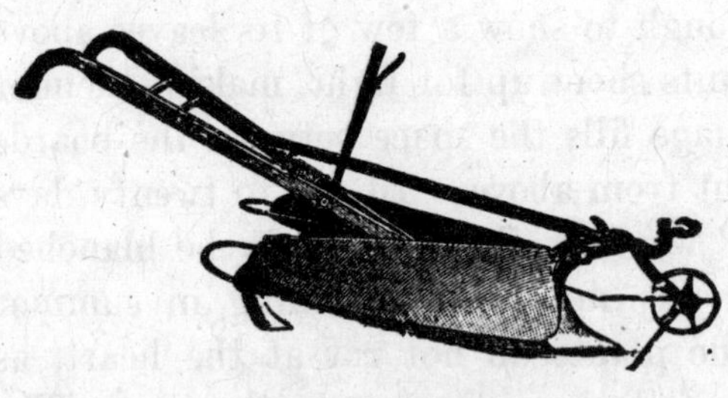

61. Celery plow.

means of celery plows, implements with very high mold-boards that throw a great quantity of earth against the plant (Fig. 61). If celery is to be blanched by the banking process, the rows are rarely less than 3½ feet apart, and if the tall-growing varieties are used, the rows are often put at 5 or even 6 feet. Double rows, 6 inches apart and the plants about 6 inches in the row, lend themselves well to earth banking, the space between these pairs of rows being 4 to 6 feet. In this case, of course, about twice the number of plants is required as in the single-row planting. The late or green (not self-blanching) varieties are grown for earth-banking.

Blanching by means of close planting was formerly known as the "new celery culture." This consists in growing the plants so close that the light is excluded and the plants blanch themselves. Plants are usually grown as close as 6 to 10 inches either way. It will be seen that this system can be used only when the soil is very rich and when there is abundant supply of moisture. Whenever the water-table is close to the surface or when one can practice irrigation, it may be considered. It is usually successful in small home gardens where one can use a hose. The self-blanching varieties are usually grown in the close-planting method.

Blanching in storage is the usual practice with late winter celery. If it is thoroughly blanched before putting in storage, it will not keep well. It is usually advisable, however, to "handle" the crop at least once in the field in order to induce a straight upright growth and to begin the blanching process. Thereafter the plants are set in pits or sheds so close together that the blanching proceeds.

Marketing; storing.

For market, celery is prepared by being thoroughly washed and usually scrubbed, so that all earth and sand are removed. The outside leaves are removed and usually the root is trimmed away, leaving a pointed base to the whole cluster, although the shape of the trimmed product differs between places. A few plants (3 to 8) are tied together to form attractive bunches. These plants are then shipped in crates or boxes, the style of box and the number to be packed in each depending largely on the market in which one sells. For high-class local markets the product is sometimes handled in attractive paper-lined baskets and hampers (Fig. 237). In all careful marketing the celery should be closely graded. The plant lends itself to such assortment.

The celery may be lifted from the field by means of a spade or shovel. In large plantations the plants are plowed out or removed by horse or power implements made for the purpose.

There are two or three methods of storing celery. Storing in outside cellars or pits is sometimes practiced. The early winter and midwinter celery, however, is usually stored in special celery houses, which are permanent sheds with windows at intervals along the roof, to supply light enough for the workmen. Wooden chimneys are provided to afford ventilation. These houses are sometimes supplied with heat by means of stoves, so that the temperature does not fall much, if any, below the freezing point. In beds in these houses the celery plants are set close together and the blanching proceeds during storage. Any celery house must be kept cool and moist. To avoid rot, only healthy sound plants should be stored, and the handling should be so thoughtful that the plants are not broken or bruised. The plants are often stored and blanched by removing them to deep spent hotbeds, which are well covered in winter.

62. Home storing of celery.

In the home garden celery is sometimes stored in trenches in the open, after the method suggested in Fig. 62 (adapted from R. W. De Baun, N. J. Extension Bulletin, 1917). A roof is placed lengthwise the line of plants, and it may be covered with straw or other material as the winter closes in. Care must be taken not to cover the tops too soon or too tight, as the plants quickly spoil if kept warm and close. This method is successful only on well-drained land.

The old method of celery culture grew the crop in a trench; in such case the blanching largely took care of itself and the covering of the row for winter was an operation of little labor. This method is still good for the skilful home gardener.

Varieties.

The varieties of celery may be ranged in the "self-blanching" and "green" classes. There are no sharp lines of demarcation between the two. The former are simply easy-blanching types developed largely by selection. Most of the celery is now of this kind. It is well adapted to blanching by means of boards. White Plume is an old favorite, but Golden Self-Blanching is now more popular. The green kinds, as Boston Market and Pascal, are usually slow blanchers, requiring banking or blanching in storage, and are of the class of good keepers.

The Celery Plant

Apium. *Umbelliferæ.* About 20 species, as the genus is usually accepted, of annual, biennial and perennial herbs widely distributed over the globe.

A. graveolens, Linn., var. **dulce,** DC. Prodr. iv, 101. 1830. (*A. dulce,* Mill. Gard. Dict. No. 5. 1768. *A. Celeri,* Gærtn. Fruct. i, t. 22, 1788.) Celery. Strong-smelling glabrous biennial (perhaps sometimes perennial): root leaves many and well developed, the petioles and rachises usually expanded: stems erect and branching, 2 to 3 ft. tall, many-grooved, with conspicuous joints: radical leaves pinnate, ovate to oblong in outline, the long petiole with an expanding base; leaflets usually two or three pairs and a terminal one, each one pinnately ternately compound and stalked, the lateral segments often again divided, the segments and divisions cuneate-ovate and more or less cut and coarsely toothed: blossoms very small, white, in small compound umbels among the leaves; first umbel sessile or nearly so and with subsequent long-stemmed umbels from the same joint, the involucels minute or wanting: flowers a dozen or more in each umbellet, on short rays or peduncles, the 5 broad petals incurved and surrounding the 5 anthers; calyx not evident: fruit ("seed" of gardeners) one of the two separable carpels, short-oblong with curved back and straight front, about 1 mm. (1/16 in.) long, smooth, brown, bearing three prominent ridges and two lesser ones on the front edge, weighing ¼ to 1 mg.; germinating longevity 5 to 8 years; sometimes the two carpels cohere in commercial samples, making a "seed" twice the bulk of the above weight; the short recurved styles, one to each carpel, are usually broken off in the commercial seed.—A plant of cultivation, grown from early times but not of ancient domestication and not greatly modified from the wild plant. The wild original, *A. graveolens*, Linn. (*Celeri graveolens*, Britt.) is wild in ditches and wet places in Europe and Asia, mostly near the sea. (The Latin word *graveolens* means "strong-smelling," whereas *dulce* is "sweet" or "pleasant," here designating the edible cultivated plant.)

Var. **rapaceum**, DC. Prodr. iv, 101. 1830 (*A. rapaceum*, Mill. Gard. Dict. No. 5. 1768). CELERIAC. A race producing a thickened turnip-like root (*rapum* is Latin for "turnip"), the leafstalks not developed. See page 193 for cultivation.

CHAPTER VI

BULB OR ONION CROPS

Onion	**Ciboule or Welsh onion**
Leek	**Shallot**
Garlic	**Chive**

All the bulb crops are hardy, require a cool season and moist rich soil with excellent surface tilth. Usually they are not seed-bed crops. They require little room and may be planted close. They are used both as main-season and secondary crops. They are propagated both by seeds and bulbs.

These crops are grown chiefly for the underground bulbs; but the leaves are often used in stews and seasonings. The onion is the only commercially important plant in the above group in this country. Garlic, leek and the others are known chiefly to citizens of foreign birth or to those who grow products for the large cities. The onion, however, is a major oleraceous crop, being grown under large field conditions as well as habitually in the home garden. These various vegetables are sometimes known as alliaceous plants, from the Latin *allium* or *alium,* the garlic; all of them belong to the genus Allium.

Seeds of these plants are grown by planting over-wintered bulbs in spring. The bulbs should be planted two or three inches deep, a few inches apart in the row. Seed-stalks soon arise, and the seeds are produced in heads on top. Some of the kinds, as garlic, seldom produce flowers and seeds.

ONION

Cool rather moist and level land, soil with the best possible surface condition and containing much quickly available plant-food, careful attention to the selection of seed, the most perfect shallow tillage, are some of the essentials in the growing of a good crop of onions. The commercial onion supply is grown from seeds, sown where the plants are to grow, the early table onions from bulbs of different kinds and to some extent from transplanted seedlings. All onions withstand considerable frost in their growing state. In the South, onions are grown as a winter crop.

Being cool-season plants, onions are sown or planted as early in the spring as the ground can be made ready. In mild climates, seed is sometimes sown in autumn. Onion seed is sown ½ in. to 1 in. deep. Sets, tops, and multipliers may be planted at intervals until steady warm spring weather comes. One ounce of seed is sown in about 150 feet of drill, and 3½ to 5 or even 6 pounds to the acre. Rows stand 12 to 16 or 18 in. apart, and the plants are thinned as they stand, so that the mature onions will not crowd. If the onions stand 3 x 14 inches, nearly 150,000 plants are required to the acre. A good crop of onions is 300 to 400 or 500 bushels to the acre, but 600 to 800 bushels are secured under the best conditions, and sometimes as much as 1,000 bushels.

ONION SMUT (*Urocystis cepulæ*).—Smut can be detected by the presence on leaves and bulbs of black pustules that rupture and expose a powdery black mass of spores. Only onions grown from seed are attacked, and these only in the very young stage. Affected plants gradually die throughout the season. *Control:* Formaldehyde solution made by adding

one pint of commercial formaldehyde to sixteen gallons of water should be applied in the furrow with the seed at the time of sowing at the rate of two hundred gallons to the acre. The application may be made in the open furrow just ahead of the coverers by means of a watering device attached to the drill. About a five-sixteenth inch flow of liquid from the tank should accomplish the full application. A properly equipped drill should discharge, when stationary, one gallon of the solution every fifty seconds.

Onion mildew (*Peronospora schleideniana*).—The disease may be recognized by the furry fungus coating on the outer surface of affected leaves. As the fungus develops, the plants yellow and finally die. The disease usually becomes evident at a few points in a field and rapidly spreads under favorable conditions of moisture. Partial recovery may occur in a dry period by the growth of new leaves, but under favorable conditions the disease will develop anew. *Control:* Burning of dead tops to prevent the over-wintering of the fungus in them and crop rotation to reduce infection from spores over-wintering in the field, are desirable. Tillage may aid the plants to outgrow the fungus. Spraying with bordeaux mixture to which has been added resin-fish oil soap is sometimes recommended; applications should begin before the disease has become established and will perhaps need to be repeated several times.

Onion thrips (*Thrips tabaci*).—Minute elongate yellowish insects, 1/25 inch long when mature, that attack the leaves, especially under the sheath at the base, causing them to turn whitish and giving the plants a dirty yellowish appearance. The tender leaves at the center become thickened, curled and deformed. Badly injured plants fall over on the ground. Most injurious in seasons of drought. *Control:* Spray early before the leaves turn down with "Black Leaf 40" tobacco extract, 1 pint in 100 gals. water in which 5 or 6 lbs. soap have been dissolved. Use the material liberally and direct the spray downward into the base of the leaves. Make three or four applications at intervals of four or five days.

ONION MAGGOT (*Phorbia ceparum*).—The parent fly lays her white elongate oval eggs on the plants near the base or in cracks and crevices of the soil. The small whitish maggots, about 1/3 in. long, work their way down the stem usually inside the sheath. Young plants are killed; later the maggots burrow into the bulbs, causing decay. *Control:* Many of the flies may be poisoned before laying their eggs by using the following formula:

Sodium arsenite	1/5 ounce
Cheap molasses	1 pint
Water	1 gallon

The mixture should be placed in tin cans cut down to a depth of about 3 in. The tins should be distributed about the field and kept filled from the time the onions first show above ground till the injury is past.

The beginner is likely to be confused by the different methods of propagating the onion; yet the various propagation-forms of the plant represent only one species. The case may be presented as follows:

A. Propagated by means of bulbs: mostly for early or spring onions.
   1. From sets, which are small onions of arrested development that resume growth on being planted the following spring.
   2. From top onions, which are bulbels or small bulbs produced on the flower-stalk in the place of flowers and seeds.
   3. From multipliers, which are bulbs that break up into two or more distinct bulbs when planted.

B. Propagated directly from seeds: main field crop and also some of the early table green onions. Crops grown from seeds are often called "black seed onions," but the name has no significance for all onion seeds are black; the contrast is with the bulb-propagated group.

Early Green Onions

The small early or spring onions, used green or fresh and usually sold in bunches, are grown from either bulbs or seeds, usually from bulbs. These bulbs, as we have learned, are of three kinds: "top onions," or bulbels that are produced on the top of the flower-stalk, in the place of flowers; "sets," which are small onions, arrested in their growth; "potato onions," or "multipliers," which are compound bulbs, each component part forming a new bulb. The top onions (sometimes called "tree onions" and "Egyptian onions") and the multipliers are distinct races or types of onions, but sets are only the partially grown bulbs of any common onion which it is desired to propagate in this way.

To raise sets, seeds are sown very thickly on a rather light and dry piece of ground. As much as 40 to 70 pounds of seed are sown to the acre. The plants soon crowd, and by midsummer the tops begin to die for lack of food, moisture and room. The bulbs should not be more than one-half or three-fourths inch in diameter. They are cured and stored as are ordinary onions. The following spring, when planted, they resume growth, and in a very short time give edible onions for the table.

63. A multiplier onion (× about ½).

The illustration (Fig. 63) shows a multiplier onion. A cross-section (Fig. 64) shows that it has three "hearts" or "cores." As these cores grow, each gives rise to a separate bulb. If allowed to remain in the ground, each part develops two or more cores; and so the multiplication continues. When planted, the parts or cores are separated and planted as if they were sets; or if they do not readily separate in the hand, the entire onion is planted and a cluster of young onions is produced.

Multiplier onions seldom produce flowers and seeds. If not harvested for green onions, the small bulb grows into a large one which again breaks up into small ones. Sometimes the multiplier onions are planted in autumn. These plants are really perennials, continuing themselves by successive division of the bulb, whereas the ordinary seed onion is usually biennial.

All green or "bunch" onions, whether grown from bulbs or seeds, may be planted very thick. Usually they stand as close as 2 inches in the row. Often the rows are wide, so that three or four bulbs may stand abreast, but this increases the difficulties of tillage and weeding; but it may be said that weeds are usually not troublesome early in the season, if the land is clean to start with.

The little onions, or "acorns," from the flower-cluster of the top onion resume growth in spring, as if they were sets, and soon give an agreeable table supply. If left in the ground, the following year they will send up flower-stalks the same as will ordinary dry onions; but instead of producing only flowers and seeds, they will bear a greater or lesser number of bulbels with the flowers. In old gardens, even in the Northern States, a row of these plants is sometimes allowed to grow at will year after year, supplying enough little bulbs to afford the table supply of green onions.

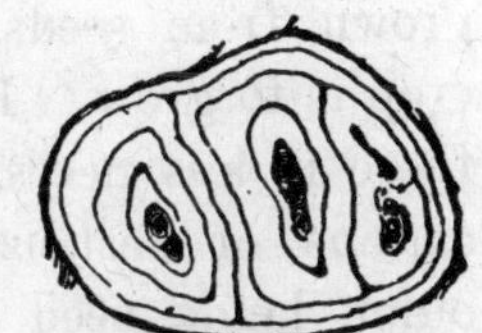

64. Cross-section of a multiplier onion, showing the cores, of which there are three in this case.

Early onions are grown to a considerable extent from transplanted seedlings. This method is sometimes known as "the new onion culture." The plants are started January, February or March in hotbeds or the forcing-house, and are transplanted to the open when the season will permit. In mild climates, as California, onion seedlings may be transplanted in spring from seed-beds sown in autumn. The large quick-growing southern types of onions, as Gibraltar and Prizetaker, may be grown to perfection in the North by this method, whereas the season may not be long enough for plants started in the open. Of course these transplanted onions may be carried through to maturity for autumn and winter use as are other onions grown from seeds, and extra quality bulbs may be produced.

Main-Crop Dry Onions

The general commercial onion supply is the crop of mature ripened bulbs, harvested and cured in autumn and sold in bulk as are potatoes. This main-season crop is grown from seeds, sown directly in the field where the crop is to grow. Earliness is not particularly desired, and there is less necessity, therefore, of making heavy applications of fertilizers which are quickly available. All onion lands need to be well fertilized, however, particularly with the materials rather rich in potash. Onions are relatively surface feeders; therefore the top of the soil should be very finely prepared, and the fertilizer should not be plowed under. Every attention should be given to preventing the soil from baking and to keeping the surface in uniformly good tilth. Fig. 65 shows the graceful curves in an onion.

Soils that become dry and hard produce a poor crop of onions. The best soils are those naturally loose and moist. Lowland areas are usually chosen for the growing of commercial onions. Reclaimed marshes, from which the roots and peat have been removed, are excellent. It is also of great advantage to have level land, as it facilitates the use of the hand tools and the finger work so essential in the growing of a good crop of onions.

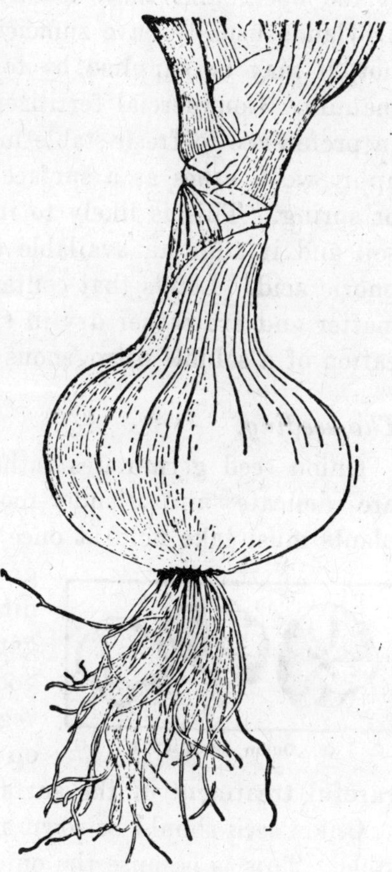

65. A globe onion (× ½).

It is customary to prepare onion land the preceding autumn. This not only insures earliness but it also allows the surface to become weathered and comminuted so that it is in perfect condition for the seeds as soon as the season opens. All clods and stones should be removed by a garden rake, horse weeder, or other fine-toothed tool. The land should have been in good cultivation for some years previous, if possible, that it may not contain seeds of weeds; for weeds are difficult to eradicate in an onion bed. Raw and coarse stable manures are rarely used for onions because they make the land rough and keep it too open, and they usually bring in seeds of weeds. Lowlands usually have sufficient humus, but if they have not, it may be supplied by top-dressings of old and fine manure. Commercial fertilizers are usually to be advised in preference to fresh stable manures. It is customary to apply wood ashes as a surface dressing either in autumn or spring. This is likely to improve the structure of the soil and it adds an available supply of potash and phosphoric acid. Lands that contain relatively little vegetable matter and are rather dry in spring may receive an application of a soluble nitrogenous fertilizer.

The seeding.

Onion seed germinates rather slowly and the plantlets are delicate and slender-rooted (Figs. 66, 67). The plants must take hold at once if they are to make a good growth. The onion-bed condition of tilth is considered by gardeners to be the measure of good treatment of land. No vegetable-garden crop raised on a large scale demands such careful treatment of the surface soil as the onion.

66. Onion seeds (× 4).

Onion seed should be sown as early in the spring as possible. This is because the onion delights in a cool season, and also because the plants should become established before the dry hot weather of summer. In garden practice, the seed should be sown thick, for there is likely to be failure of the seeds to germinate; and if the first sowing does not give a good stand it is rarely advisable to make a second sowing because of the lateness of the season. In field culture, thinning is expensive, and one must take great care to secure good and viable seed. The seed is sown with various kinds of hand seed-drills, some of which sow several rows at a time. The rows stand about 14 inches apart, varying, however, from 12 to 18 inches. In the rows the plants are thinned to 2 to 5 inches, depending on the size of the bulb in the particular variety. For field-crop onions, about 5 to 6 pounds of seeds are sown to the acre. The intervals between the rows are commonly 14 inches.

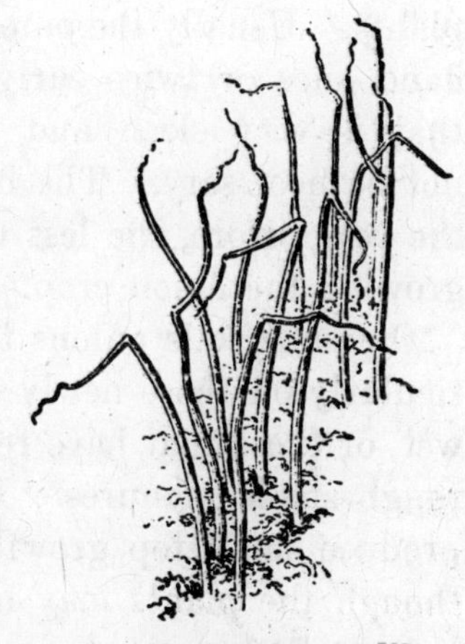
67. Onion seedlings (× ½).

The character of the crop depends very largely on the seed stock. The onion quickly runs down or deteriorates if the stock is not carefully selected and grown. Cheap onion seed is always to be avoided. Those who make a business of growing onions prefer to buy seed from parties whom they know, even though it costs twice as much as the ordinary seed of the markets. Poor seed may mean mixed varieties, lack of uniformity in the crop, the production of "scallions" or onions that do not make large bulbs.

Field practices.

Tillage is by means of hand wheel-hoes. If the land is rough, hard and uneven, these hoes cannot be worked to the best advantage. The land should be so finely pulverized that the lumps and clods do not roll on the young plants. Usually the onion patch will need to be weeded by hand once or twice early in the season, although in land that is very clean and free of weeds this expense may not be necessary. The better the preparation of the land the year before, the less will be the trouble and expense of growing the onion crop.

On some soils onions tend to run too much to top, particularly on those newly turned over from sod, or that are wet, or those that have received too great an application of rough stable manures. Dry soils and dry seasons tend to produce small top growth and a relatively large bulb, although the plants may mature so early that the bulbs do not reach the actual size they attain on moister land. If the tops are still rank and green late in August, or early in September, and show little tendency of ripening naturally, it is well to break them down to check the growth. A common way of doing this is to roll a barrel lengthwise the rows. The best onion crops, however, are those that ripen naturally. Late growth is sometimes due to the seed. If seed is from plants that have been grown for a number of years in a long-season and moist climate, as in England, the progeny tend to grow very late.*

The onion is a somewhat difficult crop to handle and to store unless the autumn season is warm and one has good facilities for handling the bulbs. The onions are usually allowed to dry or cure for a day or two before they are put into storage. If they cannot be handled in the field, they should be cured under cover, for the bulbs should be

*On this point consult Bailey, Bull. 31, Mich. Agric. Coll. 42 (1887).

dry and free from earth when they are sent to market or put into winter storage. Curing under cover is more expensive than curing in the field, but it usually gives brighter-colored bulbs and is to be advised when one caters to a special market.

The tops must be removed. It is customary to pull the onions before the topping is done. Three or four rows of onions are thrown into one, making a small windrow. After they have cured for two or three days, the tops are removed with strong shears, or usually with a shoe-knife. The tops are cut about one-half inch above the bulb. If they are cut shorter than this the bulb is likely to rot or shrivel, and if they are cut much longer it has an untidy appearance. The top should be cut off clean, leaving no ragged ends, and care should be taken not to tear the covering of the bulb itself. Some growers cut the tops from the bulbs before the crop is harvested. This may be done if the tops have died naturally. It is usually rather more expeditious than the other way. The bulbs are pulled by hand or a potato-fork; but in large areas an attachment is rigged to a cultivator to cut under the onions and lift them out.

68. Shed in which onions are stored temporarily.

If the crop is uneven, as will usually be the case, it is advisable to grade the bulbs if the best prices are to be secured. All small, inferior, misshapen bulbs are removed, and also those of unusual color. A good means of grading onion bulbs is to run them over a rack with slat bottom, or other form of grader, the slats being at such distance apart as to allow the large bulbs to pass over, but to catch all the small ones and to drop them through the spaces. The large bulbs are worked over the end of the table into baskets or barrels.

Storing.

Mature onions ordinarily will not stand freezing and thawing. Therefore, if they are stored for the winter, they must be put in a frost-proof place. They must be kept dry. Winter store-houses in the North are often provided with fire heat. Onions may be frozen with safety, however, provided they do not thaw out until spring and the thawing is then gradual. They may be stored in the loft on the north side of a building, where the sun does not strike the roof, and covered several feet thick with straw or loose hay. In the spring the straw is gradually removed and they are allowed to thaw slowly. When the winter temperature is very uniform, this method of keeping onions may be safe; but in regions of marked fluctuations in winter temperature it is not to be recommended.

69. Storage-house for onions.

Most onion-growers prefer to sell the crop in the fall. Usually it is put in temporary storage in open sheds, much as corn is stored in the crib. One of these sheds is shown in Fig. 68. There are wide spaces in the outside boarding of the shed, and the floor is raised a few inches above the ground and cracks are left in it. The eaves should project enough to carry all water clear of the sides. If the onions are dry and clean when put into storage and the tops have been carefully removed, the onions may be stored several feet deep in narrow bins or cribs of this kind.

70. Seeds of leek (× 6).

Frost-proof storage-houses are most reliable. They are provided with good ventilation, and kept near freezing temperature. Only mature well-cured onions should be stored in them, and particular attention should be given to having only rot-free bulbs. Fig. 69 represents "a good onion storage-house," drawn from M. T. Munn, Bull. 437, of the New York (Geneva) Experiment Station.

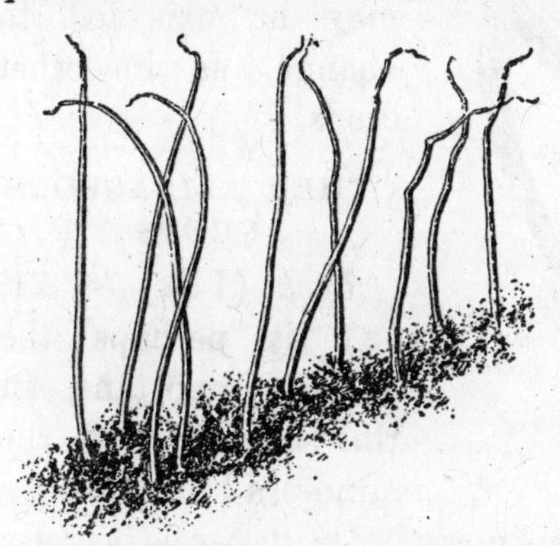

71. Seedlings of leek (× ½).

The kinds.

Varieties of onions are many. They differ in season, size of bulb, shape, color, quality, keeping ability. The yellow-skinned varieties are popular for dry onions. Danvers (Yellow Danvers) is a favorite; as also Globe Danvers and the very similar Southport Globe, Weathersfield, and others. The so-called Italian and Spanish onions are usually larger and require a longer season than most of the American types. The Bermuda and Texas onions are mostly of this European family. At present the globe type of onion is most in demand, whereas formerly the flat onions were most popular. The fashions may be expected to change, as in other crops.

72. Leek plant (× 1/7).

OTHER ALLIACEOUS CROPS

Leek (Figs. 70, 71, 72) is perhaps the most important, in this country, of the minor alliaceous plants. It should be better known. Its flavor is usually milder than that of the onion. The soft bulb, scarcely thicker than the neck, and the thick leaves are used in cookery. The plant requires the entire season in the northern parts of the country. Seeds are sown early in spring as for onions, and the plants thinned or transplanted to stand 4 to 6 inches apart, the rows being one foot or so apart. The plants are transplanted in early summer if especially good results are desired. Usually the plants are blanched for a considerable height above the crown by hilling or growing in trenches. Leeks are stored after the manner of celery, or they may be left in the ground if the climate is not very severe. In the South,

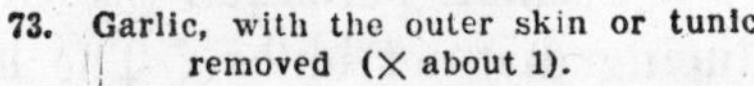

73. Garlic, with the outer skin or tunic removed (× about 1).

74. Welsh onion

the seed is sown in autumn, also sometimes in the North and the plants carried over in frames.

Garlic (Fig. 73) is grown from "cloves," which are the separable parts of compound bulbs, comparable in some ways with multiplier onions. These cloves or bulblets are planted in early spring; the compound bulbs mature in summer or early autumn, and after cured are commonly sold in bunches made by braiding the tops together. The plant rarely bears flowers.

Ciboule, or *Welsh onion* (Fig. 74), is like a common onion without the expanded bulb. It is grown for its leaves, which are used in seasoning. It is mild in flavor. Propagated from seeds as are onions.

75. A form of onion, often known as shallot.

Shallot is very like garlic in manner of growth, but the cloves are separate at maturity, whereas they are inclosed in a common skin in the garlic. They are mild in flavor. Cultivation as for garlic. Much of the stock known as shallot is only a form of onion (Fig. 75), either the multiplier type or small bunched onions in the spring.

Chive or *Chives* (Cive) is a small perennial growing in dense tufts and not producing distinct bulbs. The leaves are used for seasoning. It is perfectly hardy. It is a neat and interesting plant for a permanent edging along the garden walk. It is propagated by division of the clumps, although it sometimes seeds freely. The leaves are cut off as needed. When the vitality begins to decline, the plants are taken up, divided, and the parts re-set.

THE ONION PLANTS

Allium. *Liliaceæ.* Probably 300 species, widely distributed in the northern temperate regions of the globe, biennials and perennials, mostly bulbous. Many species are native to North America, some of them being known as leeks. The wild species often produce bulbels in the flower-cluster, as does the "top onion." The plants carry the characteristic alliaceous odor, of which the onion flavor is one of the manifestations. The black angular seeds of these plants do not long retain their vitality; 2 to 3 years, is the usual longevity. Many of the edible forms have been domesticated from prehistoric time, the onions and others being considerably modified through long cultivation. The plants have brought with them several of the ancient substantive names: *prason*, the Greek word for leek, now preserved to us in such combinations as Schœnoprasum (rush- or reed-leek), Scorodoprasum (compounded of Greek words for garlic and leek, anciently used for a kind of garlic), Ameloprasum (vine-leek); Porrum, Latin word for leek, allied to Greek *prason;* Cepa, the Latin *cæpa*, an onion.

A. Leaves fistulose (cylindrical and hollow).
- B. Plant a tufted perennial, without prominent bulbs, growing in mats. 1. *A. Schœnoprasum.*
- BB. Plant mostly biennial or plur-biennial, usually with evident bulb or bulb-like enlargement.—ONIONS proper.
  - C. The leaves large, not numerous.
    - Bulb large and prominent. 2. *A. Cepa.*
    - Bulb little thicker than the neck or crown. 3. *A. fistulosum.*
  - CC. The leaves small, awl-shaped, numerous. 4. *A. ascalonicum.*

AA. Leaves plane or flat (not fistulose).
- B. Bulb of several parts or cloves. 5. *A. sativum.*
- BB. Bulb simple, not much enlarged. 6. *A. Porrum.*

1. **A. Schœnoprasum,** Linn. Sp. Pl. 301. CHIVE. Erect glabrous perennial, 6 in. to 2 ft. high when in bloom, growing in tough clumps or tufts: stems enlarged somewhat into long slender bulbous bases; roots many and tough: leaves many in the clump, grasslike, some of them radical and others sheathing the stems, terete, hollow, long-pointed, usually equalling or surpassing the scapes: flowers rose-purple, many in a single terminal head which is subtended by the two thin

spathe-bracts; perianth segments lance-acuminate, ⅜ to ½ in. long and mostly equalling or exceeding the slender pedicels; stamens (6) included; pistil single, a long straight style arising from the summit of the emarginate ovary: fruit (capsule) splitting into 3 parts, several-seeded, the seeds black, about ⅛ in. long, oblong and pointed on either end, convex on the back, keeled on the front, weighing about 1 mg.—Native in Europe and Asia; the native plant in the U. S. is now separated as var. *sibiricum*, Hartm., or as *A. sibiricum*, Linn. It is not unlikely that more than one plant is in cultivation as chives.

2. **A. Cepa**, Linn. Sp. Pl. 300. ONION. Mostly biennial, not cespitose (not growing in tufts or sods), glabrous and glaucous: bulb large, much expanded, globular, oblong, conical, oblate, or other forms, the outside membranes thin and tunicate: leaves the first year radical, sheathing over each other at base and forming a neck, long and pointed, soft, hollow, swollen in the lower half: stem (produced usually the second year) simple, straight and erect, 2 to 4 ft. high, hollow, much enlarged, swollen below the middle, much overtopping the few or many prominently sheathing leaves, sometimes the stem-leaves disappearing before flowering: flowers numerous, lilac or whitish in a large terminal globular umbellate head subtended by 2 or 3 reflexed spathe-bracts, about ¼ in. long and borne on slender radiating pedicels ½ to 1 in. or more long; perianth segments narrowly lanceolate, acute, the stamens exserted, the filaments of the 3 inner stamens very broad at the base and lobed or toothed on either side; ovary globose or depressed-globose, smooth, with a single style: fruit dehiscing into three parts: seed black, about ⅛ in. long, nearly as broad as long, convex on the back and angled on the front, usually irregularly shrunken, weighing 3 to 5 mg. —Western Asia. Var. **viviparum**, Metz, acc. Alef. Landw. Fl. 301. 1866. (Var. *bulbellifera*, Bailey, Princ. Veg. Gard. Ed. 1, 316. 1901.) TOP ONION. Bulb small, undeveloped; bulbels borne in the flower-cluster with the flowers, and used for purposes of propagation; sometimes the cluster is proliferous,

sending out flower-bearing (and bulbel-bearing) branches. Var. **solaninum,** Alef. l. c. 300. (Var. *multiplicans*, Bailey, Princ. Veg. Gard. Ed. 1, 316. 1901.) POTATO ONION. MULTIPLIER ONION. Plant propagating by the natural division of the parent bulb: flowering stems (not often produced) short and slender, the umbel few-flowered.

3. **A. fistulosum,** Linn. Sp. Pl. 301. WELSH ONION. SPRING ONION. CIBOULE. Differs from *A. Cepa* in its more clustered or cespitose habit, more leafy and the leaves usually equaling or surpassing the stem, the bulbs little exceeding the broad soft stem-base: stem short and stout, 12 to 20 in. high, much swollen throughout its middle part and tapering to the flower-head: flowers white or hyaline, in a dense terminal head, stamens long-exserted, alternate filaments broadened at the base, perianth ⅜ in. long and about equaling or even exceeding the pedicel, segments long-acuminate: seeds onion-like, about 2 mg. in weight.—Native in Asia.

4. **A. ascalonicum,** Linn. Amœn. Acad. iv, 454. 1788. SHALLOT. Differs from *A. Cepa* in its small stature, slender awl-like leaves, and small ovate-oblong or oblong-conical gray more or less angular bulbs that break up into several distinct bulbs that cohere at the base: flowers (seldom produced) white or violet, in globose heads, the perianth scarcely exceeding the pedicel, segments spreading, oblong-lanceolate and acute. —Supposed to be Asian, but not certainly known in an indigenous state; by some writers thought to be a form of *A. Cepa*. It is doubtful whether the true shallot is in common cultivation; see page 156. The plant bears the name of Ascalon, eastern Mediterranean.

5. **A. sativum,** Linn. Sp. Pl. 296. GARLIC. A weak-growing flat-leaved plant of strong characteristic odor, producing several distinct hard parts or cloves, each with its integument, all inclosed in a silky-thin white or pink envelope comprising the compound mother bulb (the delicate envelopes sometimes decay and vanish if the mature bulbs are left too long in the ground, particularly if the season or the place is wet); planted in early spring, these cloves grow rapidly, produce another

compound bulb, and the leaves die down in summer, leaving no trace above ground: flowers seldom produced.—Southern Europe.

6. **A. Porrum,** Linn. Sp. Pl. 295. (*A. Ampeloprasum,* Linn., var. *Porrum,* Gay, Ann. Sci. Nat. 3d ser. viii, 218. 1847.) LEEK. Stout vigorous glabrous green very slightly glaucous biennial: bulb single, not much broader than the stout neck and gradually passing into it, with numerous stout roots beneath it: leaves equitant, keeled, 2 to 3 ft. long and at the base 1½ to 2 in. wide, very long-pointed: flower-stem slender, pithy and not fistulose, 2–3 ft., leafy below, the bulb more evident: flowers borne in a terminal umbellate head, subtended by a single spathe-bract, color pinkish, ¼ in. long, much exceeded by the pedicels; segments lance-ovate, acute, the midnerve usually colored; anthers exserted, the filaments of 3 of them very broad and with a slender branch on either side near the top exceeding the anther; ovary conic, the style arising within the notched top: fruit dehiscing into 3 parts: seeds black, about 1/6 in. long, onion-like, weighing 2 to 4 mg. —Not certainly known wild; considered to be an ameliorated form of *A. Ampeloprasum,* of Europe and western Asia.

A related plant is *A. Scorodoprasum,* Linn., the rocambole, sometimes cultivated for uses like garlic, native in Europe; it is a lesser plant than the leek, with smaller umbels which bear bulbels, the stamens not exserted; the ovoid bulb bears stalked offsets or bulblets.

CHAPTER VII

ROOT CROPS

Beet	**Parsnip**
Radish	**Celeriac**
Turnip	**Chervil**
Rutabaga	**Salsify**
Horse-radish	**Scorzonera**
Carrot	**Scolymus**

Root crops require a cool season and deep soil. They are grown in drills, and usually are not transplanted. They are used both as main-season and secondary crops. All are hardy. No particular ingenuity or skill is required in growing them.

The necessity of deep soil is apparent when one considers that the value of a root depends to a large extent on its straightness or symmetry. In hard and shallow lands roots are short and they tend to be branched and irregular. Fine tilth does much to insure quick growth, and quick growth improves the quality. Tile-draining and subsoiling greatly improve land to be used for root crops. The use of clover as a green-manure is also desirable, as it loosens and ameliorates the soil to a greater depth than most other green-manure crops.

Most root crops succeed best in cool soil. They thrive in the North, or in the cool season in the South. Those that do not require the entire season in which to complete their growth usually thrive best in spring and autumn.

Root crops are of two general classes as respects the purposes for which they are grown—fodder crops and oleraceous crops. The former are not intended here; neither are sugar-beets. Most of the vegetable-gardening root crops are able to secure their food from relatively unavailable combinations, and they generally use rather freely of potash, although they are also heavy nitrogen and phosphorus feeders. To start them quickly, a light dressing of available nitrogen compound is useful, particularly if the roots are needed for a particular season. These crops, as a class, are supposed to be more exhaustive of the plant-food elements than the cereals and legumes.

The earliness of the root crop in market-garden handling is likely to make all the difference between success and failure. The earliness is determined largely, according to Voorhees, "by the amount and availability of the nitrogen and phosphoric acid applied." Frequent top-dressings of soluble nitrates are advised. An application of 1,000 to 1,500 pounds of the basic fertilizer (page 383) "is frequently employed at time of seeding, followed by a top-dressing of 50 to 100 pounds of nitrate of soda to the acre once every week or ten days, for at least three or four weeks after the plants have well started."

Probably the most laborious part of the growing of root crops is the harvesting, particularly of the long late kinds. This labor is much lessened by plowing out the roots. Even if the roots are too deep for the plow, two or three furrows may be thrown from either side of the row, and the pulling is made easier. Usually, hand-pulling is unnecessary. As soon as the roots are out, the tops should be cut off about an inch above the crown, if the crop is to be stored or sold in bulk. The roots should lie in the sun until the earth is dry enough to shake from them, when they may be stored in the pit or cellar or sent to market. They are easy to keep.

The market value of a root depends largely on its looks. All strong side roots should be cut off, and branchy specimens should be discarded. Early in the year, such roots as beet, carrot, radish, and turnip are sold in bunches of 6 to 12; but as the season advances and prices fall, they are sold in bulk. When sold in bunches, care should be taken to have all the specimens in the bunch of uniform size and shape. The leaves are allowed to remain, and the bunches are tied neatly by a tape or other cord passed around the leaf-stalks. The bunches should be kept well sprinkled and away from the sun, for wilted leaves give them a stale and unattractive appearance.

Seeds of these crops are grown from roots carried over winter. Plant the roots in spring, the crown level with the surface of the ground, 2 feet or more apart. Flower-stalks are soon sent up, and seeds are usually produced freely.

The species of roots may be assembled by their botanical affinities. The beets of all kinds are allied to spinach and the pigweeds (Chenopodiaceæ). The radish, turnip, rutabaga, horse-radish are cruciferous, being members of the Cruciferæ or Mustard family and therefore allied to the cole crops. Others are umbelliferous, belonging to the Umbelliferæ or Parsley family, as carrot, parsnip, celeriac, tuberous-rooted chervil. Salsify, scorzonera, scolymus are compositous, representing the Sunflower family, Compositæ.

BEET

A loose deep rich fresh relatively cool soil and a continuous growth are the prime requisites in the cultivation of garden beet. It is usually a companion- or succession-crop in the vegetable-garden. The crop is hardy and easy to raise. The round varieties are relatively surface feeders and early in growth. The land should be kept well tilled to conserve moisture and to keep down weeds, particularly in the early part of the season.

Sow in drills as soon as the ground is ready, and thin to 5 to 8 in. apart; the thinnings may be used as greens. The drills should be far enough apart to admit of wheel-hoe tillage,—12 to 18 in. Field beets should be far enough apart for horse tillage, 20 to 30 in. Five to eight pounds of seed are required for an acre; 1 ounce sows 75 to 100 feet of drill. Seed is covered about 1 in. deep. Average crop is 300 to 400 bushels to the acre.

LEAF-SPOT (*Cercospora beticola*).—Ashen gray leaf spots, each surrounded by a reddish purple border, are characteristic symptoms. Frequently the central tissue drops out and the leaf presents a shot-holed appearance. As the outer leaves die off, new ones are formed, thus elongating the crown of the beet. *Control:* Sanitary measures in the field together with crop rotation are beneficial. A thorough application of bordeaux mixture at intervals will afford control.

SPINACH LEAF-MINER (*Pegomyia hyoscyami*).—A whitish maggot, ⅓ in. long when full-grown, that mines the leaves of beet, spinach, orach, and chard. The mine is at first thread-like but soon enlarges and becomes a blotch. Many maggots often infest the same leaf. The insect also breeds on lamb's quarters (pigweed). *Control:* Clean culture and the

destruction of its wild food plants; with some crops the injury may be avoided by growing the plants either early in the spring or late in autumn, when the insect is less abundant.

SUGAR-BEET WEBWORM (*Loxostege sticticalis*).—A yellowish white caterpillar marked on the back with three dark stripes, about 1 in. in length when full-grown, that devours the leaves and covers its feeding grounds with a slight silken web. *Control:* Spray with 3 lbs. paris green in 100 gals. of water to which 6 lbs. of whale-oil soap or 3 lbs. of lime are added, or dust the plants with paris green, 2 to 4 lbs. in 100 lbs. air-slaked lime.

HAWAIIAN BEET WEBWORM (*Hymenia fascialis*).—A small slender pale green caterpillar which skeletonizes the underside of the leaves; restricted to the Southern States. *Control:* Spray with arsenate of lead (paste), 2 lbs. in 50 gals. of water, taking care to hit the underside of the leaves.

SPOTTED BEET WEBWORM (*Hymenia perspectalis*).—A small shining green caterpillar marked with rows of small black dots that at first skeletonizes the leaves but later eats the whole leaf; restricted to the Southern States. *Control:* Same as for the preceding species.

SOUTHERN BEET WEBWORM (*Pachyzancla bipunctalis*).—A glossy dark dirty green caterpillar, about ¾ in. long when full-grown that devours the foliage, folding and webbing the leaves together with silken threads. *Control:* Same as for the spotted beet webworm.

Two general types of beets are grown for vegetable-gardening purposes: the short-season turnip varieties (Fig. 76) and the main-season long-rooted varieties. Certain oval half-long types are intermediate in season. The long-rooted varieties are less popular than formerly, for the turnip varieties may be grown in autumn for winter use, and fresh beets are to be had from the South in winter. Formerly the long blood beet was used for stock-feeding to some extent, but the mangel-wurzel has largely taken its place. The early beets lend themselves well to the intensive practices of market-gardeners.

The soil for beets, particularly for the early kinds, should be mellow and quick, on the loamy order. Hard, poorly-tilled and cloddy lands are not adapted. Fresh manure is usually avoided, but well-rotted manure is used freely, and chemical fertilizers are desirable.

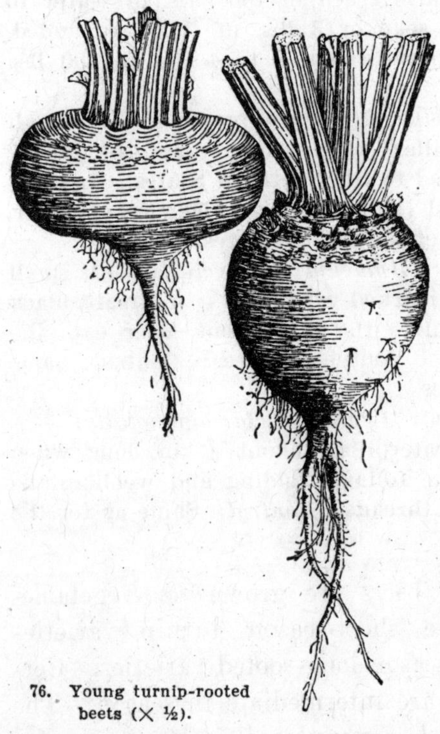

76. Young turnip-rooted beets (× ½).

The plants should be kept growing continuously. They seldom completely recover from a marked check or setback, at least not in time for a dated early market. Good frequent level tillage is required.

Beet seeds require considerable moisture to germinate. This is because the "seeds" are really fruit clusters with hard shells, each cluster containing two or three small seeds (Fig. 77). The husks or walls of the fruit are relatively impervious to water. Therefore, if sown late in the season special care should be taken to have a moist seed-bed. For the reason that the fruits rather than the seeds are sown, beets are likely to come up in little clumps, and careful thinning is essential if the best results are to be secured. Specially constructed seed-drills, or special attachments, are necessary for the proper sowing of the rough uneven-sized seeds of beets. Young plants of beet are seen in Fig. 78.

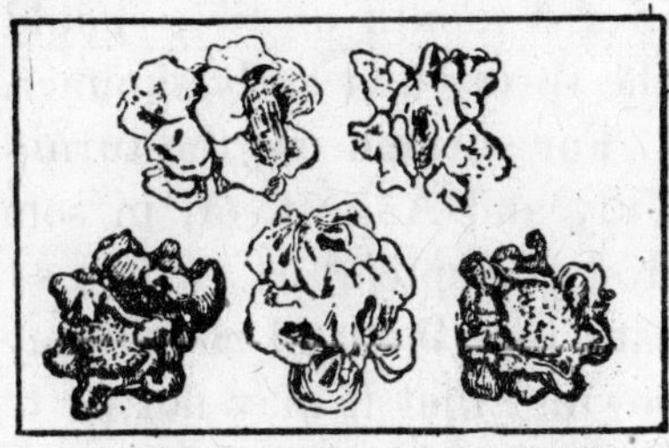

77. Fruit clusters of beet (× 2).

Vegetable-gardeners now chiefly know the early turnip-rooted varieties. These varieties may be grown either as a spring or fall crop. They mature in two to three months (60 to 90 days) and roots large enough for bunching of some of the earliest varieties may be had in six weeks to two months.

The early turnip varieties of beet may be sown as soon as the land can be worked in spring if one wishes to secure an early crop. They may be followed by a later crop, as celery, late potatoes, cabbage or cauliflower. In some cases, they are grown as a companion-crop in the rows with a main-season crop, as cabbage. For very early results, it is well to sow the early varieties in hotbeds, or cold-frames. They may be allowed to mature in the frames, or in special cases they may be transplanted into beds, although transplanting is rarely done, as it does not pay.

78. Seedlings of garden beet (× about ½).

For home use, two or three rows fifty feet long, the seeds being sown at intervals (as every fortnight) extending over a month or two, should give a sufficient supply for the spring and early summer.

For autumn use the turnip-rooted beets may be sown in July and August, or, in some places, even as late as the first of September. When sown late, however, it is important that the land should have been well tilled previous to sowing, that it may not be too dry. The firmest and best roots may be stored for winter in pits or in the cellar in boxes of earth or moss.

The long or blood beets are usually sown in early May in the Northern States, and they occupy the ground the whole season. The half-long kinds are useful in autumn and winter, and they may be sown later than the long kinds, following early peas or other crop.

Young beets are much used for greens. They are rarely grown especially for this purpose, but the seed is sown thick and the thinnings are sold in bunches or in small packages. The whole plant, root and top, is thus used as a potherb. Certain kinds of beets produce thick leaves rather than roots; these are essentially leaf crops and are discussed under that head. See Chard, page 59.

Early beets are usually sold in bunches of about six, before the roots are full grown, but the later crop is sold in baskets, crates, and barrels. The price depends much on the earliness and freshness of the product.

Good early and mid-season beets are Egyptian, Bassano, Eclipse, Bastian, Detroit Dark Red, Crimson Globe, Columbia, Edmand. A standard winter variety is Long Blood. There are many other good varieties.

The Beet Plant

Beta. *Chenopodiaceæ.* Perhaps a half dozen species of herbaceous plants, biennial and perennial, on the coasts of Europe, Asia and Africa.

B. vulgaris, Linn. Sp. Pl. 222. (*B. esculenta*, Salisb. Prodr. 152. 1796. *B. vulgaris* var. *esculenta*, Guerke, in Richter-Guerke, Pl. Eur. ii, 127. 1897.) Cultivated Beet. Biennial (rarely annual), glabrous, smooth, the growing parts often red, yellow or metallic-green (particularly midribs and petioles: taproot thickened into a single downright tuber, in many sizes, shapes and colors: stem produced the second year, one from the top of the tuber and sometimes a few small supplementary ones, slender and grooved, erect but falling with the load of fruit, much branched and leafy, the main stem 2 to 4 ft. tall: leaves in a tuft from the crown, the blade ovate to oblong-ovate in outline, truncate or semi-cordate or abruptly tapering at base, obtuse or muticous, the margins undulate and entire or irregularly sinuate-dentate, the slender petiole usually exceeding the blade; stem leaves petioled, smaller, the lower ones of similar shape to the radical leaves, those in the inflorescence passing into linear spreading bracts: flowers greenish, very small, sessile, in long paniculate racemes, the plant producing great numbers on its many slender branches, usually about 2 or 3 flowers together, with minute bractlets beneath the perianth, which has 5 incurving parts, on the inside of which parts are the 5 stamens; ovary 1, sunk in a disc or hypanthium, the styles usually 3 and with blunt or ovate stigmas; the perianth and disc are persistent, inclosing the single seed in a hard case bearing corky protuberances which are the thickened and modified perianth-parts, the 2 or more flowers in the cluster growing together by their bases and forming the very irregular fruit-mass known as the "seed" of commerce; this fruit-mass weighs 5 to 50 mg., and on the faces of it one is able to make out the 5 prominences of the different flower-parts; longevity of seed about 5 or 6 years.—Unknown wild; regarded as an ameliorated form of *B. vulgaris* var. *perennis*, Linn. (B. *maritima*, Linn. *B. vulgaris* var. *maritima*, Koch), of the sea-coasts of western Europe, a very different looking plant, perennial, with long hard thick-branched root, smaller leaves, and many prostrate or decumbent stems. The evolution of the beet is a remarkable example of modification, in which the whole habit and habitat of the plant have been changed. The sugar-beet (*B. vulgaris* var. *saccharifera*, Alef.) is part of this modification. The mangel-wurzel, or mangold of English and American writing, is another form of it. In North America the beet is thought of in relation to its thick edible root, but another race is developed in its leaves rather than in its roots. We may therefore distinguish the leaf-beet and the root-beet; in England the latter is known as beet-root; in France the leaf-beets are known as poiré.

Var. Cicla, Linn. Sp. Pl. 222. LEAF-BEET. Root downward, not developed into a fleshy edible part, sometimes branched: leaves much developed, usually larger and broader than in the common beet, sometimes 2 ft. long, the midrib usually broad and often fleshy. Here belong the ornamental-leaved beets and also the Swiss chard. The word Cicla refers to Sicily.

RADISH

Quick and continuous growth, carefully selected seed, rather cool weather for the early bunching kinds, protection from the root-maggot—these are prime considerations in the growing of radishes. The radish is a partial-season crop. It is easy to grow on light fertile land.

Radishes are usually sown as early in spring as the ground is fit, even before the frosts are past. Sow in rows 6 to 12 in. apart, or farther apart if a wheel-hoe is to be used. Cover ½ to ¾ in. Thin 1 to 3 in. apart, depending on variety. For family use, sow at intervals of 7 to 10 days. As the season advances, choose a cooler site, as a northern exposure.

Usually the sowings are discontinued from the last of June until late August. One ounce of seed sows 100 feet or more of drill; 8 to 10 lbs. are required for an acre.

There are no prominent diseases of the radish.

CABBAGE ROOT-MAGGOT (*Phorbia brassicæ*).—See detail under Cabbage. Radishes may be raised free from maggots by screening the beds with cheesecloth.

FLEA-BEETLES.—Screening the beds with cheesecloth will prevent injury.

SPINACH APHIS (*Myzus persicæ*).—See under Spinach. The first pair of leaves sometimes becomes badly infested on the underside. Spray with "Black Leaf 40" tobacco extract, 1 pint in 100 gals. water in which 4 or 5 lbs. soap have been dissolved. Do not use with bordeaux mixture on young radish plants as it will stunt them.

In North America the radish is known mostly as a spring crop, although it is sometimes grown in autumn. In the Old World, however, it is known also as a summer crop, but the varieties grown in the hot weather are usually unlike those raised in the spring and autumn. In the Orient (particularly Japan) it is a winter and spring vegetable, extensively eaten.

There are three general types of radish roots: the ordinary small spring or autumn radish, usually light red or clear white (Fig. 79); the large turnip radishes, useful for summer cultivation, white, gray or black; the winter radishes, that make a long hard red, white or black root. The winter radishes are relatively little grown here, although deserving to be better known. They are usually sown late in the season, as are late turnips (July and August) and the roots may be kept over winter as other roots are stored. Some of them make roots 12 to 20 inches long and several inches in diameter. The flesh is solid, quite different from the little table radishes eaten as delicacies; these winter radishes supply numbers of people in other countries with substantial food.

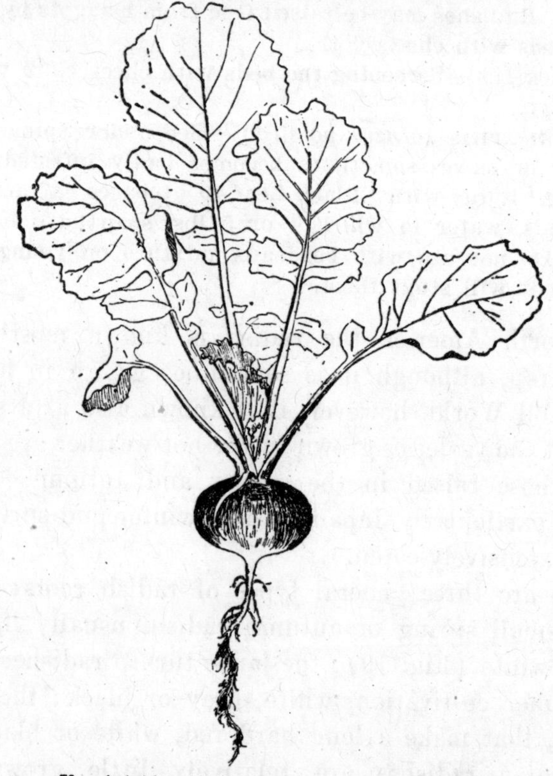

79. A table radish. Raphanus sativus (× ½).

Radishes are usually treated as a companion-crop when grown in the open field. They may be sown in drills between the rows of cabbages, peas or other later-maturing vegetables. Sometimes they are sown directly in the drill with the other vegetables. The seeds are quick to germinate and thereby break the crust and mark the row (Fig. 80) and thus facilitate tillage, and the roots may be harvested before the other crops need the space. For family use, radishes are often grown in beds by themselves. In clean friable land they are sometimes sown broadcast. They may be forced in winter, and grown for very early spring use in hotbeds and later in coldframes. Better roots and a more uniform crop are secured by sowing only the large seeds (Fig. 81). The small ones may be sifted out with a hand screen.

80. Radish seedlings (X about ½).

If the land is loose and rich, the spring radishes should come to edible maturity in four to six weeks. The roots are of better quality when they are relatively small and crisp. When growth ceases the roots become stringy, bitter, and often hollow, and the plant runs to seed (as it does also from too late sowing). Sow at frequent intervals for a succession. Radishes do not come to their full perfection in hard and dry land. The roots are so

81. Radish seeds (X 4).

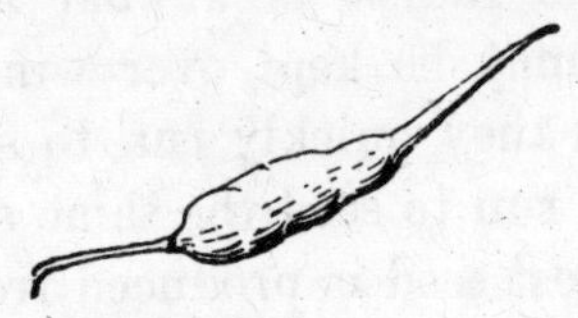

82. Pod of radish (X 1/3).

small and short that the plants are essentially surface feeders.

If radishes are to be grown in hot weather, the land should be as cool as possible and supplied with abundance of moisture to keep them growing continuously. It is well to grow the regular summer radishes, as Strasburg, but as there may not be a market for them, the small spring radishes may have to be carried into the summer.

83. Leaf of long-pinnate radish. Raphanus sativus var. longipinnatus. (× ⅛).

For the market, radishes are washed and tied in bunches of 4 to 10, with the tops left on. They should be kept moist until sold. If the tubers are graded to size, shape and color they make very attractive produce.

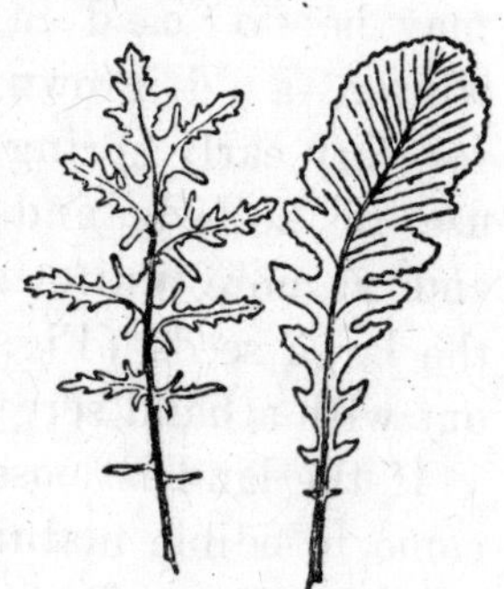

84. Leaf-forms of small-pinnate radish. R. sativus var. parvipinnatus. (× 1/3.)

The radish is annual and biennial. Roots maturing late may be kept over winter and planted in the spring, when they quickly run to seed. Spring and summer radishes run to seed the same season if left in the ground, but the best seed is produced from plants that are transplanted when young. Little radish seed is grown in North America, probably largely because of the high price of labor.

Probably the most popular variety is French Breakfast,

and various forms of the same type. Other good kinds are Olive-shaped, Scarlet Short-top, Wood Early Frame, White Box. For summer, good varieties are Icicle, Chartier, Lady-finger, White Naples, White Vienna, Strasburg, Stuttgart. For winter, Scarlet Chinese, Celestial, Black Spanish, White Spanish may be mentioned.

The Radish Plant

Raphanus. *Cruciferæ.* Probably 8 or 10 species, annual, biennial, perennial, Europe to the East Indies.

R. sativus, Linn. Sp. Pl. 669. Common Radish. Annual and biennial: root thickened, white to pink to purple to nearly black, short and globular to conical to oblong to spindle-shaped and extending into a long taproot that bears most of the feeding rootlets: stem stout, erect, 2-3 ft. high, long-branching in flower, usually falling when laden with fruit, striate or grooved, more or less glaucous, glabrous or bearing few scattered straight stiff colorless hairs: lvs. very variable in size, shape and division, all petiolate, sometimes smooth but usually with scattered sharp stiff colorless hairs on both surfaces, strongly veined; radical ones 3-6 in. long and 1-2 in. broad, obovate or short-oblong in outline, usually lyrate-divided, the terminal part large and the lateral divisions becoming very small along the petiole, the margins irregularly crenate or crenate-dentate; cauline lvs. large, mostly strongly lyrate-pinnatifid and long-petioled, terminal lobe very large and mostly rounded (sometimes acute!) and more or less shallowly lobed, the inferior divisions few or several, the upper lvs. passing into lanceolate or linear undivided bracts in the inflorescence: flowers white to red-veined to lilac, slender-pedicelled, on long branches; petals 4, long-clawed, the oblong or obovate obtuse blade spreading usually at right angles in full anthesis; 4 narrow sepals about as long as the claws of the petals; some or all of the 6 anthers exserted in the throat of the corolla, as is also the single style with its globular stigma: fruit an indehiscent spongy pod, 1 to 3 in. long

(Fig. 82), with a long beak and 1 to 6 seeds in the thickened part: seeds brown, variable in size and shape, globular-angular, large ones about ⅛ in. long and weighing 8 to 10 mg., the small ones only about half as heavy; longevity about 5 years.—Probably Asian, but known only as a cultigen (in cultivation and frequently escaped). Thought to be a development from *R. Raphanistrum*, Linn., the charlock, a weedy plant with slender taproot, yellow flowers fading to white or violet, and slender furrowed pods with marked constrictions between the seeds: this plant is now widely spread, and is an introduced weed in North America. The radish is variable in the size, season, shape and color of its tuberous roots (the word radish is connected with the Latin *radix*, root), and botanical groups are usually defined in terms of these characters; better botanical characters, however, reside in the leaves and pods. There are marked groups in the pinnate division of the leaves, and one group in which the leaves are undivided.

Var. **longipinnatus,** Bailey, Gent. Herb. i, 25. 1920. Plant large and stout: radical leaves elongated and narrow, sometimes 2 ft. long, the leaflets 8 to 12 or more pairs: root large and long, usually a winter radish.—Apparently most of the oriental winter radishes belong here (Fig. 83).

Var. **parvipinnatus,** Bailey, l.c. Plant slender, with large root: leaves small, sometimes with very slender divisions and sometimes merely lobed: pod slender, nodose, with a very long beak.—India and Japan; apparently not cult. in this country (Figs. 84, 85).

Var. **nonpinnatus,** Bailey, l.c. Leaves entire, the radical ones obovate and on the stem oval or oblong, the margins entire or obscurely crenate-dentate.—China, not recognized in this country (Fig. 86).

Var. **caudatus,** Alef. Landw. Fl. 258. 1868. (*R. caudatus*, Linn. Mant. i, 95. 1767.) RAT-TAILED RADISH. Pods rather than root greatly developed, sometimes more than 1 ft. long, curved and sometimes twisted (Fig. 87).—The young pods are the edible parts, sometimes pickled and sometimes eaten raw as are radishes. Only now and then grown in this country, as a curiosity.

TURNIP AND RUTABAGA

The turnips of all kinds are cool-season crops of quick germination and rapid growth. They are partial-season plants, usually following early crops. They grow long after tomatoes, corn and many other crops are killed by frost. Seeds are usually sown where the plants are to stand. The soil should be loose and fertile.

For garden use, particularly for the early season, turnips are sown in drills 10 to 18 inches apart. In drills, 1 ounce of seed may be used for every 200 to 300 feet, or 1 pound to the acre; broadcast, 2 to 3 pounds to the acre. The plants should be thinned to stand at first 3 inches apart; and then, as some of the young roots are removed for eating, until the main crop allows a foot of space for the development of each full-sized tuber. The late or fall crop is often sown broadcast, particularly if it is to be used for stock-feeding. Better results are secured, however, when the plants are grown in rows. For general field purposes, the rows are placed 18 to 30 inches apart, to allow of wheel-hoe or even horse-hoe tillage. Seeds are sown ½ to ¾ inch deep. Yields run from 600 to 1000 bushels to the acre.

The diseases of these plants are black-rot and club-root; in insects the turnip aphis may be troublesome: see the discussions for Cabbage, page 71. Flea-beetles often infest turnips and rutabagas: see the account of this insect on page 435.

85. Pod of small-pinnate radish (× ½).

The true or "flat" turnips usually have flattened or very oblate roots, soft white flesh, and green rough leaves. They do not require the full season in which to mature, and are therefore grown as a spring or autumn crop. The herbage is very hardy, withstanding considerable frost without injury. They are grown somewhat for stock feed, but not so largely as the rutabaga; only the vegetable-garden use of them is intended in this writing.

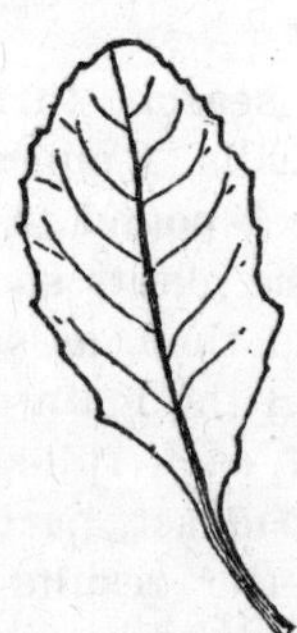

86. Leaf of simple-leaved or non-pinnate radish (× ¼).

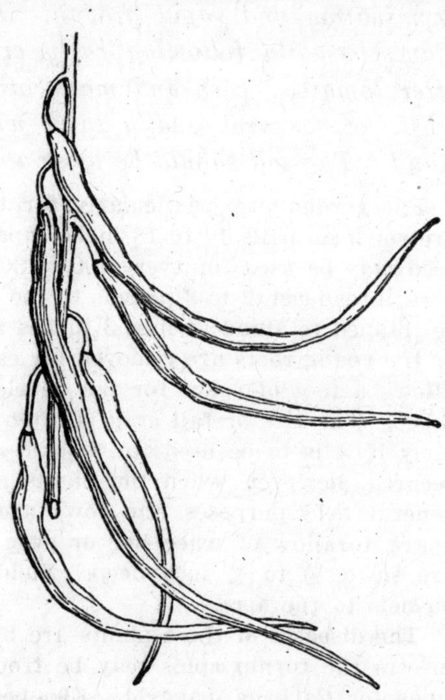

87. Pods of rat-tailed radish, Var. caudatus (× 1/3).

For early use, turnips are sown as soon as the land can be prepared in spring. They should give roots large enough for the table in six to ten weeks. For the fall crop, seeds may be sown in the Northern States as late as the last week in July, and in the Central States as late as the middle of August. The plants will grow until heavy freezing weather, at which time they may be pulled and stored as are other roots. The roots will not stand hard freezing.

The value of the turnip as an article of food lies very largely in its tenderness and succulence. If the plant grows slowly, it is woody, stringy and bitter. To secure a quick growth, the land should be rich and moist, and in fine tilth. If the plants are raised in broadcast seeding, the land should be in excellent condition and free from weeds, as no subsequent tillage is possible.

88. Seeds of turnip (× 8).

The turnip is one of the easiest plants to grow, except that it is often seriously attacked by the root-maggot. This pest can be kept in check by injecting bisulfide of carbon into the ground about the plants, but this labor is usually more than the turnips are worth. It is better, therefore, to grow turnips on land that has not been infested; or, if there is no such land on the premises, it is advisable not to grow turnips until the insects are starved out.

Early turnips are sold in bunches, like early beets, the tops usually removed. The main crop is sold by the bushel or the barrel. Roots are stored for winter like potatoes.

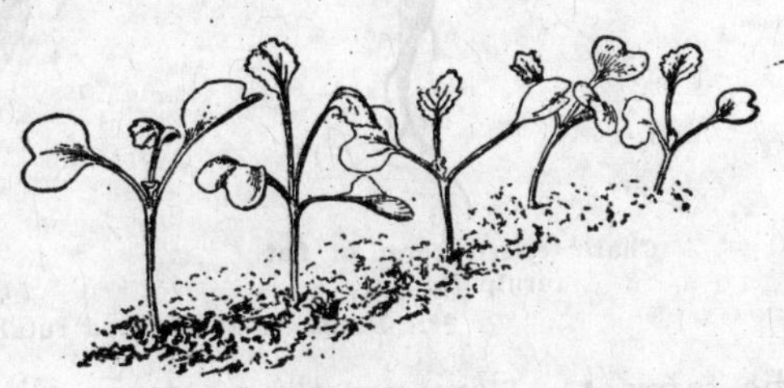

89. Seedlings of turnip (× ½).

Standard varieties of turnip are Milan, Snowball, Strapleaf Flat Dutch. Figs. 88, 89, 90 show the turnip.

Rutabaga

The requirements for the rutabaga are the same as for turnips, except that the plants require a month to six weeks' longer time in which to mature. It is not raised as a spring vegetable.

Rutabaga differs from the turnip in having a denser and mostly yellow-fleshed root, which is rounded or elongated and not distinctly flat, the leaves glaucous-blue and not hairy, the crown long and leafy, the roots arising from the under side of the tuber as well as from

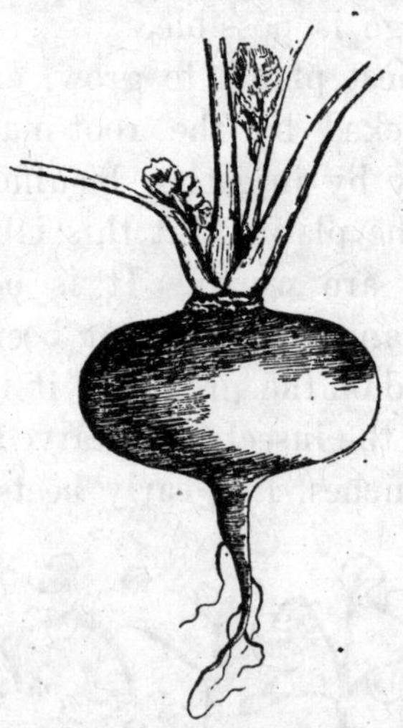

Characteristic form of flat turnip (× ¼).

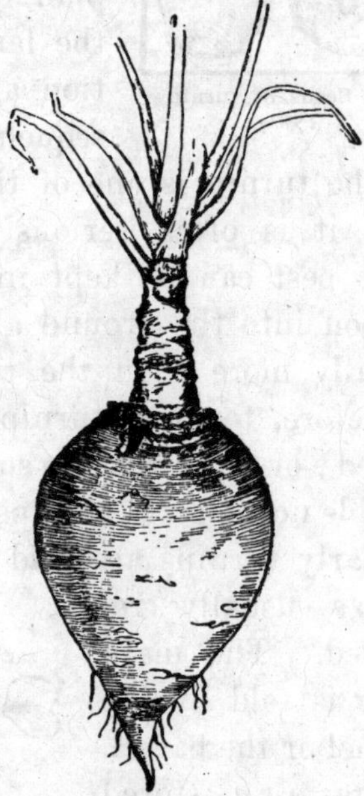

Rutabaga (× 1/5).

90. Forms of turnip and rutabaga.

the taproot. Compare the roots in Fig. 90. It is a richer vegetable than the turnip. It is grown either as a spring or autumn crop. As in the case of the turnip, the product grown for stock is raised from summer-sown seeds. For the main crop, the seeds are usually sown as early as the first of July or the middle part of June in the Northern States.

For the botanical account of turnips and rutabagas, see the discussion of brassicaceous plants in Chapter IV (pages 95, 96).

HORSE-RADISH

Horse-radish is a perennial grown commercially as an annual, propagated by root-cuttings (sets). It is perfectly hardy. Grown usually as a combination-crop and succession-crop, occupying the land completely late in the season, when it makes its principal growth. It requires a very deep and fertile soil. The grated or shredded root is used as a piquant sauce and relish.

Cuttings of the side roots are employed for propagation directly in the field, and the plants stand 10 to 18 in., more or less, in rows far enough apart for good tillage, which is usually 3 to 4 ft. if the plants are started between other crops. The commercial yields are 3 to 5 tons to the acre, varying less or more.

HORSE-RADISH FLEA-BEETLE (*Phyllotreta armoraciæ*).—A black strongly convex flea-beetle about ⅛ in. long, having each wing-cover yellowish except a narrow black stripe along the outer margin and a wider one on the inner margin. The eggs are laid in clusters on the petioles of the young leaves. The larvæ burrow in the petioles. The beetles are more destructive early in the season and the larvæ later. *Control:* Spray the plants with bordeaux mixture containing 4 to 6 lbs. arsenate of lead (paste) in 50 gals. Several applications may be necessary. Change the location of the beds from time to time in order to avoid the beetles.

SPINACH APHIS (*Myzus persicæ*).—See under Spinach.

HARLEQUIN CABBAGE BUG (*Murgantia histrionica*).—See under Cabbage.

Sharp distinction is to be made between the home-grown supply of horse-radish and the commercially-raised product. It is the same plant; but in the home premises it is usually allowed to persist year after year, often as a weedy plant, and is dug in spring as wanted. It is customary to plant the old crowns, and sprawling crooked roots are the result. These roots are good enough for home use, but they would not sell on the market. For commercial purposes, a clean straight shapely root is desired (Fig. 91); and to obtain this root, careful propagation, good land and thorough tillage are essential. In some parts of the country the growing of horse-radish is an important industry.

91. A good root of horse-radish (× 1/6).

As a commercial crop, horse-radish is grown as an annual, being propagated from cuttings of the small side roots. These cuttings are made from the trimmings when the roots are dressed for market in autumn. A good cutting should be the size of a lead pencil up to that of one's little finger (Fig. 92). It is usually made 5 to 8 inches long, and the lower end is cut slanting to designate the right end up when planting (Fig. 93). These cuttings or sets are tied in bundles and stored in the cellar or pit, as are other roots.

92. Horse-radish sets (× 1/3).

Sets may be planted at the first opening of spring, but since the plant makes the larger part of its growth late in the season, it is customary to hold them rather late and to plant them with some other crop. They are often planted in the rows of early cabbages or beets. When the cabbages are off, the horse-radish takes the land. The sets are dropped right end up in furrows or holes, which are made with a strong-pointed stick or crowbar or a dibber. They are usually placed in a somewhat slanting position although the upright position is probably as good. The top of the cutting usually stands 3 to 5 inches below the top of the soil. This deep planting delays the appearing of the plants and thus prevents interference with the combination-crop. The rows are far enough apart to allow of horse tillage, and the plants should stand 10 to 16 or 18 inches in the row.

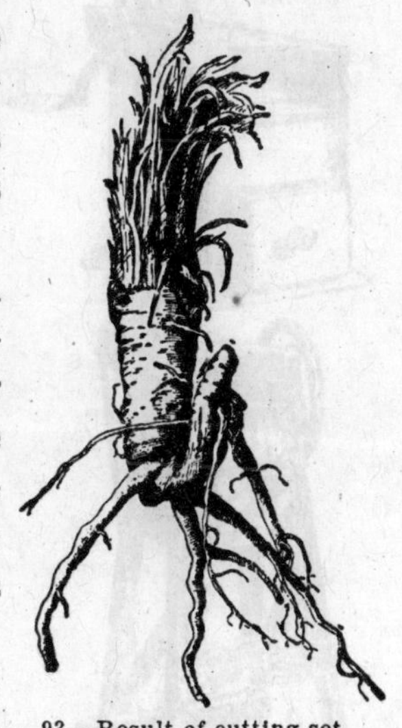

93. Result of cutting set wrong end up.

The plant will stand much abuse. If it grows so rapidly as to interfere with the cabbages or other plants with which it is planted, the tops may be cut off two or three times early in the season. After the other crop is removed, the land is given good surface tillage.

Sometimes horse-radish is made the main crop, and other crops are grown incidentally. In this case, it is planted in rows 3 to 4 feet apart on ridges, and spinach, early beets or lettuce are grown on the sides of the ridges. The crop will grow until freezing weather.

It is best to plow out the roots in autumn and to store or sell them. As horse-radish is likely to become a bad weed, it is necessary that all the small roots be taken out of the land. When the crop is harvested, therefore, all the loose roots are picked from the furrow and destroyed. If these furrows are left open until spring many more of the roots will be exposed, and they may then be removed. Subsequent plowing and dragging will often expose still others. It is usually impossible to get all the roots out of the land, but if the ground is occupied with other crops and is kept in good tillage, the horse-radish should not become a nuisance.

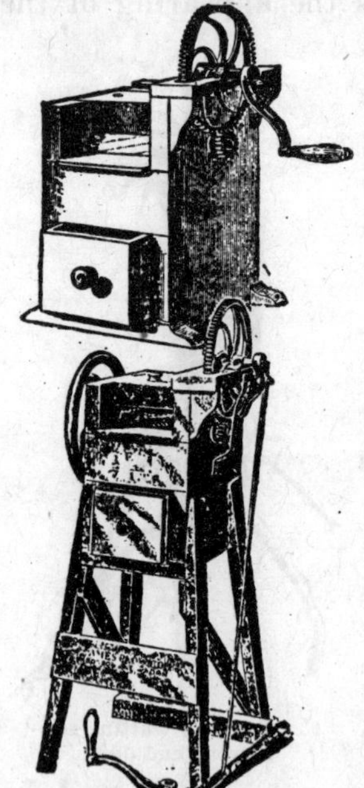

94. Two kinds of hand-power horse-radish graters.

The roots are washed and trimmed before they are sent to market. For special trade, the roots may be tied in bunches of 6 or 8, but the crop is generally marketed in barrels or in bulk. As the roots must be grated (Fig. 94) before they are used, it is necessary that they be long, symmetrical, uniform and as large as possible in order to fit the grating machines. Small and branchy horse-radish can scarcely be sold at any price. From 3 to 5 tons (or more) should be raised on an acre, the latter quantity when the ground is deep and rich and when the plants do not suffer for moisture.

The Horse-Radish Plant

Armoracia. *Cruciferæ.* A few species of herbs in Europe and Asia. The horse-radish has an involved synonomy, due (1) to different interpretations of generic limits, as to whether it should go in one genus or another; (2) to the nomenclature tangle in which the former genus Nasturtium is involved. For botanical and nomenclatorial reasons, it is here separated in the genus Armoracia. The plant has no immediate relation to the radish; and the word *horse* was probably originally used in this connection in the sense of "coarse" or "large."

A. rusticana, Gærtn. Mey. & Scherb. Fl. Wett. ii, 426. 1800. (*Cochlearia Armoracia,* Linn. Sp. Pl. 648. *Nasturtium Armoracia,* Fries, Fl. Scan. 65. 1835. *Roripa Armoracia,* Hitch. Spring Fl. Manhattan, Kans. 18. 1894. *Radicula Armoracia,* Robinson, Rhodora, x, 32. 1908.) HORSE-RADISH. Stout glabrous perennial with dock-like leaves: root branching, long, hard and deep: lower leaves of two kinds, mostly oblong or oblong-ovate and undivided, long-petioled, margins crenate-dentate, but sometimes lobed or even pectinate both from the root and on the lower part of the stem; main and upper stem leaves mostly sessile or tapering to a petiole-like base: stem erect, 18 to 36 in. high, branched above: flowers white, ⅛ in. or more across, in panicled racemes, the petals obovate: pods (sometimes not forming) ovoid to short-oblong, ⅛ in. or more long, slender-pedicelled, with very short style and large stigma, 2-celled with seeds in 2 marginal rows in each cell: seeds seldom maturing, never sought for propagating the cutivated plant, cordate-orbicular.—Southeastern Europe, by some writers thought to be possibly a form of another species; in this country it has run wild in moist land and along ditches, where its abundant white flowers are conspicuous in late spring. (The word *Armoracia* is an old substantive in Latin—from the Greek—designating the horse-radish.)

CARROT

Very clean and mellow land, particularly soil that will not "bake" over the seeds, and close attention to surface tillage, are requisites for the culture of carrots. Seeds are slow to germinate and they are sown where the plants are to grow. The crop is half-hardy. It is easy to grow after the plants are well established. It is mostly a succession-crop.

Carrots are sown in drills from 10 to 18 inches apart, depending largely on the variety and the method to be employed in tilling. The early crop is thinned to 4 or 5 inches in the row, and the late large varieties to about 6 or 8 inches. Rows are 10 to 16 inches apart, or twice this distance for horse tillage. If it is not desired to plant the late varieties for autumn use, one may choose the early varieties for that purpose, sowing the seed late in July or even the first of August. Unless the soil is in very fine tilth and moist, however, it is difficult to secure a stand as late in the season as this. Carrot seed should always be sown thickly to allow for any failure in germination. It is sown about ⅓ or ½ inch deep. For an acre, 2 to 3 lbs. of seed are required; for 300 feet of drill, 1 oz., if the seed is fresh. Good crops run 200 to 400 bu. to the acre, and in special cases more than this if the very large kinds are grown.

STORAGE ROT (*Sclerotinia libertiana*).—Frequently carrots in storage show a soft rot over which there later appear white felts of mycelium containing hard black fungous bodies. These black bodies or sclerotia serve largely to carry the fungus over winter. *Control:* Carrots should not be grown on land infested with the organism. The removal of affected plants in a field is desirable to eradicate the fungus. Thorough drying of the roots in the field, careful sorting out of decaying carrots, and storage under cool dry conditions are important.

CARROT RUST-FLY (*Psila rosæ*).—A slender straw-colored maggot, $\frac{3}{10}$ in. long when mature, that burrows in the root of carrots. Fortunately in this country serious attack is not likely to continue in the same locality for more than one or two seasons in succession. No satisfactory control is known.

CARROT BEETLE (*Ligyrus gibbosus*).—A reddish brown beetle resembling a June beetle, about ½ in. long, that feeds mostly underground, gnawing out holes in the roots and underground stems. *Control:* Clean farming and a short rotation of crops.

Carrots are grown for human food and also for livestock. In the former utilization, which is the only part of the subject under consideration here, there are two leading types: those grown for spring or early summer use, and those grown as a main crop and used in the winter. The main-season carrots are not cultivated very extensively as a vegetable-gardening crop. Young fresh carrots may be shipped from the Southern States so cheaply that there is relatively little need of storing the roots for market. The carrot does well as a hotbed crop.

95. Fruits ("seeds") of carrot (× 5).

Light quick fertile land is essential for the growing of tender sweet carrots. In such lands the germination is also more certain and uniform. The carrot is a fairly hardy plant, and the early varieties may be sown as soon as the land is fit in the spring. The late varieties may be sown as late as the middle of June in the Northern States. Carrots mature rather slowly, and even the early varieties require 2 to 2½ months to bring them to edible size, unless they are aided in their growth by a covering of sash. On land to be used for late carrots, it is well to sow some early stuff in spring, as radishes, and to keep the ground clean until it is needed for the carrots. The early weeds will then be killed, and

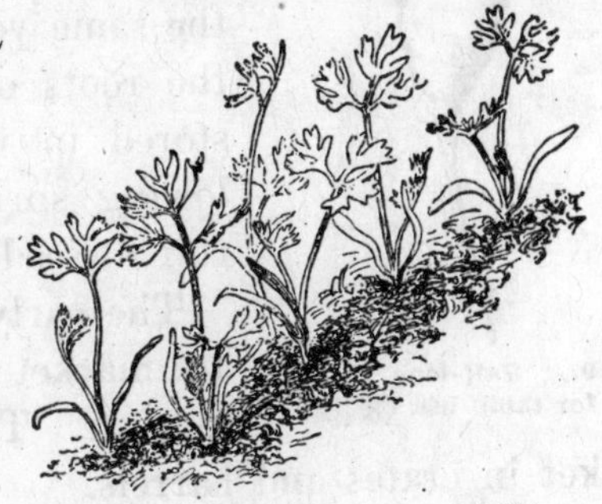

96. Seedlings of carrot (× ¾).

the young carrot plants will have an opportunity to grow. Special care must be taken to keep down weeds. In their early stages, carrot plants are shallow-rooted and delicate, and the tillage should be very careful. A late crop may follow early carrots, and an early crop may precede the late ones.

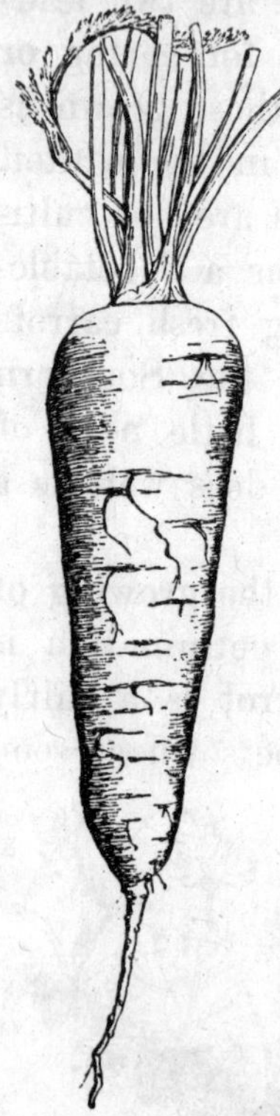

97. Half-long carrot, for table use (× 1/3).

The seeds of carrots are small (Fig. 95) and germinate slowly (Fig. 96). Unless the soil is in good condition and free of weeds the young plants are likely to suffer. It is well to sow seeds of radishes, turnips or other quick-germinating things with the carrots to mark the row and to break the crust.

The carrot is annual and biennial. The early varieties send up flower-stalks the same year if left in the ground; but the roots of the late varieties must be stored in winter, and set out the following spring, when they will quickly run to seed.

The early short and half-long carrots are marketed in small bunches, with the tops on. The main crop is sent to market in crates and barrels.

Varieties of carrots are either yellow-fleshed or white-fleshed. They are also of severai forms. The stump-rooted or half-long varieties (Fig. 97) are popular for garden work. These are early or mid-season varieties fit for using either early in the season or late in summer. The

Early Forcing (or similar varieties) is one of the best for growing in hotbeds or coldframes, or in autumn for home use. The Half-long Danvers is one of the reliable mid-season varieties. For late or main-season crop, the Long Scarlet is excellent; and for stock-feeding the Long Orange and Long White or Belgian are used. These latter types are also good for home use, although when they are allowed to reach their full size they are likely to be somewhat coarse in texture.

The Carrot Plant

Daucus. *Umbelliferæ.* About 60 species in many parts of the world, including several native in North America, very few known to cultivation.

D. Carota, Linn., var. **sativa,** DC. Prodr. iv, 211. 1830. Cultivated Carrot. More or less hairy annual and biennial, with fern-like foliage: taproot single, much thickened and forming the carrot of gardens: leaves sparsely bristly-hairy, mostly long-stalked, the base of the petiole expanded; blade pinnately decompound, the many ultimate segments nearly linear and acute: stem erect, 2 to 3 ft., bristly-hairy, grooved, much branched, bearing showy compound many-rayed umbels on the ends of long branches, the involucre bracts leaf-like and cleft into linear divisions: flowers small and numerous, in globular umbellets, the whole umbel more or less globular, the outer flowers with unequal petals and usually on longer pedicels or rays; petals 5, obovate and obtuse or emarginate; anthers exserted; style short and stout: fruit ("seed") one of the two separable carpels, oblong, about ⅛ in. long, convex on the back and bearing 3 ridges and intermediate spiny or wavy ribs, flat and 2-ribbed on the front or face, crowned with the short style-beak (which may be broken off in commercial seeds), weighing 1 to 2 mg.; longevity 4 or 5 years.—Cultigen; derived from the wild carrot (*D. Carota,* which is native in Europe, N. Africa and Asia, and introduced and extensively spread in North America. There are apparently points of difference between the domesticated and wild plant aside from the thickened root of the former. The flower-head of the garden carrot is likely to be globular, as are also the umbellets, rather than flat or saucer-form, as in the wild plant. The foliage, particularly in virgin plants, seems to have peculiarities between the two. (The word *Carota* is Latin for carrot, and from which the English word is derived.)

PARSNIP

A cool very deep rich open soil and one that does not "bake" over the seeds and a full-length season are requisites for parsnip-growing. Seeds are sown where the crop is to stand. The plant is hardy.

The seeds of parsnips germinate slowly, and retain their vitality only a year or two; therefore they should be sown thickly. Seeds are usually sown in drills far enough apart to allow of wheel-hoe or horse tillage, and the young plants are thinned to stand about 6 to 8 in. in the row. In gardens, the rows may be 14 to 18 in. apart; in field culture with horse tillage, 24 to 30 in. The seed is covered ½ in. to 1 in. with earth. One ounce of fresh seed is used to 200 to 250 feet of drill; 4 to 6 lbs. are generally sown to the acre. A good crop is 500 to 600 bushels to the acre, but more than this is obtained under the best conditions.

There are no menacing diseases of parsnips, and the insects are mostly those of carrots (which see). The larvæ of the black swallow-tail butterfly sometimes attack parsnips; see under Celery; also carrot rust-fly and beetle.

Parsnip webworm (*Depressaria heracliana*).—Small greenish yellow caterpillars, that web together and devour the unfolding blossom-heads of parsnip and celery, greatly decreasing the seed crop. The parent moths hibernate under flakes of bark, and on emerging deposit their eggs on the plant near the flower-heads. *Control:* Spray or dust the flower-heads with arsenate of lead after they have opened. So doing will kill many of the caterpillars.

The parsnip occupies the land the entire season. The seeds are sown in spring as early as the ground is fit. As they germinate slowly, it is well to plant radishes or other quick-growing seeds with them to break the ground and mark the row; of course these other plants must be quickly removed, and this may not be practicable in a large area. The crop is sometimes grown for live-stock.

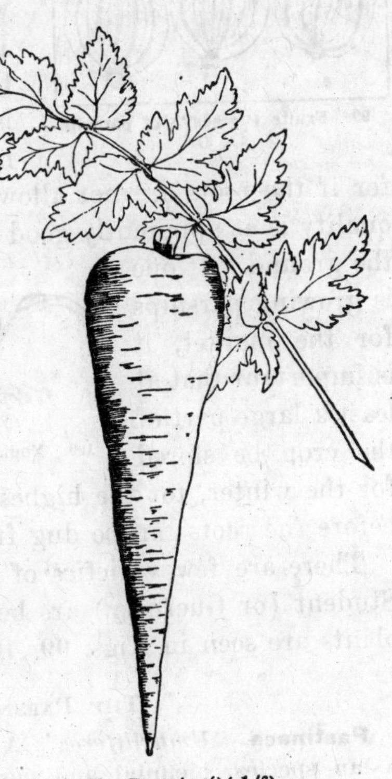

98. Parsnip (× 1/6).

The plant makes a long-cylindrical tapering root (Fig. 98): therefore the ground should be deep. Much of the value of the parsnip as a market crop is destroyed when the roots are branchy and forking. Land that is shallow and lumpy tends to make such roots. Good parsnip roots should be 1 foot long, and straight, clean and comely.

Parsnips are rarely sold before the end of the season. They are sent to market in crates, boxes and barrels. They are stored in the same way as beets and turnips—in bins in the cellar, and in pits.

The roots may be harvested in autumn and stored in the cellar or in pits, or they may be left in the ground until spring. The hard freezing of winter does not injure them. In fact, many persons think that the quality of the roots is improved by freezing. This notion is probably unfounded, for if the roots are not allowed to shrivel in winter, their quality is as apparently good as when allowed to remain in the ground. If one is growing parsnips for the market, it is important that at least a large part of the crop be stored for the winter, for the highest prices are usually obtained before the roots can be dug from the field in spring.

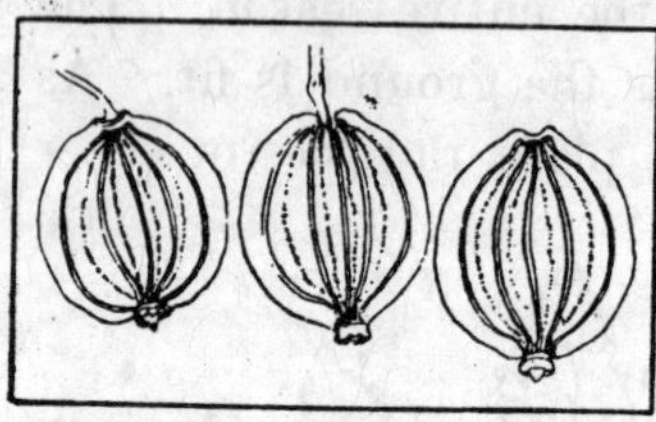

99. Fruits ("seeds") of parsnip (× 2).

100. Young plants or seedlings of parsnip (× ½).

There are few varieties of parsnip. Hollow-crown and Student (or Guernsey) are best known. Seeds and young plants are seen in Figs. 99, 100.

The Parsnip Plant

Pastinaca. *Umbelliferæ.* A dozen or so European and Asian species, biennial and perennial.

P. sativa, Linn. Sp. Pl. 262. CULTIVATED PARSNIP. Tall stout mostly glabrous strong-scented biennial (rarely annual): taproot single and enlarged to form the parsnip of gardens: leaves long and rather narrow, odd-pinnately compound, long-stalked with petioles expanded at base, leaflets ovate to oblong, sessile or short-stalked, more or less irregularly lobed, the margins toothed or cut: stem erect, strongly grooved and angled, 3 to 4 or 5 ft. high, branched: flowers greenish, small, in compound umbels that are mostly devoid of involucre and involucels; umbels enlarging in fruit, the rays sometimes 6 in. long; petals obovate, clawed, incurved; stamens exserted; styles 2, spreading or recurved: fruit ("seed" of commerce) one of two closely appressed but separating carpels, very thin, flat, oval, about ¼ in. long, wing-margined, strongly ribbed on the outside and less so on the inner face, weighing 2 to 5 mg., holding germinating power only a year or two.—Europe and Asia; in var. *sylvestris*, DC., extensively spread in this country as an introduced weed.

CELERIAC

The celeriac, or turnip-rooted celery, has a short, thick, tuberous crown-base, from which many roots arise. This tuber is the edible part, being used either as salad or a cooked vegetable (Fig. 101). It has the celery flavor. The plant is dwarf; it requires no blanching, being generally grown only for the root. Sometimes the seeds are sown where the plant is to grow, but as they are as slow to

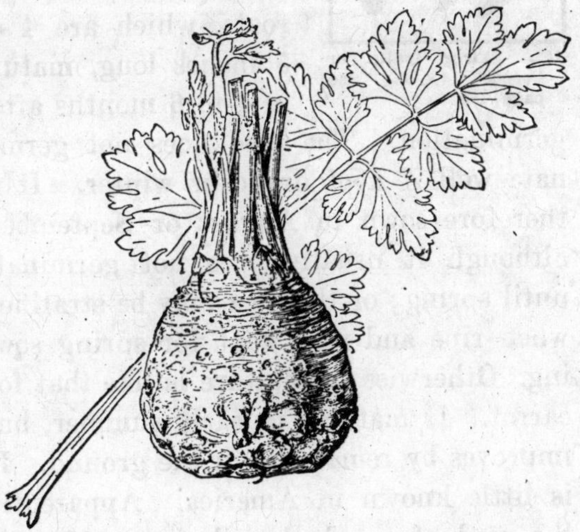

101. Celeriac, trimmed root and leaf (× 1/3).

germinate as those of celery it is advisable to start in a seed-bed and transplant. The plants are allowed 6 or 8 inches in the row, and the rows may stand at 12 to 20 inches. The roots may be stored in winter as are other roots.

For a botanical account of celeriac, see page 139.

TURNIP-ROOTED OR TUBEROUS CHERVIL

The chervil is a small-rooted plant, something like carrot and of similar utility, but that the roots are gray or nearly black and of different flavor. The roots, which are 4 or 5 inches long, mature in 4 to 6 months after germination. The seed does not germinate well if kept dry over winter. It is therefore sown in August or September, although it usually does not germinate until spring; or the seed may be stratified when ripe and thus kept for spring sowing. Otherwise the culture is like that for carrot. It matures in early summer, but improves by remaining in the ground. It is little known in America. Apparently the seed of salad chervil (page 124) is sometimes sold for this plant.

102. Skirret fruits, or "seeds" (× 6). See page 195.

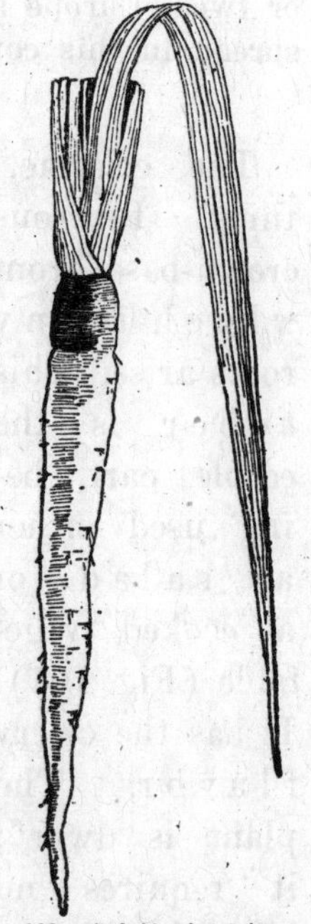

103. Salsify (× 1/6).

Tuberous or turnip-rooted chervil is **Chærophyllum bulbosum**, Linn. Sp. Pl. 258, native in Europe. It is an upright

branching more or less hairy biennial, 2 to 3 ft. tall, with ternately decompound leaves, the ultimate segments being linear rather than ovate or fernlike, as in salad chervil (*Anthriscus Cerefolium*), producing underground spindle-shaped tubers 2 to 4 in. long: fruit ("seed") nearly linear, about ¼ in. long and more or less curved, not long-tapering as in the anthriscus, plane and unmarked on the front, convex on the back and with 4 dark-colored furrows on the back and sides, weighing 2 mg.

SKIRRET

Seeds of skirret (Fig. 102) are sometimes offered by American seedsmen, but the plant is little known in this country. It is raised for the thick but small prongy clustered roots, which are used in the same way as salsify and parsnip. The plant is perennial (but commonly treated as annual) and roots may be left in the ground over winter, being harvested as wanted. If seeds are sown in spring, good roots should be had in autumn. Sometimes the small roots and side prongs are used for propagation, the same as seeds. The plants are usually spaced 6 to 8 inches in the row and the rows may be 12 to 15 inches.

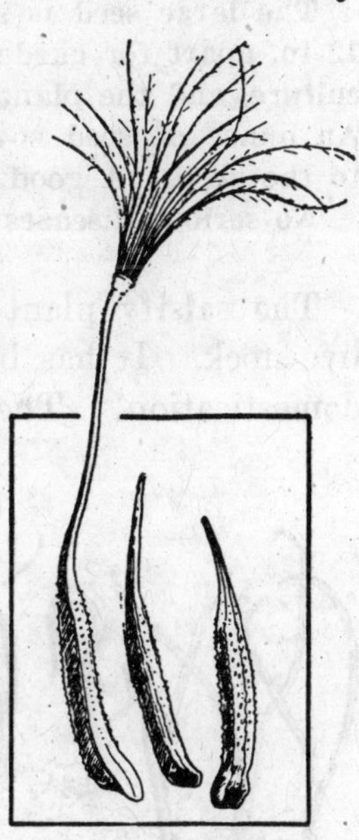

104. Seeds (properly fruits) of salsify, the one at the left with the beak and pappus remaining (× 1½).

Skirret is *Sium Sisarum*, Linn., native in Asia, one of the Umbelliferæ or Parsley family. The plant grows 1 to 3 feet tall, with odd-pinnate leaves and one to three pairs of lanceolate pointed toothed leaflets; flowers small, white, in terminal compound involucrate and involucellate umbels; fruits ("seeds") more or less curved, the ribs usually 3 on the back and 1 on either edge (Fig. 102).

SALSIFY

Deep rich cool soil and the full-length season are required for the production of good salsify. It is not transplanted. Hardy and easily grown.

The large seed is sown about 1 in. deep in drills or rows 12 in. apart for garden culture and sometimes 18 in. for field culture, and the plants are thinned to stand 3 to 5 in. apart. An ounce of seed sows about 70 feet of drill; 8 to 10 lbs. to the acre. A good yield is 200 to 300 bu. to the acre.

No serious diseases or insects are reported on salsify.

The salsify plant is grown for cooking only, not for live-stock. It has been comparatively little improved by domestication. There is a relatively large-rooted form known as the Mammoth Sandwich Island, and another called the Improved French. Even of the largest varieties, the roots are small, rarely more than 2 inches in diameter at the crown (Fig. 103). Because of its flavor of oysters, it is commonly known as the oyster plant or vegetable oyster.

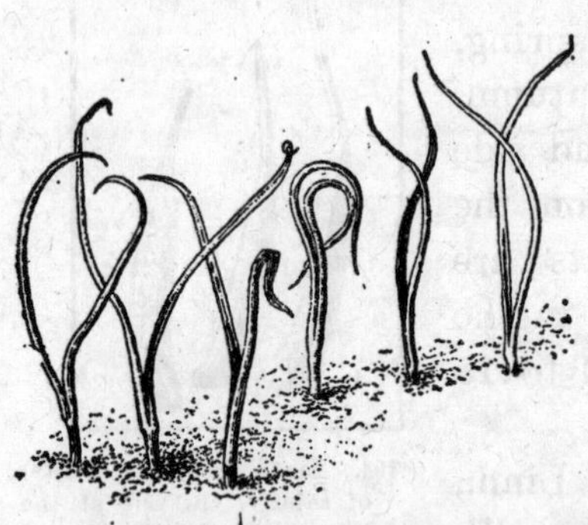

105. Seedlings of salsify (× ½).

The seed (Fig. 104) is sown in drills as soon as the

ground is ready in spring and the young plants (Fig. 105) thinned as they stand. The plant is perfectly hardy and the roots may be left in the ground over winter, as they are not injured by frost. If one desires to use the plant in winter, however, or wishes to find the best markets, a large part of the roots should be stored in the cellar or in pits. The seeds germinate readily; they are long and stick-like, and are rather difficult to sow with the seed-drill.

Sometimes salsify is bunched in autumn, but usually it is sent to market in crates or other receptacles.

The Salsify Plant

Tragopogon. *Compositæ.* Between 30 and 40 species of annual, biennial and perennial herbs, in Europe and Asia.

T. porrifolius, Linn. Sp. Pl. 789. Salsify. Oyster Plant. Vegetable Oyster. Stout erect glabrous biennial, with milky juice, the slender thickened long taproot constituting the salsify of gardens: leaves many, alternate, grass-like or garlic-like (*porrifolius* means "leak-leaved"), ¾ in. and less wide near the base, very long and long-pointed, the base broad and clasping, margins entire: stem 3 to 4 ft. tall, usually forked: heads solitary and showy, 2 to 3 in. across when expanded, terminal on long naked branches or peduncles that are enlarged and fistulose at the summit, the involucre of many linear acuminate green bracts (in a single series) that equal or exceed the purple rays, flowers closing at midday: flowers many in the head, all perfect and ligulate, the rays 5-toothed: fruit 1 to 1½ in. long, comprising the ripened carpel and a slender beak or stalk of greater length, the outer fruits in the head having upwardly serrate lines and the slenderer inner ones nearly or quite destitute of them; on the beak is borne the tuft of soft plumose pappus: the "seed" of commerce is the stick-like brown or gray fruit from which the pappus and more or less of the beak have been broken, ranging about ½ to ¾ in. long, angular, grooved and roughened, tapering above into the beak, the pieces weighing 10 to 25 mg.; longevity about 2 to 3 years.—Mediterranean region; an introduced weed in North America and other countries, along roadsides and in waste places, in such cases not producing the thickened roots of the cultivated plant.

SCORZONERA OR BLACK SALSIFY

The cultivation of this plant is in all ways like that of salsify, except that it should be given much more room. It is perennial, however, and the roots continue to enlarge without becoming inedible if left in the ground for more than one year.

It has a long black root, yellow flowers, light-colored seeds, and broader leaves than salsify. It is used in the same way as salsify. The plant is little known in North America (Figs. 106, 107).

Black salsify is **Scorzonera hispanica**, Linn. Sp. Pl. 791, of the *Compositæ*, closely related to Tragopogon. It is perennial, with milky juice, bearing many slightly pubescent keeled leaves 12 to 18 in. long, the mid-blade 1½ in. wide, lanceolate and tapering gradually into a long sharp point and below into a long-winged petiole: taproot thickened like that of salsify: stem erect, 2 ft. or more, the leaves with clasping bases: heads single, terminal, the involucre bracts in two or more series and not leafy, the flowers yellow: fruit nearly white, angular, grooved, the inner ones in the head smooth and the outer ones slightly serrate, bearing a long beak and tuft of pappus; the commercial "seeds" lack the beak and pappus, the former disarticulating, the remaining part ½ to ¾ in. long and weighing 10 to 15 mg.—Central and southern Europe.

SCOLYMUS OR SPANISH SALSIFY

This plant is cultivated like salsify, and the roots are used for the same purposes. It makes a root much like salsify, except that it is lighter colored and considerably longer. Its flavor is less pronounced, but when carefully cooked it possesses a very agreeable quality somewhat intermediate between that of the salsify and parsnip. It is adapted to all the methods of cooking employed for those vegetables. The particular value of the vegetable, aside from affording a variety in the kitchen-garden, is its large size and productiveness as compared with the

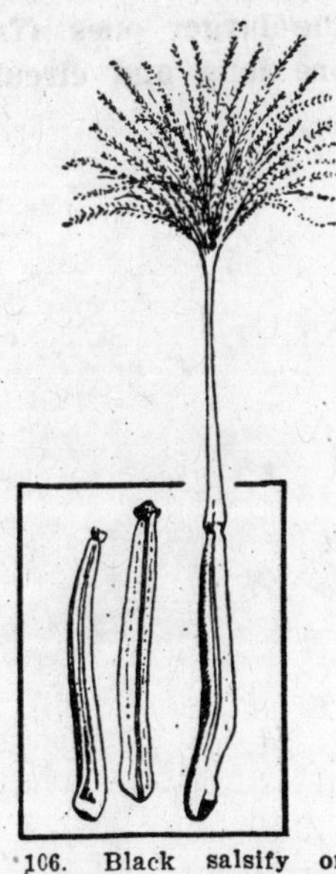

106. Black salsify or scorzonera (× 1/3).

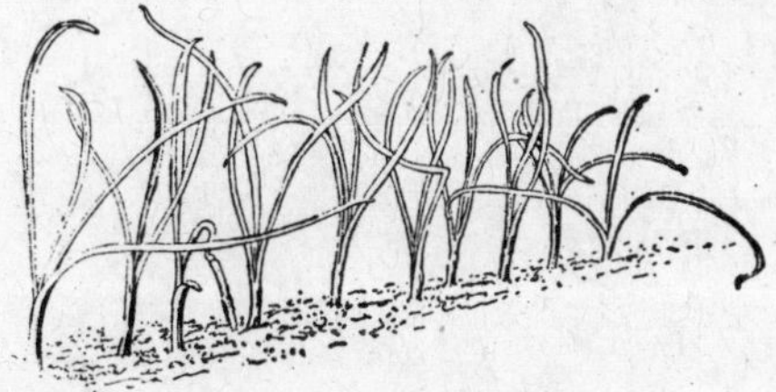

107. Young plants of scorzonera (× ½).

salsify. Almost twice the crop can be grown on a given area. The seeds are much easier to handle and sow than those of the salsify. It can be dug either in the fall or spring. Perhaps the greatest disadvantage of the plant is the very prickly leaves, which may make it unpleasant to handle. It is worth more attention in American gardens, but it may not survive the winter at the North and produce seed, although edible roots are made the first year. The plant is sometimes called golden thistle.

The Spanish salsify belongs in the Compositæ, being one of the three species of Scolymus, **S. hispanicus**, Linn. Sp. Pl.

813, native in central Europe. It is a very spiny thistle-like biennial with milky juice and long pinnatifid shiny hairy green leaves that have lighter colored ribs and veins: root single or branched: stems 1 to 2 ft. or more high, spreading, bearing many stiff spine-tipped clasping and decurrent leaves: flowers yellow, in sessile axillary heads: fruits (seeds) chaff-like, wing-margined, variable in size, the larger ones (from the outside of the head) ¼ in. or more long and circular-oblong. (*Hispanicus:* Spanish.)

CHAPTER VIII

THE POTATO CROPS

Potato **Sweet potato**

The potato crops are major horticultural products requiring not only choice and preparation of land but forethought in the arrangement of rotations and in the assembling of equipment and supplies. They are heavy products and require the use of good machinery and vehicles. The grower must prepare for the supply of labor, horse or other power, manures and fertilizers, good seed, insecticides, fungicides and sprayers, graders and handling conveniences, and must look long in advance into the transportation and market facilities. The outlay for growing heavy crops on any important scale—which is the only profitable way—is so considerable that the man should be ready and well prepared at the start. The potatoes are "money crops," and are likely to consume a large proportion of the man's time and plan.

The potato crops are two, the common or Irish potato, and the sweet potato. The former is staple in the North and the latter in the South. The two are so unlike in cultural requirements that it is not expedient to endeavor to state principles that apply to both. Yet they are usually associated in the public mind and may be brought together for comparison if not for agreement. What is known as "potato" in the South is the batatas or sweet potato; in the North it is solanum (Irish or round potato). Potatoes are tuber crops grown underground, and similar types of tools are required as well as good knowledge of the heavy handling of land.

POTATO

Deeply pulverized cool soil holding much capillary moisture and rich in potash, carefully chosen seed tubers that are also free from disease, deep and early planting, level culture, frequent surface tillage to conserve moisture, careful and persistent attention to the many diseases and insects: these are requisites of the best potato culture. The potato is propagated by divisions or cuttings of tubers. It thrives best in a relatively cool climate: in the South, it is successful only as an autumn to spring crop, for the midsummer season is too continuously hot. The potato is not tender to light frosts.

Potatoes are planted in drills or continuous furrows, 3 to 3½ feet apart. Single pieces of tubers are dropped at intervals of 12 to 18 inches. If the pieces are cut to one strong eye and dropped at above distances, 8 to 10 bushels are required to plant an acre. Usually the pieces are cut to bear about two good eyes or buds. Many planters use too little seed. The "seed" is covered 3 to 5 inches deep, the latter depth only in light or loose soil. The yield of potatoes averages about 75 bushels to the acre, but with forethought and good tillage and some fertilizer, the yield should run from 200 to 300 bushels, and occasional yields much exceed the latter figure. In large-area operations potatoes are planted and harvested by machinery, or by specially made plows. There are various devices for sorting and grading them.

Late blight (*Pytophthora infestans*).—The appearance of water-soaked areas on the leaves and in wet weather the occurrence on their under surface of a white mildew are characteristic of late blight. The disease spreads rapidly, the blighted plants giving off a disagreeable odor. Irregular discolored lesions, which later become somewhat sunken, appear on the tubers. It is in these diseased tubers that the fungus lives over winter. *Control:* Spray with bordeaux mixture 5–5–50, beginning when plants are six inches high and repeating every ten days to two weeks throughout the season. Insecticides may be added directly to the bordeaux. Potatoes in blighted fields should not be dug until the vines are dead and dry.

EARLY BLIGHT (*Alternaria solani*).—Irregular dark brown spots that show concentric rings develop on the leaves, and premature death of the foliage may result. *Control:* Thorough spraying with bordeaux mixture 5–5–50 will afford control. Applications should be begun early.

SCAB (*Actinomyces chromogenus*).—The disease is due to a parasitic bacterium that attacks the skin of the potato tuber, causing rough corky areas. The organism not only overwinters on the tubers but also in the soil and manure. *Control:* Uncut tubers should be soaked for 1½ hours in a solution made by adding 4 ounces of powdered corrosive sublimate to 30 gallons of water; spread them out where they will dry quickly. It is important to use wooden containers for the solution, and tubers should preferably be treated before sprouts have developed to any great extent.

RHIZOCTONIA (*Rhizoctonia solani*).—The most easily recognized symptoms are black scurf on the surface of affected tubers, reddish brown cankers on young sprouts, dwarfing or rosetting of vines, and the production of numerous small ill-shapen potatoes. *Control:* The treatment of seed tubers with corrosive sublimate as recommended for potato scab is advisable. Crop rotation is important.

BLACK WART (*Synchytrium endobiotica*).—Rough warty outgrowths are produced on the tubers, especially at the eyes, and may occur on other underground parts of the plant as well. The warts are at first brown, but later become black and show decay. The disease is very destructive. *Control:* Report the disease promptly to the State experiment station and receive recommendation for control.

MOSAIC.—Affected leaves are frequently wrinkled and present a mottled appearance, light green or yellow areas alternating with the normal green of other parts of the leaf. *Control:* Plant seed from mosaic-free fields.

LEAF ROLL.—A rolling of the leaves, beginning with the lower ones, accompanied by a change in color to a pale green is characteristic of this disease. The development of the plant is checked and the yield is greatly reduced. *Control:* Seed tubers should be obtained from fields free from disease.

POWDERY SCAB (*Spongospora subterranea*).—In the early stages of this disease small blisters are formed on the skin of the tubers. Later these blisters rupture, exposing a dark powdery mass and appear as raised pustules surrounded by the torn skin of the potato. *Control:* The use of disease-free tubers, seed treatment and crop rotation are important.

FUSARIUM WILT (*Fusarium oxysporium*).—Rolling and wilting of the leaves, together with yellowing of the foliage and premature death of the vines, are characteristic. Stems of affected plants show a blackening of the sap tubes. *Control:* The planting of field-selected or certified seed is advisable, and crop rotation is important.

BLACKLEG (*Bacillus phytophthorus*).—This is a bacterial disease carried on the seed tubers. It may cause rot of the seed tuber and thus occasion an uneven stand. The stems of affected plants become black at the base. Diseased plants show lack of vigor and usually die without setting tubers, although they may become diseased after the tubers are formed. *Control:* Affected tubers should never be used for seed. Seed from disease-free fields should be employed and seed disinfection is advisable.

COLORADO POTATO BEETLE (*Leptinotarsa decemlineata*).—The adult is a convex black- and yellow-striped beetle about ⅜ in. long, which passes the winter in hibernation in the ground.

The elongate oval orange eggs are deposited in small masses on the underside of the leaves. The larvæ, known as "slugs," are about 3/5 inch long, red, with the head, legs and two rows of spots on each side black. When mature, the larvæ enter the ground and pupate. There are usually two broods annually. *Control:* Spray with paris green, 1 lb. in 50 gals. water to which 2 lbs. lime should be added to prevent burning of the foliage. Paris green may be applied also in the form of a dust, 1 lb. in 20 lbs. air-slaked lime, or use arsenate of lead (paste), 3 or 4 lbs. in 50 gals. water. It is best to apply the poison in bordeaux mixture except when it is not necessary to use this fungicide for the control of diseases. In the home garden the beetles may be hand-picked into a pan containing a little kerosene. This insect is the familiar "potato bug."

THREE-LINED-POTATO BEETLE (*Lema trilineata*).—A yellow leaf-beetle, about ¼ in. long, marked on the wing-covers with three black stripes. The eggs are laid in clusters on the underside of the leaves. The grubs are yellowish, with the head and legs black and about ⅓ in. long when full grown. *Control:* Spray with arsenicals as for the Colorado potato beetle.

POTATO APHIS (*Macrosiphum solanifolii*).—Plant-lice, some of which are green and others pink. They attack potatoes, causing the leaves to curl and turn brown; in some cases the death of the plants may result. *Control:* Spray with ½ pint "Black Leaf 40" tobacco extract in 50 gals. bordeaux mixture. In case bordeaux mixture is not needed for the control of diseases, use the "Black Leaf 40" in water, adding 3 or 4 lbs. soap. The spraying should be done with great thoroughness, using plenty of material, and care should be taken to hit the underside of the leaves. Begin early, before the plants become too badly infested.

APPLE LEAF-HOPPER (*Empoasca mali*).—A small pale yellowish green leaf-hopper, ⅛ in. long, that sometimes attacks potatoes. The eggs are inserted into the tender parts of the potato plant and the nymphs feed on the underside of the leaves, which turn brown and the edges roll up and die. *Control:* Keep the foliage protected by thorough spraying with bordeaux mixture alone or in combination with arsenate of lead, taking care to hit both surfaces of the leaves.

POTATO STALK-WEEVIL (*Trichobaris trinotata*).—A bluish gray snout beetle, 1/5 in. long, which lays its eggs on the stalks of potato. The larva is a grub, yellowish white, legless, 1/4 in. long when mature. Its presence is indicated by a wilting and dying of the leaves. *Control:* Practice clean farming and collect and burn the vines after harvesting the crop. Destroy all solanaceous weeds.

COMMON STALK-BORER (*Papaipema nitela*).—A caterpillar, 1 1/4 in. long when mature, that bores in the potato stalks in gardens of small patches; not usually found in large commercial fields except along the edges. Until the last moult it is grayish brown with a white dorsal stripe and two white stripes on each side, the later stripes being broadly interrupted toward the front. The parent moths lay their eggs in the fall on the stems of such weeds as ragweed, pigweed and dock. The eggs hatch the following spring and the larvæ at first attack weeds, migrating later to potato. *Control:* Clean cultivation and the destruction of weeds around the potato patch.

POTATO TUBER MOTH (*Phthorimæa operculella*).—A serious potato pest in Texas and California. The parent insect is a yellowish brown more or less spotted moth. The eggs are deposited in the field early in the spring. On hatching the larva enters the leaf, producing a blotched line and then bores down the petiole into the stalk, causing the branch to wilt. Reproduction is continuous throughout the season. Some of the caterpillars migrate from the stalks to the tubers and where the soil is loose the moths may lay eggs on the tubers. At digging time, if the tubers are left exposed in the field during the afternoon or night, the moths will lay eggs on them. The larvæ burrow through the potatoes in all directions, causing decay. In storage the insects continue to breed as long as the potatoes are in condition to serve as food.

Control: Injury to the vines is not serious. The greatest loss comes from infested tubers. Plant deep and keep the vines carefully hilled so as not to allow any of the tubers to become exposed. When harvesting, do not leave any of the potatoes exposed overnight. When potatoes are found infested in storage, fumigate with carbon bisulfide at the rate of 2 lbs. to 100 cu. ft. space, allowing the fumigation to continue for 48 hours. Repeat at intervals of a week in summer or two weeks in winter. Do not plant potatoes after potatoes; destroy all solanaceous weeds.

POTATO FLEA-BEETLE (*Epitrix cucumeris*).—A small black flea-beetle, $\frac{1}{16}$ in. long, that riddles the leaves with holes. *Control:* Keep the plants thoroughly covered with bordeaux mixture. When an arsenical is added for the control of the potato beetle many of the flea-beetles are also killed.

The potato is such an important article of food and commerce that much study has been given it and an extensive literature has developed. To the books and bulletins the reader is referred if he intends to make anything like a specialty of the crop. Although potatoes will grow practically anywhere, within reason, yet real success in the cultivation of them is a question of good soil and location and of thoughtful experience. At this time only the simplest advice may be given; and of course this book has in mind the vegetable-garden handling of the crop.

The early potato crop, for market-gardening use, is secured by (1) choosing "early" soil and site; (2) by preparing the land the fall before, either by means of special plowing or by growing a late-tilled crop; (3) by using quickly available concentrated fertilizers; (4) by choosing early varieties; (5) by sprouting the potatoes in a warm place before planting (before the tubers are cut), allowing the sprouts to become 3 to 6 inches long. It is a widespread practice to raise the early crop from northern-grown seed stock.

Land and tillage.

The land should be loamy rather than heavy, well-drained, working up deep and mellow. The potato crop is good to follow sod and to prepare the land for other crops. Not only is the land well prepared for the crop and well tilled, but the digging amounts to another tilling and cleaning of the land.

In most cases a heavy yield of potatoes is largely a question of moisture, as well as of fertility. If planted late, the crop loses the benefit of much of the winter precipitation. Planting on ridges or hills wastes the soil moisture in most cases. "Hilling up" is often necessary, however, because the land is not deep enough to allow the tubers to grow well below the surface; and in market-garden operations the practice may conduce to earliness by exposing the soil more fully to sun heat. The ground should be such as to allow the tubers to be planted at least four inches beneath the level. If the potatoes are dropped in a deep furrow, the earth is plowed over them, and the surface may be harrowed two to three times before the plants are up, thus conserving moisture and destroying weeds.

The land should be fertile, for the tonnage of the product is heavy. Raw heavy stable manure is usually avoided, or it may be applied on the sod the preceding autumn and plowed under. Well-rotted or old manure is often used. The potato responds specially well to commercial fertilizer, and brands rich in potash are preferred.

The very heavy continuous yields of potatoes are largely a question of the proper soil.

Five to eight light surface tillings are required during the season to save the moisture. Even after the vines have begun to spread and to cover the ground, tillage may be necessary in a dry year.

Seeding.

The size in which pieces of the seed tuber should be cut has been the subject of much controversy, but the question is easy of solution if careful and comparable experiments are made. Arthur long ago showed (Proc. Soc. Prom. Agr. Sci., 1891, p. 11; Bull. 42, Purdue Univ.) that the unit in such tests should not be the number of eyes to the piece, but the size of the piece. The piece contains food. The more food the stronger the initial growth of the plant; and the stronger the initial growth, the better the crop, other things being equal. But if the piece is too large it contains so many eyes that there will be too many stalks to appropriate the food and to struggle with each other. The pieces on the tip or "seed end" may contain several eyes, but those from the other parts of the tuber usually should contain only one or two eyes. Seed should not be cut any considerable time in advance of planting unless it is rolled in plaster to prevent excessive drying.

The character of the crop depends greatly on the breeding. Seed tubers should be taken only from productive hills showing the qualities of the particular variety. Choosing good-looking tubers from the bin is not a form of plant-breeding; the selection should always consider the pedigree. Breeding for resistance to disease is important.

The grower who is not a potato-breeder should purchase seed of quality from persons who give it special attention.

In the Southern States, the common or Irish potato (also called "round potato" and "white potato") is a minor crop in general farm operations. The crop must be grown either early or late in the season to avoid the long hot summer. It is then difficult to keep the potatoes from the spring crop until the next spring, or even until it is time to plant the second crop in August (in the Gulf States). "Seed" is commonly secured from the North, and only a spring crop is grown for the Northern market.

Harvesting and storing.

In small areas, potatoes are dug by hand, a potato hook or fork being used. In field operations, various horse-drawn diggers are employed. The implements cut under the row and lift out the potatoes, or turn them out as from a furrow. Usually there are rear fingers on the

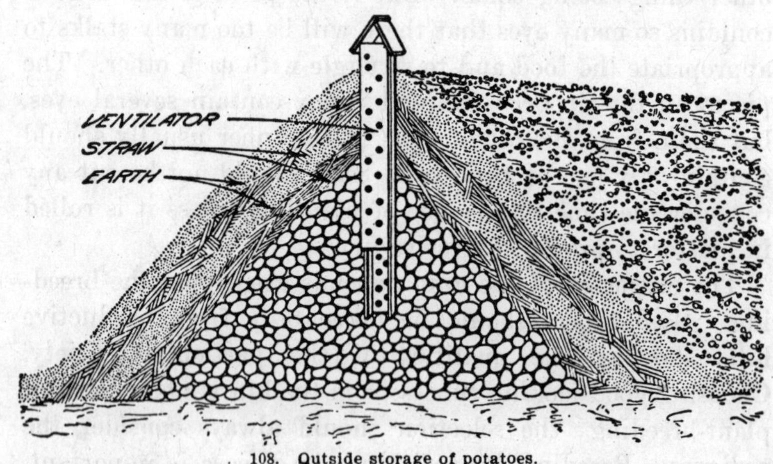

108. Outside storage of potatoes.

implement to sift out the earth and rattle the tubers clean. In other styles, there is a carrier that takes the potatoes to the rear of the machine and drops them there. The potatoes are allowed to lie in the sun for an hour or more, so that the earth will dry and shake off; then they are taken to shed, cellar or to regular storage.

109. Potato storage-house in Maine.

The potatoes should be graded for good market results. This may be accomplished by the workmen as they pick up the potatoes in the field and deposit them in the baskets or crates. Several kinds of mechanical graders are now in use for the sorting of the commercial crop. Potatoes are marketed in bushel crates, sacks, ventilated barrels, and sometimes in bulk.

110. Potato cellar.

Growers commonly prefer to dispose of the potato crop before winter, as it is heavy and bulky to store and shrinkage is likely to be heavy. If the crop has been grown free from disease and well matured, however, it keeps well, in ordinary cool unheated cellars, in pits or in houses constructed for the purpose. Houses should be certainly frost-proof and capable of maintaining a temperature of approximately 40 degrees. Some of the forms of storage pits are shown in Chapter XIX. The illustrations show types of storage structures as described by William Stuart in Farmers' Bulletin 847, U. S. Department of Agriculture. Fig. 108 is a cross-section of a potatc pit insulated with layers of straw and earth, showing the perforated ventilator in position and the potatoes piled in inverted V-shaped fashion. Fig. 109 is Maine type of potato storage-house, with central driveway into the basement part. Fig. 110 is a good outside potato cellar.

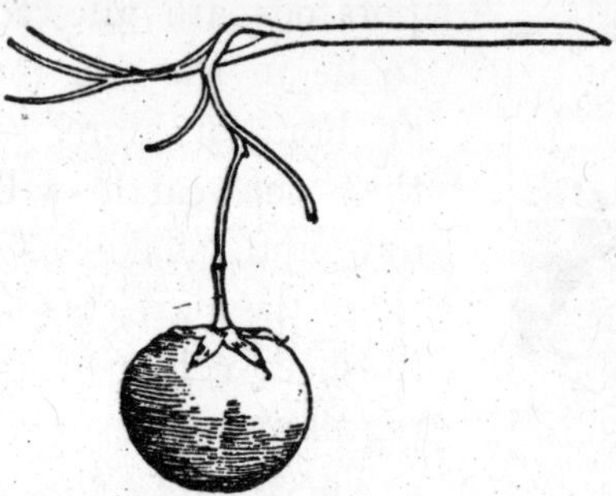

111. Fruit or berry of potato (× ¾).

112. Cluster of potato flowers (× 2/5).

On keeping potatoes in the South from the spring crop to the fall crop, McKay makes the following discussion (in Bull. 54, Miss. Exp. Station): "If exposed to the hot sun a few hours Irish potatoes will become blistered. To prevent this, dig on cloudy days or else arrange to remove to a shady place or cover in some way shortly after they are dug. Several methods of keeping potatoes during the hot summer months are practiced, and with varying success. Upon examination it will be found that, as a rule, those left in the field, scattered through the soil, keep better than those that are carefully housed. Taking this lesson from nature, we have tried the method of bedding the potatoes in the field, somewhat after the usual plan of bedding sweet potatoes for growing slips, and with good success. We are careful to see that the potatoes are covered to the depth of 6 or 7 inches with dirt, and that the bed is well drained. We have practiced the same method of bedding

113. Good sample of seed potatoes.

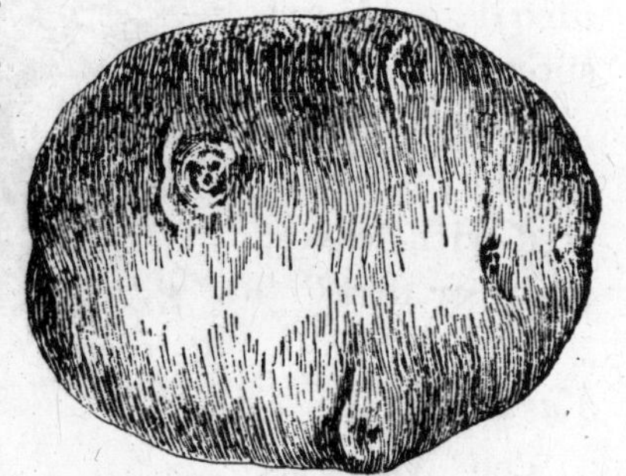
114. The round type of potato.

the potatoes in the shade of spreading trees, and on the cellar floor. A cool, shady situation is better than the open field. We have had much better success with potatoes covered with soil than with those spread out in open air in the cellar, or under trees covered with leaves. In no event should the potatoes be piled or heaped together, so long as warm weather continues. If potatoes intended for the table are exposed to the light for any considerable length of time they will turn greenish in color and become unwholesome for food. If not spread in a dark place they should be covered with leaves, straw or dirt."

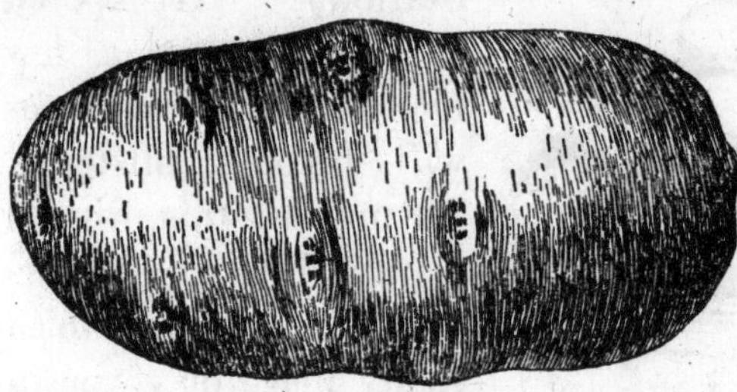

115. The oblong type.

116. A potato plant.

Varieties.

The varieties of potatoes are numerous and poorly defined, and it is not worth the while to enumerate them here. Because of variation and inattention to selection, varieties of potatoes run out. New kinds are easily grown from the true seed, but the seed-balls (Fig. 111) are not often produced in the highly-bred potatoes, probably because of insufficient pollen supply. Seeds are taken from the balls or berries and kept in the same way as from tomatoes. These seeds are sown the following spring, and the small tubers produced the first season are planted the second season, when potatoes of full size will be obtained. If the seed is started early in hotbeds or greenhouse, however, and the seedlings transplanted two or three times, full-size tubers may be obtained the same year. Every seedling may be considered a new variety. The results are likely to be much more certain if the seed is of selected parentage from hand-pollinated flowers. Fig. 112 shows the potato flower. A "sample of certified seed potatoes" is seen in Fig. 113, adapted from J. G. Milward, Bull. 252 Wisconsin Experiment Station.

The desirable forms of potatoes are the short (round) and the oblong (Figs. 114, 115). The eyes should be shallow, so that the wastage in paring is reduced to the minimum. Fig. 116 shows a "hill" of potatoes, in section.

The Potato Plant

Solanum. *Solanaceæ.* A vast genus, comprising probably 1,200 species, on many parts of the globe, particularly in tropical America; herbs, shrubs, twiners, trees.

S. tuberosum, Linn. Sp. Pl. 185. (*Lycopersicon tuberosum,* Mill. Gard. Dict. No. 7. 1768). Potato. Spreading soft-stemmed usually pubescent perennial, persisting by means of tubers borne on the ends of white underground stems (rhizomes) that arise from the stalk above the seed tuber, with the roots: stem angled, branching, 2 to 2½ ft. high: leaves dull dark green, rugose, alternate, oblong-ovate in outline, the petiole flattened or grooved above, odd-pinnate; leaflets 3 or 4 pairs and smaller ones between, the larger ones 1 to 2 in. long, ovate to oblong-ovate, pointed, unequal at base, stalked, margins undulate but entire, the intermediate leaflets ¼ to ¾ in. long, sessile or essentially so: flowers white to bluish, in stalked forking clusters, about 1¼ in. across; corolla gamopetalous, with 5 broad acute shallow lobes; calyx about one-third the length of the corolla, the 5 lobes linear-lanceolate; stamens 5, the long colored anthers connivent about the single pistil and conspicuous in the center of the flower, the very short filaments attached on the base of the corolla; ovary globular, sutured on either side, 2-celled with many ovules on central placentæ; style long, exceeding the anthers, the protruding stigma capitate: fruit (corresponding to the tomato fruit) a 2-celled or 3-celled berry ½ to ¾ in. diameter, green or yellowish, the calyx not enlarging.—Temperate Andes. Anciently cultivated by aborigines in South America.

SWEET POTATO

A warm sunny climate, long season, loose warm fertile soil, liberal supply of moisture in the growing season and a less supply when the tubers are maturing, careful attention to diseases and insects, are some of the requirements of a good sweet potato crop. The plant is tender to frost. It is propagated by means of its tubers, usually from the slips or shoots that arise when the tubers are planted in beds or frames; also by cuttings of vines.

One bushel of ordinary sweet potatoes will give 3,000 to 4,000 plants, if the sprouts are taken off twice. The plants are usually set in drills 2½ to 3½ feet apart. The plants stand 12 to 18 inches apart in the drill. At 18 x 36 inches, 9,680 plants are required for an acre. These should be produced by 2 to 3 bushels of "seed" tubers. The average yield of sweet potatoes is about 100 bushels an acre (65 to 170 bu.); but 300 to 400 bushels and more can be secured.

STEM-ROT (*Fusarium batatis* or *Fusarium hyperoxysporum*).—Leaves of affected plants become dull in color and yellow between the veins. Diseased vines wilt and the interior of the stems is blackened. *Control:* Disease-free potatoes obtained by selecting tubers from healthy plants in the field should be used for propagation. It is also advisable to treat the tubers just before planting five to ten minutes in a solution made by dissolving one ounce of corrosive sublimate in eight gallons of water. Treated potatoes should be rinsed in water then dried in the sun. Careful sanitary measures in the hotbed are important. After being used once in the hotbed, soil should be removed and the framework and earth surrounding it drenched with a solution of formaldehyde or copper sulfate. Crop rotation is important since infection may take place in the field if plants are set in contaminated soil.

BLACK-ROT (*Sphæronema fimbriatum*).—Sunken black spots somewhat circular in shape appear on the surface of affected potatoes or other underground parts of the host. The lesions frequently enlarge and extend up the stem to the surface of the soil. The stem often rots off. *Control:* Selection of disease-free tubers for use in propagation is essential. Crop rotation and non-infected soil in the seed-bed, as recommended for the control of stem-rot, is important.

TORTOISE BEETLES (*Cassida bivittata, C. nigripes, Coptocycla bicolor, C. signifera,* and *Chelymorpha argus*).—Small convex beetles of changeable color that feed on the foliage in both the adult and larval stage. The larvæ have the peculiar habit of retaining their cast skins and excrement in a mass on a fork composed of two spines which extends forward over the back. *Control:* Spray with arsenate of lead (paste), 2 lbs. in 50 gals. water, taking care to hit the underside of the leaves.

SWEET POTATO WEEVIL (*Cylas formicarius*).—A slender snout-beetle, ¼ in. long; the head is black, prothorax and legs reddish and the wing-covers bluish black. The larvæ burrow in all directions through the tubers, causing decay. The life-cycle is completed in about a month, one brood following another as long as food is available. *Control:* Rotate crops and do not plant sweet potatoes near infested fields. Slightly injured tubers may be fed to stock, but those badly infested may be buried deeply. Do not introduce the insect into uninfested localities by means of infested tubers used for seed.

SWEET POTATO FLEA-BEETLE (*Chætocnema confinis*). — A small flea-beetle, $\frac{1}{16}$ in. long, pitchy black with faint bronzy reflections. It hibernates as an adult under rubbish. The beetles do not eat holes in the leaves but cut channels on the upper surface, causing the leaf to turn brown and die. The larvæ feed on the roots of bindweed. In New Jersey the attack is confined chiefly to plants in the field. Farther south the beetles injure the plants in the seed-bed. *Control:* Spray with arsenate of lead; dip the plants before transplanting in arsenate of lead (paste), 1 lb. in 10 gals. water.

The sweet potato is one of the leading crops of the South, and it is extensively grown as far north as the light lands of New Jersey. In more Northern States it is often grown in a small way on ridges in the garden. Sweet potatoes are shipped to all parts of the country, being one of the common foods in northern parts. They are also valuable for live-stock. They are little known to the people of central Europe. The largest quantities are grown in the Carolinas, Georgia, Texas, Alabama, Mississippi, Virginia, New Jersey. Certain varieties of sweet potatoes are called yams in the Southern States, but the word "yam" properly be-

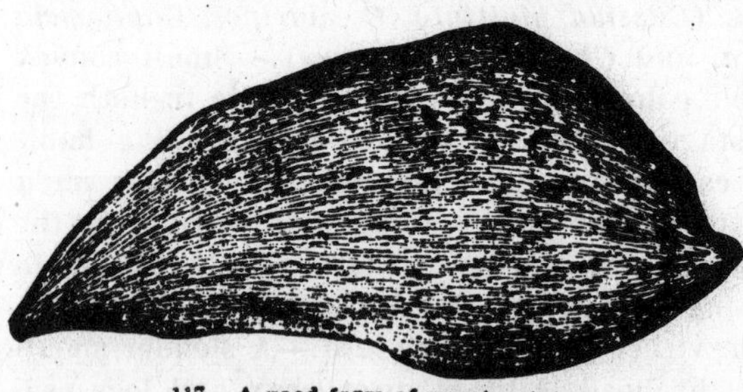

117. A good form of sweet potato.

longs to a very different kind of plant, the Dioscoreas. Fig. 117 shows "a well-proportioned sweet potato" as illustrated by R. G. Hill, of the N. Car. Extension Service (Ex. Cir. 30). Note also Fig. 118.

118. A good lot of sweet potatoes.

The sweet potato requires a deep well-drained sandy loam. The soil should be liberally supplied with well-rotted manure. Wood ashes is often a most excellent fertilizer. Commercial fertilizers are extensively employed. The soil should be well prepared before the slips are set, to avoid the necessity of cultivating close to the roots. Clean tillage should be practiced until the ground is thickly covered by the vines. After this, large weeds should be removed with hand tools. Rows are about 3 feet apart, and the slips themselves are usually 15 or 18 inches. In level culture, 24 to 30 inches either way are good distances, requiring 8,000 to 10,000 plants to the acre. In ridge culture, 30 to 40 inches between rows and 12 to 18 inches in the row constitute good practice; 8,000 to 12,500 plants go on an acre.

On the fertilizing of sweet potatoes, R. W. De Baun writes as follows in Circ. 114 of the New Jersey Experiment Station: "Well decomposed stable manure applied in the row is beneficial in producing maximum yields, but when manure is used in the drill, the surface of the sweet potato is more likely to be disfigured with black marks known as scurf, and spoken of as 'soil stain,' 'mottling,' 'rust,' etc. This spoils the appearance of the sweet potatoes and reduces their keeping qualities. A fertilizer containing a moderate amount of nitrogen and relatively high in phosphoric acid and potash gives a splendid yield. Fertilizer is easy to apply and is not likely to cause the development of scurf. The amount used to the acre varies from 600 pounds to a ton. The most economical results are probably obtained by using 1200 pounds to the acre of a 3-9-6 fertilizer. Following a clover cover-crop the percentage of nitrogen in the fertilizer may be somewhat reduced. Lime is of direct benefit to the sweet potato crop, especially where the soil is quite sour. If it is to be used, light applications are recommended. Wood ashes which have never been wet are particularly beneficial for sweet potatoes, because they usually contain 5 to 6 per cent of potash and 30 per cent of quick-acting lime."

Propagation.

It is the custom to grow all varieties from shoots or cuttings, although the Spanish variety may be cut and planted like the Irish potato. The slips are grown in beds and transplanted to the field. Many growers prefer to plant only a small part of the field with the slips and the remainder with the prunings from the growth of these slips. Propagation is usually accomplished by means of (1) slips, and (2) cuttings.

(1) Slips are the sprouts that arise from tubers when they are planted or buried. Tubers of medium size are laid on a mild hotbed and covered two inches deep with loose earth or leaf-mold. In the extreme South the tubers are sometimes "bedded" in loose warm earth, without bottom heat, but unless the weather is settled they are likely to rot and the vegetation is slow. When the shoots are 3 to 5 inches high they are broken off next the tuber and set in the field. Roots will have formed while they were still attached to the tuber. Two to four crops of "slips" or "draws" may be taken from one tuber. The tuber is usually planted whole; but large and sound tubers may be cut in two lengthwise and the cut side laid downwards, although this treatment invites decay. Six to eight bushels of seed potatoes produce sufficient plants to set an acre if "drawn" once, or half that quantity if "drawn" three or four times; 4,000 plants is a large yield from a bushel. About two months are required to produce salable plants for setting.

119. Sweet potatoes ready for covering.

(2) Cuttings are made from the ends of vines. They are taken from the earliest-planted or most vigorous vines; sometimes a few vines are set very early for the particular purpose of securing plants for the remainder of the field. The cutting is usually 10

to 12 inches long. The leaves are removed, except at the tip, and the cutting is buried directly in the soil where it is to grow permanently, being laid in a nearly horizontal position, with only an inch or so of the tip projecting. Cuttings are very desirable to avoid the spread of tuber diseases. "Seed" selection is also very important in controlling diseases.

Harvesting and storing.

Immediately after the first frost the potatoes are gathered. A common method is to clear away the vines and

120. Banks or pits of sweet potatoes.

then to plow up the potatoes with a "hill sweep" (2-winged furrowing-plow. Special diggers are on the market. Potatoes are gathered into small piles, where they remain until removed from the field.

The common method of storing is to bank in a cone-shaped pile. This pile is then covered with hay, and this is thatched with cornstalks, or covered shingle-like with pine bark. It should be kept dry and should be on a slightly elevated place.

The storage of sweet potatoes presents special problems, however, and further consideration may be given it with illustrations and comments adapted from H. C. Thompson in Farmers' Bulletin 970. Fig. 119 is a pile of sweet potatoes ready to be covered with cane-tops and earth; a ventilating hole or shaft is provided at the top. Fig. 120 shows a row of sweet-potato banks, with the ventilator openings at the top. Pits or banks are advised only when storage buildings cannot be provided. "Storage pits should be located where the drainage is good. In making a pit a little of the surface soil is thrown back to form a level bed of the size desired. It is a good plan to dig two small trenches across the bed at right angles to each other, to provide for ventilation at the bottom. Lay boards or place troughs over the trenches, and at the point where the trenches cross set a small box on end to form a flue up through the pile of potatoes. The earth floor of the pit is covered with 4 or 5 inches of straw, hay, leaves or pine needles, and the potatoes are placed in a conical pile around the flue. A covering of straw, hay or similar material is put on the pile and

121. Storage of sweet potatoes on a farm in the South.

over this a layer of soil. The covering of soil should be only a few inches thick at first but increased as the weather gets cold. Keep the ends of the trenches and flue open until it is necessary to close them to keep out the frost. It is better to make several small pits rather

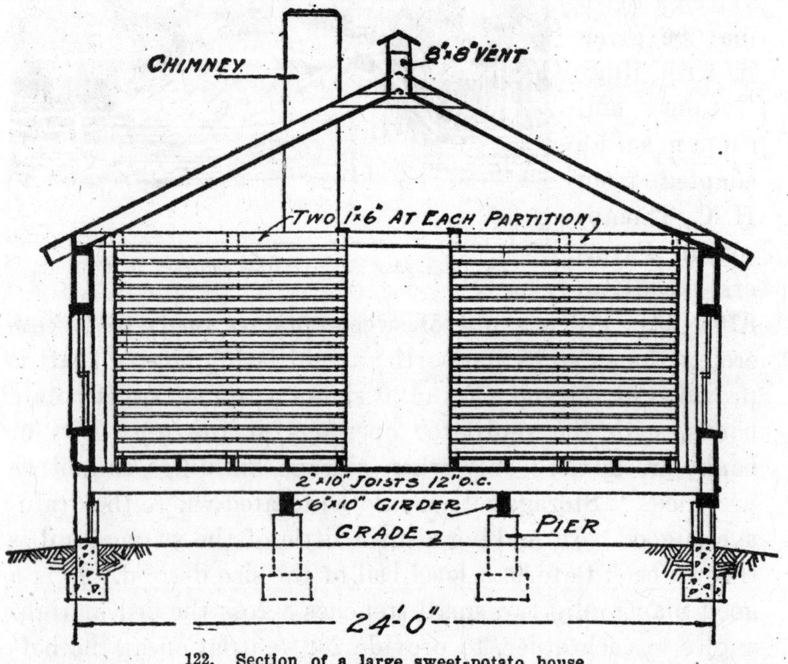

122. Section of a large sweet-potato house.

than one large one, because it is best to remove the entire contents when the pit is opened."

The illustration (Fig. 121) shows a crude but serviceable type of outdoor cellar sometimes used in the South for sweet potatoes. It should have openings near the top and bottom and through the top for ventilation. Regular "sweet-potato storage houses may be built of wood, brick, hollow tile, cement, or stone. Wooden houses are preferable, because they are cheaper and easier to keep dry than the other types. It is difficult to keep moisture from collecting on the walls of a cement, stone, or brick house. Where such houses are built for sweet-potato storage they should be lined with lumber, so as to keep the air in the house from coming in contact with the masonry walls. It is best to build sweet-potato storage houses on foundations that allow a circulation of air under them." Fig. 122 is a cross-section of a 24 x 60-foot sweet-potato house, and Fig. 123 of a 12 x 16 house. "To keep sweet potatoes in good condition they must be (1) well matured before digging, (2) carefully handled, (3) well dried or cured after being put in the house, and (4) kept at a uniform temperature after they are cured." The Figs. 122 and 123 illustrate cross-sections of the structures. The horizontal lines indicate the construction of bins, which are slated on both

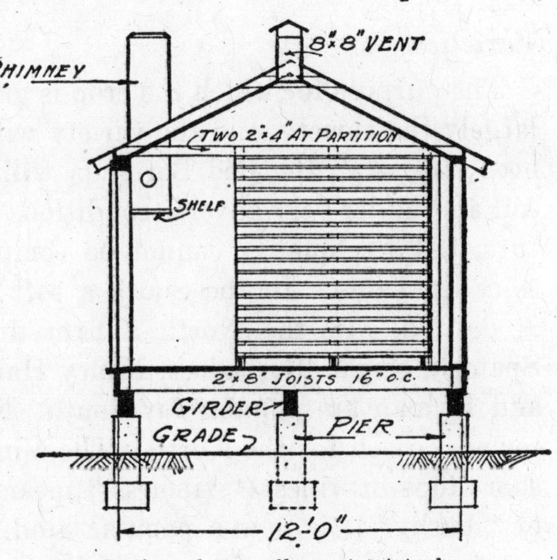

123. Section of a small sweet-potato house.

sides of 2 x 4 supports, with 1 x 4-inch material to provide air space between the bins; circulation is also provided by starting the bins 4 inches above the floor.

Varieties.

The purpose for which the crop is grown determines very largely the variety, and the variety will determine the care necessary; *e.g.,* the Red Bermuda will grow in almost any soil and under very adverse conditions of climate and moisture, but the quality cannot be compared to that of the so-called yams. In the South a soft sugary sweet potato is desired. In the North a firm dry tuber is wanted. Spanish, Sugar, Barbadoes, Nancy Hall, Triumph, Dooley, and Hyman are popular far South. Nansemond and Jersey are prized for the North. The Vineless, a variety with short tops or vines ("vineless" meaning "not running," or "bushy"), is now a popular kind. As with the Irish potato, careful attention should be given to breeding.

In the United States, the sweet potato rarely sets seed. In fact, it does not often bloom, although blossoms may appear late in the season under favorable conditions. The production of new varieties depends on tuber-selection and the appearing of mutations or sports.

The Sweet Potato Plant

Ipomœa. *Convolvulaceæ.* Twining herbs, shrubs, even trees, largely of tropical countries, of about 400 species.

I. Batatas, Poir. Encyc. vi, 14. 1804. (*Convolvulus Batatas,* Linn. Sp. Pl. 154. *Batatas edulis,* Choisy, Convolv. Or. 53. 1834.) Sweet potato. Tuberous-rooted perennial with long running tops, juice milky: stems prostrate, slender, extending many feet, rooting, angled, mostly with sparse thin hairs: leaves alternate, long-stalked, thinly hairy or glabrous, exceedingly variable in shape, usually ovate to round-ovate in outline, cordate or truncate at base; blades entire and the margin merely wavy, or sometimes angled and notched, or deeply 3- to 5-lobed and the basal lobes again lobed: flowers few or several terminating axillary peduncles of varying length (much shorter or considerably longer than the petiole), light violet with a darker center, like the flower of a morning-glory; corolla about 2 in. long, obscurely obtusely 5-lobed; calyx about ½ in. long, deeply parted into unequal cuspidate lobes which are sometimes ciliate; stamens 5, the sagittate anthers and the slightly 2-lobed capitate stigma usually not half the length of the corolla; ovary ciliate, sitting in a 5-angled yellow cup or disc.—Unknown wild, but supposed from historical and geographical considerations to be native of the western hemisphere; by some botanists thought to be a probable derivative of *I. fastigiata*, Sweet, of tropical America. It was early distributed in the islands of the Pacific and apparently was in China at least soon after the beginning of the Christian era; but the Polynesians were great navigators, and they may have got it from America. It was probably anciently cultivated on the American continent. (*Batatas* or *batata* is an aboriginal American name for the sweet potato, from which the word " potato " is derived.)

CHAPTER IX

PEAS AND BEANS

Peas and beans are usually closely associated in the public mind, and they are in fact closely related botanically; yet they have few points in common from the cultivator's point of view, since peas are hardy cool-season plants and beans are tender warm-season plants. Both are leguminous crops, and are therefore capable of using atmospheric nitrogen by means of their root nodules and the bacteria in them. As garden crops, however, they may need applications of nitrogen to secure a quick start, particularly if an early crop is desired. "It is frequently the wiser economy," as expressed by Voorhees, "to apply nitrogen, particularly if they are raised upon land which has not been previously planted with these crops, and thus may not possess the specific nitrogen-gathering bacteria." The peas and beans, of divers kinds, constitute the pulse crops.

The basic formula recommended on page 383, if applied to pulse crops at the rate of 500 to 600 pounds to the acre, will usually furnish sufficient nitrogen, and may, if necessary, be supplemented by the application of amounts of superphosphate and potash salts which will add from 20 to 30 pounds of phosphoric acid, and 60 to 75 of potash (Voorhees, Fertilizers, rev. ed. 297).

One of the important attributes in distinguishing species in these plants is the nature of germination. In the garden pea, the "pea does not come up," but remains under ground, while the bean "comes up"; that is, the cotyledons or seed-leaves (resting in the seed) remain below ground, in which case the germination is said to be hypogeal ("beneath the earth"), or they are liberated and appear above ground, in which case the germination is epigeal. Deep planting is safer with the hypogeal seeds. The common beans are epigeal, but the broad beans and multifloras, and some others, are hypogeal. The distinctions between some of the species of beans (Phaseolus) are very marked in their method of germination and in the character of the first leaves following the seed-leaves. The pictures in this chapter are interesting in this connection. In Fig. 124, the peas have remained in the ground, as also the bean in Fig. 140; but in Figs. 132, 135, 143, the beans have come out of the ground and cling to the stems.

124. Young plants of pea; cotyledons hypogeal (× ½).

PEA

Peas are a partial-season crop, requiring cool climate and a soil not over rich; seed is sown where the plants are to stand; grow in drills; hardy and may be sown very early.

Peas are usually sown in two rows 6 to 8 in. apart. If tall varieties are grown, one row of brush or chicken-wire

(the wire is better) answers for both rows; if the dwarf kinds are grown, one row will help to support the other. Between each two pairs of rows a space should be left wide enough for convenient tillage. The plants should stand 3 to 4 inches in the row. One pint of seed of the small-seeded varieties will sow 100 to 125 ft. of single drill. In drills, 1 to 2 bushels will sow an acre; broadcast, 2 to 3 bushels. Early peas are usually planted deep, 3 in. and more. Table peas may be had in about 70 days from the sowing of first-early varieties. Green peas in the pod yield about 100 to 150 bu. to the acre.

Blight (*Ascochyta pisi* and *Septoria pisi*).—These are two distinct blights, yet the symptoms and control measures are in general the same. Circular gray to dark brown spots occur on the leaves and sunken spots of a similar nature on the pods. Sometimes the small black fruiting bodies of the fungus are evident in the affected areas. *Control:* Seed from disease-free plants is necessary since the fungus may be carried in the seed, and crop rotation should be practiced. Diseased pea refuse in manure carries the organisms, but after fermentation in a silo it may be safely used.

Root-rot (*Fusarium* sp. and *Pythium sp.*).—A dry rot at and beneath the surface of the ground is caused by either of these organisms. *Control:* Crop rotation, care in the use of infested manure, and the development of resistant strains seems to afford the only possible control.

Pea weevil (*Bruchus pisorum*).—A small brownish beetle, mottled with gray, white, and dark brown, that deposits its eggs on pea pods in the field. On hatching, the grub burrows through the pod and enters a seed where it reaches maturity in about 40 days. It then cuts a smooth round hole to the surface of the pea, leaving only the outer hull intact. In the South many of the beetles emerge in the fall and hibernate, but in the North they do not usually emerge till spring. Only one weevil is found in each pea. There is but one brood annually. Peas are not reinfested in storage as is the case with beans.. *Control:* Do not use infested seed for planting nor is it a good plan to use seed in which the weevils have been killed, as such seed produces only weak plants. Seed peas may be held over to the second year, by which time the weevils will have emerged and died.

PEA APHIS (*Macrosiphum pisi*).—A moderate sized pea-green plant-louse that often attacks peas in great numbers, causing the plants to take on a sickly yellowish appearance and die. Infested blossoms are blasted and injured pods are stunted and rendered worthless. The pea aphis passes the winter on clover, in the South principally on crimson clover. *Control:* Peas grown in rows about twenty inches apart are less likely to be injured than when sown broadcast. When grown in rows the lice may be controlled by spraying with "Black Leaf 40" 11 oz. in 100 gals. water in which 10 lbs. fishoil soap have been dissolved. Applications should be made at weekly intervals. Avoid loss by raising the main crop early in the season for the cannery before the lice become abundant.

PEA MOTH (*Laspeyresia nigricana*).—A small slightly hairy yellowish black-headed caterpillar about 1/4 in. long, that in the Northern States and Canada sometimes causes great damage by infesting pea pods, where it feeds on the unripe seeds. In Wisconsin the moths begin laying eggs about the middle of July, which hatch in a week or ten days. *Control:* Both very early and late varieties of peas are less liable to injury. Adopt a crop rotation in which peas do not follow peas nor are planted in fields adjourning those interested the previous year.

We may distinguish three uses or purposes for which peas are grown: as picked peas, the pods being gathered by hand and the product sold directly in the market; as a canning crop, whereby they are grown under much less intensive methods, mown with a mowing-machine, transported by wagon-load or truck-load, and shelled by running vines and pods through machinery devised for the purpose; as a general field crop, often in connection with oats, for forage.

Garden or picked peas are of the easiest culture. They thrive best in spring rather than in summer, but they also thrive in autumn from late-sown seeds. In summer they are very liable to mildew and to injury by heat. Peas and onions are the first vegetables to be sown in the open ground. Even before freezing weather is past, peas may be planted. It is customary to plant them 3 to 5 inches deep: the roots are then deep enough to be in cool and moist soil. Early peas are frequently planted more shallow.

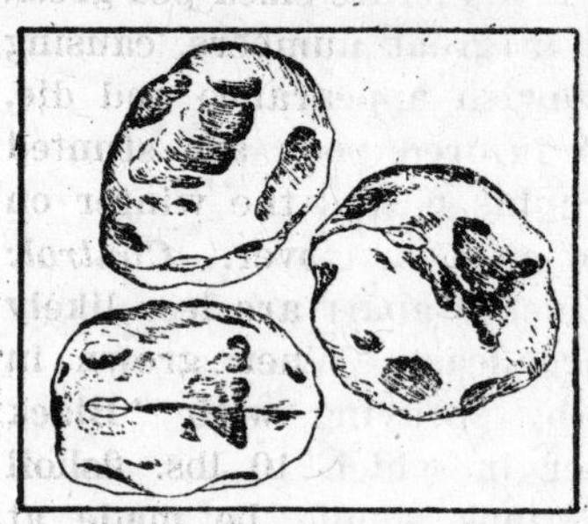
125. Wrinkled pea (× 2).

A light soil is chosen when earliness is desired; but for the main crop the clay loams are excellent. A very rich soil tends to make the plants run to vine and to delay the crop. Successional sowings should be made at intervals of six to ten days.

For early use, the dwarf varieties should be chosen. For the main or late crop the tall or climbing sorts, which are more productive, are preferred. Pinching-in the excessive growths tends to make the tall varieties somewhat earlier. Early in August in the Northern States dwarf varieties may be sown for fall use. The first sowings in spring are usually of the "smooth" peas, as they are less likely to rot in the ground than the wrinkled kinds. The very

126. Smooth pea (× about 2).

early dwarf peas are productive in proportion to the size of plant, but the actual yield is not large. Most of them are harvested in one or two pickings. The early pea does not compare with the string bean in productiveness, and allowance must be made for this fact in planning the home garden, if one is fond of green peas.

As a canning crop peas are sown broadcast or by a grain drill. Usually the crop is not tilled, as it is off the ground in June or July before the land gets very weedy. The crop is harvested with a mowing-machine, gathered with a hay-rake, and hauled to the factory where the threshing is done. The straw is used as sheep feed and is valuable as manure. In central New York canning peas are planted May 1 to 15; the crop is off in July; 1½ tons to the acre is an approximate yield (in the pod).

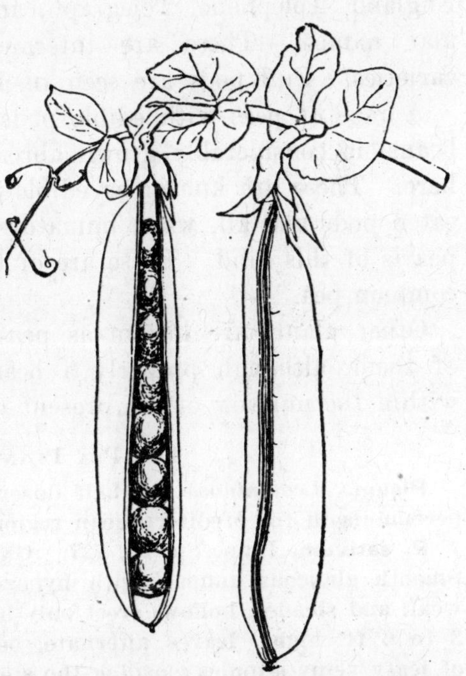

127. A legume,—the pod and seeds of pea (× ½).

Peas are of two kinds as to character of seed: the seed wrinkled and the seed smooth (Figs. 125, 126). The wrinkled are the better in quality. There are dwarf and tall varieties of both the wrinkled and smooth types. For very early there are many popular strains, as Alaska, Gradus, Thomas Laxton, Surprise, Eclipse, First-of-All, Philadelphia, Daniel O'Rourke, American Wonder, Little Gem, Blue Peter. For late, Marrowfat, Champion of England, Telephone, Telegraph, and Stratagem are popular names. These are intermediate or second-early varieties. Full pods are seen in Fig. 127.

A race of peas with edible pods, comparable to string beans, is considerably grown abroad but is little known here. These are known as edible-podded, or sugar peas, eaten pods and all, when immature. The Melting Sugar pea is of this kind. These are of the same species as the common pea.

Other plants are known as peas. The cowpea is one of them, although properly a bean. This plant is not within the purview of the present volume.

The Pea Plant

Pisum. *Leguminosæ.* **A half dozen species of annuals and perennials in the Mediterranean region and western Asia.**

P. sativum, Linn. Sp. Pl. 727. Garden or Culinary Pea. Smooth glaucous annual, with hypogeal germination: stems weak and slender, hollow, erect only by means of the tendrils, 3 to 6 ft. high: leaves alternate, odd-pinnate, with a pair of leafy veiny stipules clasping the stem; leaflets 2 to 6 pairs. of which the first 2 or 3 pairs are regular foliage blades and the remainder tendrils; expanded leaflets oval, oblong, elliptic to nearly circular, sessile, the apex rounded, emarginate or cuspidate, the margins entire, irregularly serrate or toothed; tendril-leaflets simple (not branched): flowers 1 to 3 terminating a long axillary peduncle, white, sometimes violet, papilionaceous; calyx large and green with 5 deep acute lobes; corolla about twice the length of the calyx; standard erect

or the sides inflexed or reflexed, orbicular and emarginate; wings closely appressed over the upwardly curved keel; stamens monadelphous, 9 and 1, the tube inclosing most of the smooth green shining ovary; style bent upward, not coiled, bearded on the inner face below the stigma: fruit a several-seeded dehiscent pod 2 to 4 in. long, nearly straight on the back and knife-shaped on the front, beaked at the apex, the sides more or less reticulated, the remains of the calyx persistent below its base: seeds 2 to 10, mostly whitish or greenish, even or wrinkled, globular or angled, ¼ to ½ in. diam., weighing 300 to 400 mg. and more, and retaining vitality 3 to 5 years.—Native in Europe and Asia, and cultivated from earliest times. Var. **humile**, Poir. in Lam. Dict. v, 456. 1804. (*P. humile*, Mill. Dict. No. 2. 1768.) DWARF PEA. Low, a few inches to about 2 ft. tall, the pods small, plant early: the early garden pea. Var. **macrocarpon**, Ser. in DC. Prodr. ii, 368. 1825. (Var. *saccharatum*, Hort., not Ser.) EDIBLE-PODDED PEA. Pods lacking the stiff lining, soft and edible, not dehiscent, often very large (sometimes 5 to 6 in. long and 1 in. broad), but frequently not larger than in other peas.

Var. **arvense**, Poir. in Lam. Dict. v, 456. 1804. (*P. arvense*, Linn. Sp. Pl. 727.) FIELD PEA. Flowers colored, the standard usually pinkish or light violet and the wings purple, keel often greenish: peduncles usually shorter, often little exceeding the stipules: leaves sometimes spotted with gray: pod and seeds mostly small.—Grown for forage, often with oats and other grain.

BEANS

Garden beans represent several species, but all the common kinds in North America are very tender to frost and require a warm season and sunny exposure; soil should be open and light, but fertile; seed is sown where the plants are to grow; usually grown in drills, except the tall kinds; the common bush beans are partial-season plants.

Bush string (snap) beans are sown in drills, the rows being 18 to 30 in. apart to allow of easy tillage. The plants should stand 4-8 in. in the row. Plant 1 or 2 in. deep. One pint will sow from 75 to 125 ft. of drill, depending on the variety. In drills, 1 bushel to 5 pecks are sown to the acre. One hundred bushels, more or less, is a fair acre-yield of string beans, and 200 bushels are frequently reported. The tall or pole beans are usually grown in hills 3 or 4 ft. apart.

ANTHRACNOSE (*Colletotrichum lindemuthianum*).—This disease may be recognized by the presence of black spots on the stem, leaf-stalk and leaf-veins, and black sunken cankers on the pods. Affected seeds show discolored areas on their surface. *Control:* Clean seed obtained from disease-free plants or pods should be used for planting. The Wells Red Kidney and the White Imperial are resistant, and breeding work now being conducted promises to yield other resistant types. Spraying at intervals with 4–4–50 bordeaux mixture is sometimes recommended, but it is of doubtful practicability except in small garden planting.

BACTERIAL BLIGHT (*Bacterium phaseoli*).—Water-soaked to brownish splotches on leaves and pods are characteristic. Affected seeds may show yellowish discolored areas. Field and garden varieties and lima beans are affected. *Control:* Seed from disease-free plants should be chosen. The kidney type among the field beans has proved to be very susceptible.

MOSAIC.—Alternate light and dark green areas and cupped swellings on the young leaves especially are indicative of this disease. No causal organism has been discovered. The disease is carried over in the seed. *Control:* Seed from disease-free fields should be planted. Marrow and Yellow-Eye beans are nearly free from the disease. The Red Kidney is somewhat resistant. The Michigan Robust pea-bean is a high-yielding strain apparently unaffected by mosaic. Other pea-beans and medium-beans become severely diseased.

DRY ROOT-ROT (*Fusarium* sp.).—The fungus affects the stem beneath the surface of the ground, causing a dry rot. *Control:* Plant on land free from the organism and avoid the use of bean straw or manure on uncontaminated soil. Long rotations and shallow cultivation are desirable. Experimental breeding promises to yield commercial strains resistant to the fusarium.

BEAN WEEVIL (*Bruchus obtectus*).—This is a small light-brown beetle, having the wing-covers about ⅛ in. in length, mottled with light brown, gray and black. The eggs are laid in the pods in the field and the grubs develop in the seeds and transform to beetles within cavities just under the integument. In emerging the beetle cuts out a circular lid in the seed-coat. Several beetles may develop within a single seed. The number of generations that may develop annually in the field depends on the temperature and length of the season. In the North there is only one brood but in the South there may be six or more. In storage, breeding may be continuous if the temperature is sufficiently high and the beans may be reduced to a powdery mass. *Control:* Weevils in the beans may be killed by fumigating with carbon bisulfide at the rate of about 1 ounce to each bushel of seed. A container as near air-tight as possible should be used and the fumigation continued for twenty-four to thirty-six hours. It is not advisable to use weevil-infested beans for seed since the germination is poor, and weak plants are produced.

BEAN LEAF-BEETLE (*Ceratoma trifurcata*).—This small beetle, about ⅓ in. in length, is yellowish to reddish and has the wing-covers marked with six black spots. The beetles feed on the underside of the leaves and riddle the foliage with holes. The eggs are laid on the ground at the base of the plants and the grubs attack the roots. From one to three broods occur annually, depending on the length of the season. *Control:* Spray the plants with arsenate of lead, 4 pounds of paste or 2 pounds of powder, at the first appearance of the beetles, taking care to hit both the under surface and the upper surface of the plants.

BEAN LADYBIRD (*Epilachna corrupta*).—In the semi-arid regions of the Southwest this ladybird beetle is a serious enemy of beans. It is yellowish to brownish orange, about ⅓ in. in length, and has the wing-covers marked with 16 small black spots arranged in three transverse rows. The beetles riddle the leaves with holes and attack the pods and blossoms. Eggs are laid on the underside of the leaves and the larvæ skeletonize them. There are one or two generations annually, depending on the length of the season. *Control:* In the home garden the beetles may be handpicked or the larvæ brushed off on the hot ground, where they will perish before regaining the plant. In larger fields the plants may be protected by spraying with arsenate of lead, 8 pounds of paste or 4 pounds of powder to 50 gallons of water, adding 4 pounds of lime to prevent burning of the foliage. Care should be taken to hit the underside of the leaves. It is sometimes advisable to plant the crop either early or late to avoid the insects, when they are numerous.

BEAN THRIPS (*Heliothrips fasciatus*).—In the Far West beans are sometimes seriously injured by a minute thrips. The insect is only about 1/24 in. in length and is black. The insects rasp and puncture the tissues, causing the leaves to turn yellowish or white, dry up and die. *Control:* Early planting and thorough cultivation will produce a rapid growth and help the plants to outgrow the injury. In the garden the plants may be sprayed with "Black Leaf 40" tobacco extract, 1 part in 800 parts of water, to which enough soap has been added to give a good suds.

BEAN APHIS (*Aphis rumicis*).—This black plant-louse passes the winter in the egg stage on evonymus, syringa, snowball and deutzia, from which it migrates in the summer to many vegetables and several common weeds. *Control:* On beans the lice may be controlled by spraying with "Black Leaf 40" tobacco extract, 1 part in 100 parts of water, to which enough soap is added to give a good suds.

STRIPED GREEN BEAN CATERPILLAR (*Ogdoconta cinereola*).—Bean vines are sometimes stripped of their foliage and pods by a pale green looping caterpillar striped with whitish and yellowish longitudinal lines. When mature, it is almost an inch in length. *Control:* The caterpillars may be poisoned by spraying with arsenate of lead (paste), 5 pounds in 100 gallons of water. On snap beans tobacco dust may be used to drive the caterpillars from the plants.

GREEN CLOVER WORM (*Plathypena scabra*).—While the more usual food plant of this insect is clover, it sometimes becomes very destructive to beans. The caterpillar when full grown is nearly an inch in length and striped lengthwise with whitish lines. *Control:* The caterpillars may be poisoned by spraying with arsenate of lead (paste), 2 pounds in 50 gallons of water. On string beans, where the poison would be obectionable, the tobacco dust may be used.

SEED-CORN MAGGOT (*Phorbia fusciceps*).—It sometimes happens, especially in cold backward seasons, that seed beans in the ground are attacked by a small whitish maggot that either entirely destroys them or so injures the bud that when the plant comes up no leaves are produced. Much of the injury may be avoided by planting the seed rather shallow.

As the beans are of so many kinds and types, we must state the main situation at the outset:

1. Broad bean, the bean of history, a hardy plant little raised in this country and very different from any of the following.—*Vicia Faba.* Figs. 128, 129, 130, all representing Broad Windsor.

128. Seeds of broad bean.—Broad Windsor (× ¾).

2. Common bean of North America, kidney bean of the English, haricot of the French.—*Phaseolus vulgaris:*

a. Snap or string beans, in which the green pod and

its contents are eaten, developed mostly on bush or non-climbing plants.

b. Dry field beans, for the general market, the ripe product of bush varieties for the most part.

c. Shell beans, in which the nearly full grown but unripe beans are shelled and eaten, the produce for the most part of pole or running varieties.

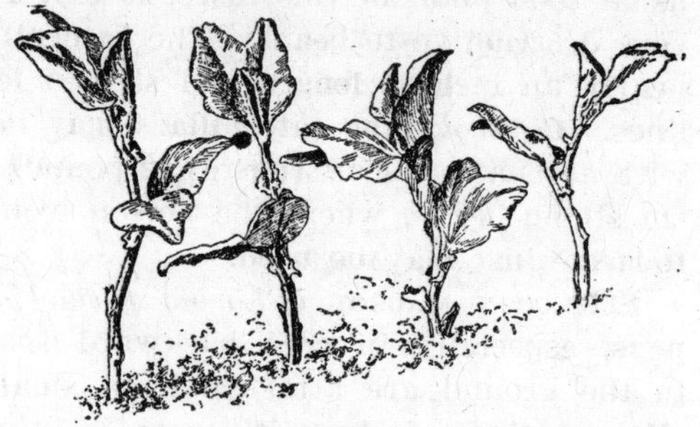
129. Seedlings, or young plants, of broad bean (× ½).

3. Multiflora beans, grown as snap or shell beans but mostly known in this country as ornamentals, particularly the Scarlet Runner; mostly pole beans.—*Phaseolus multiflorus.*

4. Sieva and lima beans,

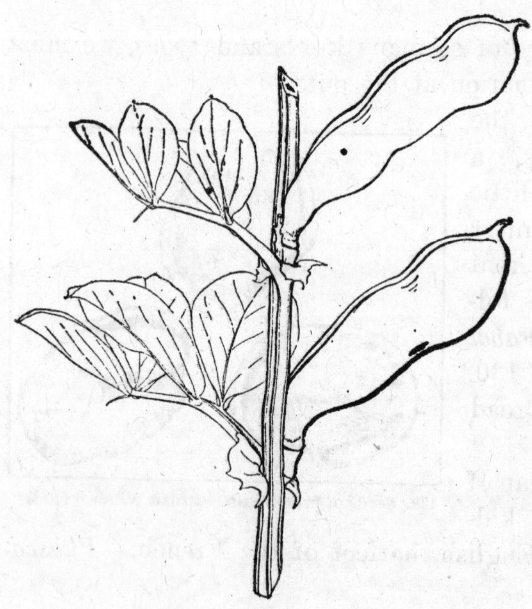
130. Broad Windsor bean (× 1/3).

grown as shell beans or for the ripe dry product; both bush and pole varieties.—*Phaseolus lunatus.*

5. Tepary, grown as ripe beans in the dry far Southwest, and for land improvement, annual indigenous bush beans, with a viney or semi-twining habit on good land.—*Phaseolus acutifolius* var. *latifolius* (Figs. 131, 132).

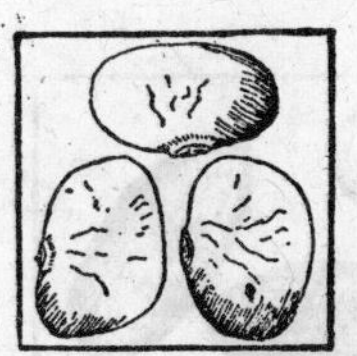

131. Tepary bean (× 1 1/3).

6. Metcalfe bean, an indigenous long-running perennial species introduced in the far Southwest for forage.—*Phaseolus Metcalfei* (*P. retusus*).

7. Various oriental beans, mostly bush, grown for the dry seeds, yet little known in this country but likely to attract attention. Among them Adzuki, Urd, Mung, Moth, Rice beans, all represent different species of Phaseolus.

8. Soybean, grown mostly in this country for forage and soil improvement. —*Glycine Soja.*

9. Cowpea and related beans, cultivated chiefly for forage and green - manuring. —*Vigna* species.

10. Velvet beans, planted far south for forage. — *Stizolobium* species.

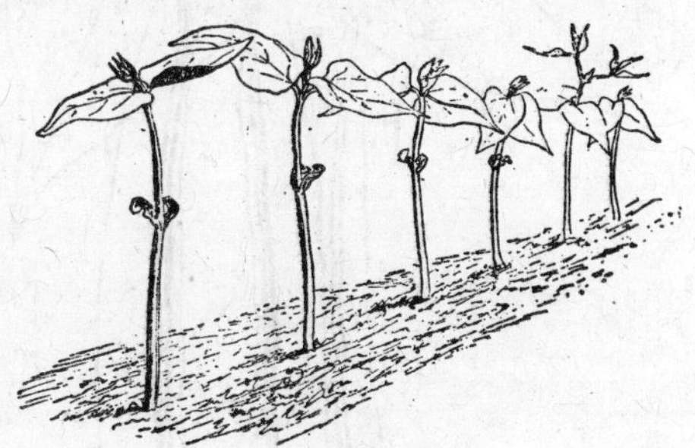

132. Tepary (× ½).

In this book, only the first four groups are discussed. There are other types of cultivated beans, in other species and genera, but so little grown in this country that they do not require listing here.

Broad Bean

The broad bean is a stiff erect plant, as hardy as peas, grown in Europe for food, either the green or dry beans being used. It is not commonly known in this country, apparently because the summers are too hot; and the winters in the North are so severe that it cannot be planted in autumn and carried over, as in the milder parts of Europe. In the southernmost States and on the Pacific Coast the crop may be seeded in September to November. The beans are large. They make a rather coarse but nevertheless very excellent dish. They are sometimes grown for stock feed, and for green-manuring. Broad beans appar-

133. Black Wax bean (× about 1).

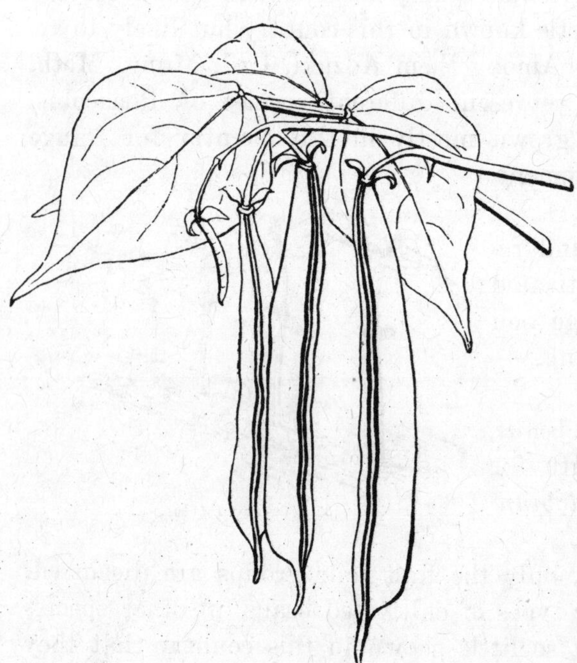

134. Black Wax (× ½).

ently thrive best in regions tempered by the sea. Inoculation of the land is desirable, with soil in which a good crop has been grown.

The amateur may start plants under glass and transplant to the open; but usually the seeds are sown where the plants are to stand, at the earliest moment in spring. If to be used as green or shell beans, care must be taken that the pods do not become hard.

135. Seedlings of wax bean (× ½).

The rows may be 2 to 3 feet apart; the plants may stand 4 to 6 inches. The varieties mostly known with us are Broad Windsor, Mazagan, Sword Long-pod. The beans should be ready for use in late spring and summer. There are many varieties, differing greatly in size of pod and in size and shape of seed. These plants as a class are sometimes known as "horse beans."

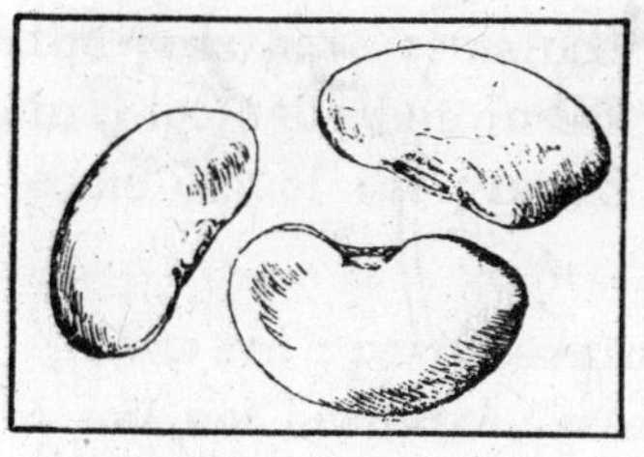

136. Dutch case-knife bean (somewhat enlarged).

The Common Garden Bean

The common bean is grown in two general types: the bush bean and the pole bean. In North America the bush bean is by far the more important since its growing obviates the labor and expense of providing support on which the plants may climb. Bush beans are grown both as a field crop and a garden crop. As a garden crop they are used mostly as "string" beans, the pods being picked when they are two-thirds grown, the pod and beans together being eaten. There are certain strains of bush beans particularly adapted to this use. They are such as have thick and fleshy pods, with very little fibrous tissue in the sutures. The pods of a good string bean have no "strings." The pods break cleanly in two, and this gives rise to the common name, "snap" beans. The snap beans are again of two groups—the green-podded, represented by the flageolets, and the yellow-podded or wax beans (Figs. 133, 134, 135), the more popular in this country.

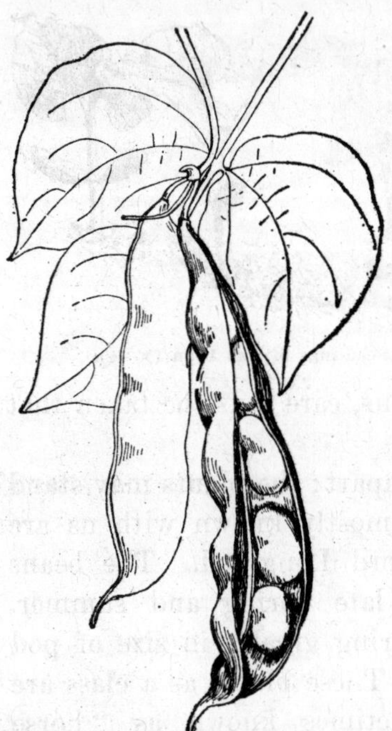

137. Dutch case-knife (× about ½).

138. Flowers of multiflora bean.—Scarlet Runner (× ¾).

In order that string beans may be of the best quality,

they should make a rapid and continuous growth. The soil should be rich and in excellent tilth. Plant only after the weather has become thoroughly settled. A succession may be had all summer. Although beans are nitrogen-gathering plants, it is nevertheless advisable to apply a little nitrogen at the start on land that is not well supplied with humus or in which beans have not been grown within a year or two.

139. Common wax bean (× 1).

For canning as string beans, the Wax and Refugee are grown. In central New York the crop is planted about May 15, and the harvest is August 1 till frost. An acre yields approximately 5,000 pounds.

String beans are productive, and if the ground is frequently tilled and the beans picked before they get hard, the yield will continue for a considerable time, in this respect differing from peas. They are picked by hand. All broken, imperfect and diseased pods should be discarded when marketing. They are sent to market in baskets and hampers.

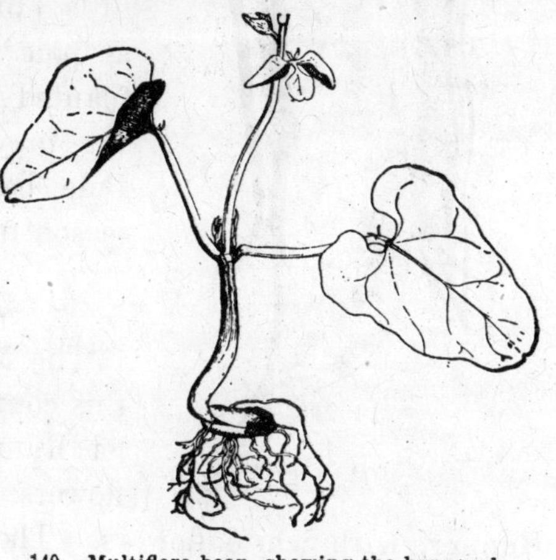

140. Multiflora bean, showing the hypogeal cotyledons (× ¼).

Other kinds of garden beans are used as "shell beans." The large soft seeds are gathered just before they begin to harden, and the pods are not eaten. Some of the best of these shell beans are pole or running varieties, the Cranberry or so-called Horticultural Lima, White Creaseback, Kentucky Wonder or Old Homestead, Dutch Case-Knife (Figs. 136, 137, sometimes erroneously referred to the Multiflora Group), being amongst the most popular. Pole beans require that the plants stand farther apart in the row, usually 1 foot or so, and the rows 2½ to 3 feet, for intensive cultivation. It is usually recommended that they be planted in hills 3 or 4 feet either way, with a pole to each hill. When planted in rows, wide wire fencing may be used for support. Pole beans require the entire season in which to make a crop.

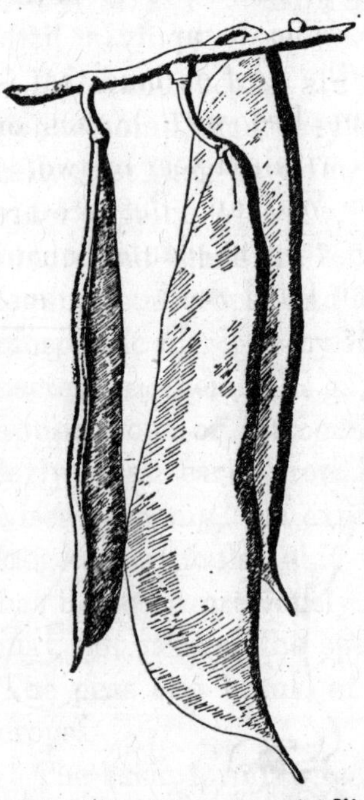

141. Pods of sieva bean (× ¾).

Multiflora Bean

The Multifloras are known in this country mostly by the Scarlet Runner, with bright scarlet flowers, and the White Dutch Runner, with white flowers. The pods may be eaten as snap beans, but usually they are grown to the shell-bean stage. These varieties are high climbers, making good screens. They may be planted along fences or lattices as are other pole beans, or in hills 3 or 4 feet apart. There are bush varieties, but little known in gardens.

These beans are perennial, and the thick roots live over winter if they do not freeze, the plants then coming into bearing early. Sometimes the roots are lifted in autumn and carried over winter in sand in the cellar. Commonly, however, the plants are treated in all ways as annuals, as are other pole beans.

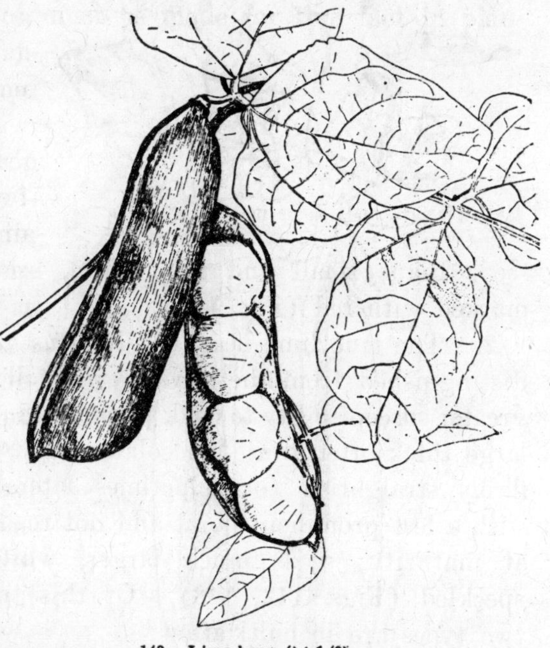

142. Lima bean (× 1/3).

Growers ordinarily do not distinguish sharply between the multifloras and the common garden beans. Aside from the duration of the plants, differences in germination, unlike flower - clusters, the plants differ also in flowers, as seen in Figs. 138 and 139; note the size, shapes, and also the calyx-bracts and the bracts at the axils. Figs. 135 and 140 may also be compared.

Sieva and Lima Beans

The limas are beans of high quality. They may be thrown into the following classes:

1. The sieva or Carolina bean (*Phaseolus lunatus*), a

relatively small and slender grower, early and comparatively hardy, apparently annual, with thin, short and mostly broad (ovate-pointed) leaflets, numerous small papery pods much curved on the back and provided with a long upward point or tip and with the habit of splitting open and twisting when ripe, discharging the seeds beans small and flat, white, brown, or variously marked with red (Fig. 141).

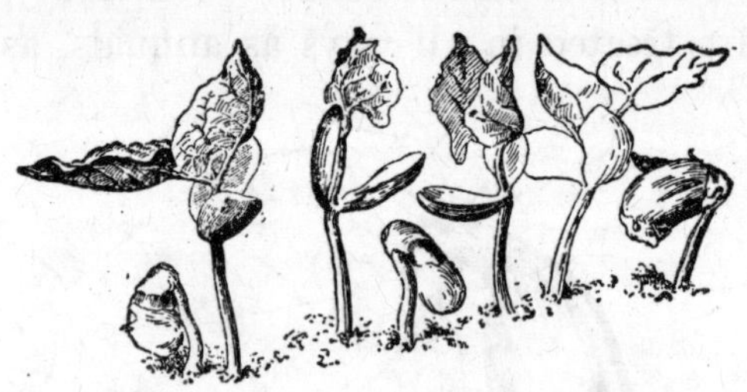

143. Germination of lima bean (× ½).

2. The true lima bean (*P. lunatus* var. *macrocarpus*), distinguished from the sieva by its tall growth, lateness, greater susceptibility to cold, perennial in tropical climates, large thick often ovate-lanceolate leaflets, and fewer thick fleshy straightish (or sometimes laterally curved) pods with a less prominent point and not readily splitting open at maturity; seeds much larger, white, red, black or speckled (Figs. 142, 143). Of this true or large lima two types are in cultivation:

(a) The Flat or Large-Seeded limas, that have large very flat and more or less lunate and veiny seeds, very broad pods with a distinct point, and broad ovate leaflets (Fig. 144).

(b) The Potato limas, with smaller and tumid seeds, shorter and thicker pods with a less prominent point, and long-ovate leaflets tapering from a more or less angular base into a long apex. There are dwarf forms.

Lima beans demand a long season and continuous growth, particularly the tall or true lima varieties. Very often the flowers are blasted by the hot dry weather of midsummer. It is well, therefore, to get the plants established as early as possible that some of the fruit may set before the hottest weather. It is important that the earliest and quickest soil be chosen and that quickly available fertilizers be applied when the seeds are planted. Light and sandy lands are usually preferable. In these, plant-food acts quickly and the plant secures a good and very early start.

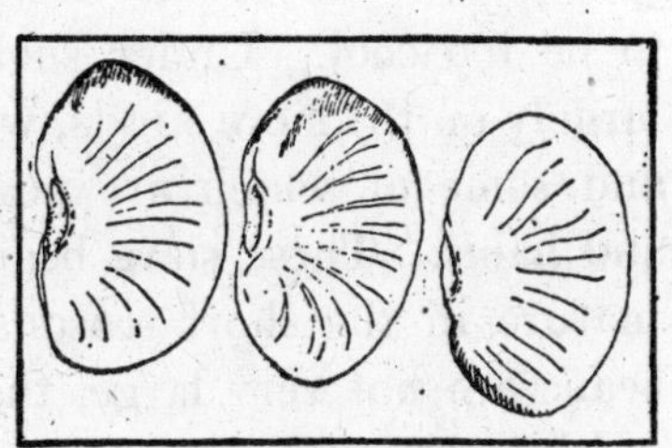

144. Large flat lima bean (nearly natural size).

The tall varieties must have strong supports. When poles are scarce, it is a good plan to set rather strong stakes 10 to 12 feet apart and to run wires or heavy cord from pole to pole, one strand near the top and one within a foot or so of the ground, and then to connect these horizontal strands with perpendicular cords. Sometimes several plants or hills of lima beans are planted in a semicircle around one strong stake, and strings are run from the top of the stake to the ground, making a cone. This is a very good plan for the home garden, since the vines are well exposed to the sun, but is too laborious for general market cultivation. In commercial plantations, one bare pole is ordinarily provided for each hill; and the hills are 3 to 5 feet asunder, sometimes as much as 6 feet. The beans are planted 2 to 3 inches deep, with the eye downward; 3 to 5 beans are left in a hill.

In the Northernmost States, it is usually inadvisable to attempt to grow the large late pole lima beans unless one's soil is particularly quick and the exposure is very warm. The seasons are usually too short, and the nights are likely to be too cool. Under such conditions it is best to rely largely on the sieva kinds, which are not very high climbers and some of which are nearly or quite "bush" in form and habit. These sieva beans are very heavy croppers and mature in the short seasons of the North. Although the beans are not very large, the quality is good.

Lima beans are more tender than the common garden beans, and are planted later.

The dwarf limas are excellent for northern gardens. Some of them are heavy croppers, and they retain the excellent quality of the pole varieties. They may be planted as close as 6 to 10 inches in the row, and the rows may stand only 2 feet asunder.

The Bean Plants

Vicia. *Leguminosæ.* A genus of wide distribution on the globe, comprising more than 100 species of annual and perennial herbs.

V. Faba, Linn. Sp. Pl. 737. (*Faba vulgaris*, Mœnch, Meth. 150. 1794.) Broad Bean. Strong erect simple or sometimes branched plant, 1 to 6 ft. high: germination hypogeal: stem glabrous, angled and grooved: lvs. many, all cauline, pinnate, petioled, with clasping stipules; leaflets 2 to 6, semi-opposite or alternate, entire, various in size and shape, 1½ to 4 in. long, obovate, elliptic to lance-ovate, blunt or very obtuse or even retuse and usually short-cuspidate, the terminal one usually represented by a rudimentary tendril: flowers few in short axillary clusters, narrow, papilionaceous, 1 to 1½ in. long, dull white with purple markings on the standard and purple wings or purple spot on them; calyx unequally

5-toothed, less than half the length of the corolla; standard folded over the much shorter wings and the wings longer than the keel which incloses the 9-and-1 stamens and the upwardly bent style which is bearded just back of the stigma: pod variable, large, 1 to 4 at a joint, 3 to 6 in. long, 1 in. or less broad, at first erect but usually becoming declined or pendent with weight, flattened or circular in cross-section, beaked at the summit: seeds 2 to 8 to the pod, nearly globular to flattened-angular, usually heavier than lima beans.—Probably W. Asian and N. African in origin, but the indigenous form unknown; cultivated from the earliest times. (*Faba* is Latin name for bean.)

Phaseolus. *Leguminosæ.* Perhaps 150 species of warm-country annual and perennial herbs, mostly twining.

P. vulgaris, Linn. Sp. Pl. 723. COMMON POLE BEAN. KIDNEY BEAN. Tall-twining pubescent annual: germination epigeal: stems very slender, branching, angled: lvs. pinnately 3-foliolate, petiole long, stipules small and acute, often falling early; stipels present; leaflets broad-ovate in general form, acuminate or acute, margins entire, the terminal one prominently stalked, the lateral ones short-stalked and unequal-sided, the lower side being the larger: flowers few on an axillary peduncle shorter than the petiole, white fading to yellowish, violet or lilac, the floral bracts (at the base of the pedicels and of calyx) green and broad-oval, ¼ in. long more or less; calyx a small shallow cup covered or subtended by the 2 bracts, about ¼ length of the corolla, obscurely 5-toothed or angled with the longer projection on the under side; corolla papilionaceous, the standard bent abruptly upward at the middle, broad and sometimes hooded, retuse, the wings projecting and between which is the upwardly coiled keel; stamens 9 and 1; style within the coil, bearded toward the end: pod long and narrow, 4 to 8 in. long and rarely exceeding ¾ in. across, curved, the sides nearly parallel, the beak slender pointed and curved: seeds 3 to 8, very various in size, form, weight and color, lending themselves well to the giving of names as if the variations represent species, usually less than ¾ or ⅘ in. long, those of the Horticultural Pole variety weighing between 400 and 600 mg. and even more; germinating vitality 3 or 4 years.—Nativity entirely unknown, with the probability of American origin; apparently anciently cultivated by aborigines. Var. **nanus,** Aschers. Fl. Prov. Brandenb. 170. 1864. (*P. nanus*, Linn. Amœn. Acad. iv, 284. 1788.) BUSH BEAN. Plant low, compact, not climbing: comprises all the dwarf string beans and field beans.

P. multiflorus, Willd. Sp. Pl. iii, 1030. 1800. (*P. vulgaris* var. *coccineus*, Linn. Sp. Pl. Ed. 2, 1016. 1763. *Lipusa multiflora*, Alef. Landw. Fl. 26. 1866.) MULTIFLORA BEAN. Distinguished from *P. vulgaris* in being perennial with a thickened root, having hypogeal germination, flowers showy and many on peduncles that much exceed the petioles or even the leaves and prominently pediceled, floral bracts lanceolate or linear-lanceolate, pod mostly larger as well as the seed, which usually exceeds ¾ or ⅘ in. long.—Tropical America. There are bush or dwarf varieties. The flowers are very showy in the Scarlet Runner, as also in the White Dutch Runner. Seeds are often highly colored.

P. lunatus, Linn. Sp. Pl. 724. SIEVA OR CIVET BEAN. Probably annual, slender; germination epigeal: leaflets short, broad-ovate, acuminate or acute, varying to lanceolate or even linear-lanceolate: flowers many, small to medium (smaller than those of *P. vulgaris*), on short or long racemes, nearly sessile or short-pediceled; flower bracts small and not conspicuous, narrow: pod 2 to 3½ in. long, broad and flat, usually broadening toward the apex, with a long sharp beak, the sides splitting away and twisting when ripe: seeds flat and thin, not large, with lines radiating from the hilum or scar.—Tropical America.

Var. **macrocarpus,** Benth. Fl. Bras. xv., pt. 1, 181. 1859-1862. LIMA BEAN. A stronger stouter later plant, with large thick mostly angled broader leaflets: pods larger, very broad and flat, heavy, the point blunt or at least not very prominent: seeds very large.—South America. Probably specifically distinct, and the synonymy needs study. Both the sievà and lima are twiners, but there are bush forms.

The tepary, now offered by seedsmen, is **P. acutifolius,** Gray, var. **latifolius,** Freeman, Bull. 68, Ariz. Exp. Sta., 589. 1912. It is a slender annual, native in the southwestern U. S. and Mexico, grown in hot semi-arid regions as a drought-resisting dry shell bean, long cultivated by the Indians and Mexicans. On poor land it is dwarf or bush, but on more fertile land it makes a long twining vine: germination epigeal: as grown in New York from commercial seeds it matures in 3½ to 4½ months, is semi-bush or tall twining, with thin slender stems and small leaves: flowers small, white or nearly so, 1, 2 or 3 together on very short axillary peduncles: pod small, 2½ to 3 in. long and ⅜ in. broad, curved, with nearly parallel sides and slender sharp beak: beans about 5, white, much like the Navy pea-bean. (The word "tepary" originated from the name of this bean among the Papago Indians.)

CHAPTER X

SOLANACEOUS FRUITS

Tomato	**Pepper**
Eggplant	**Husk tomato**

Tomatoes and eggplants are hot-season plants. They require nearly or quite the entire season in which to mature. Usually they grow until killed by frost, at least in the North, as they are perennials or plur-annuals, and the production of a heavy crop depends largely on securing an early start. They are seed-bed crops, and they need abundance of quick-acting fertilizers applied relatively early in their growth. They are grown in hills; that is, not in continuous drills.

These plants are here called solanaceous fruits because they belong to the family Solanaceæ. To this family also belongs the potato, which is a tuber. Here belong also fruits of minor importance, as the tree tomato (*Cyphomandra betacea*) and the morelle. The former is a semi-woody bush 5 to 10 feet high, bearing egg-shaped tomato-flavored fruits about 2 inches long, the second and third years from seed; in warm countries grown out of doors and in northern parts sometimes raised under glass. The latter (morelle) is a form of *Solanum nigrum* or black nightshade, a plant in the wild without edible fruits. The cultivated plant has berries larger than a large pea. It is an annual of simple culture. In this country it has been known mostly as the garden huckleberry and wonderberry. A related plant but with egg-shaped large attractive fruits 4 to 6 inches long is the pepino (*Solanum muricatum*) propagated by cuttings and fruiting the first year in the North if started very early.

TOMATO

Important points in the culture of tomato are: long warm season; "quick" soil with available fertility and one that retains moisture; frequent, or at least two or three transplantings to obtain stocky and continuous-growing plants, particularly at the North; early fruiting to mitigate loss from fruit-rot and to secure a heavy crop before frost; planting in hills.

Tomato plants are usually set about 4 to 5 feet each way in rich garden soil. In field conditions, they are usually set 3 to 4 feet. On light and early lands they are sometimes planted 3 x 3 feet. Sow about ¼ inch deep. From 1 ounce of seed, about 2,000 to 2,500 good plants should be obtained. At 3 x 4 feet, an acre requires 3,630 plants. A large yield is 12 to 16 tons to the acre; the average is much below this; 1,000 to 1,200 bushels are reported, but this is unusual; 500 bushels are frequently produced, but yields in general field culture for canning run perhaps 100 bushels an acre.

Leaf-spot (*Septoria lycopersica*).—Circular grayish black areas dotted with small black fruiting bodies develop on the leaves. As the disease progresses, the affected spots dry while the leaves yellow and drop off. *Control:* Spraying every ten days after the plants are set in the field is advisable. Bordeaux mixture to which has been added three pounds of resin-fishoil soap to fifty gallons of solution is recommended. Applications should be thorough enough to cover all parts of the vine. Spraying is generally thought to delay ripening.

EARLY BLIGHT (*Alternaria solani*). LATE BLIGHT (*Phytophthora infestans*).—*Control:* Spray as in the case of septoria leaf-spot.

LEAF-MOLD (*Cladosporium fulvum*).—Irregular areas on the leaves bearing on the under surface a purplish green mold are characteristic. This is primarily a greenhouse disease. *Control:* Care in ventilating and watering is important. Spraying with bordeaux, though usually recommended, is of doubtful value.

BLOSSOM-END ROT.—Affected fruits are marked before maturity with a sunken blackened area at the blossom end. *Control:* Unfavorable environmental conditions are thought to occasion this trouble. Maintenance of a uniform water supply may reduce its occurrence somewhat.

MOSAIC.—A mottled appearance of the leaves due to alternate light and dark green areas is characteristic. Affected leaves may be curled and abnormal in shape. *Control:* Diseased seedlings should be discarded as soon as they appear, care being taken in pruning and handling not to go from affected to normal plants, as the disease is communicable. Control of insects is important, as they may serve in disseminating the virus.

TOMATO WORMS (*Phlegethontius quinquemaculata* and *P. sexta*).—Large green or brownish caterpillars, 3 or 4 in. long, provided with a sharp horn at the hind end of the body. The adult is a large ash-gray or brownish gray moth marked with irregular brown and black lines. The pupa is dark brown, about two inches long, and the sucking tube of the future moth is enclosed in a separate case and resembles the handle of a pitcher. *Control:* Hand-picking is the most practicable means of control. Spray with arsenate of lead (paste), 2 or 3 lbs. in 50 gals. water. Spray early, while the caterpillars are still small. There is no danger in spraying tomatoes till the fruit is half grown. The poison may also be applied in the form of a dust.

COLORADO POTATO BEETLE (*Leptinotarsa decemlineata*).—See

under Potato. Use arsenate of lead instead of paris green as it is less likely to injure the foliage.

POTATO APHIS (*Macrosiphum solanifolii*).—See under Potato.

The tomato is a universal favorite and in the United States it is a regular and staple crop. It is a major forcing crop. In most parts of the territory, it is grown out of doors with the greatest ease. The soil should be rich but not over-supplied with nitrogen, particularly if it becomes available late in the season. It is commonly said that very rich soil is not to be advised for the tomato. This is probably true as respects the heavy application of stable manure; it usually gives up its fertility somewhat slowly and tends to keep the plant in vigorous growth and to delay fruiting. If, however, the soil has been made rich by previous application of manure, or of available commercial fertilizer early in spring, the best results may be expected. Experiments at Cornell showed that a rather light single application of nitrate of soda about the time the plants are set, gives better results than twice that amount applied at intervals as late as the middle of August (page 382).

145. Seeds of tomato (× 4).

Voorhees gives the following as "a mixture very generally used in New Jersey" for tomatoes: 100 lbs. each of nitrate of soda, sulfate of ammonia, dried blood (16% AM), ground fish, ground bone; 1,100 lbs. acid phosphate; 400 lbs. sulfate of potash. "This is undoubtedly an excellent mixture which may be used with safety in almost any quantity. The usual practice is to use from 1,000 to 1,200 pounds to the acre. Many farmers claim that the sulfate of ammonia causes some injury to the tomato and prefer to double the quantity of blood and fish. Whether there is any ground for this claim has never been definitely determined."

The obtaining of a good crop of tomatoes in the North depends very largely on having vigorous and stocky plants well in advance of the season, and a warm quick soil. The plants should be set in the field as soon as the weather is settled. Thereafter they need no special care except to keep the land well tilled.

Starting the plants.

Plants are usually started four to eight weeks before they are transplanted to the field. For the home garden it is well to handle the young tomato plants in pots; but in commercial operations this is scarcely practicable. The custom is to grow them in small flats not more than ten or twelve inches square and that hold about two inches of soil. In some cases, even smaller flats or boxes are used. In these boxes the plants are displayed in the stores for sale to amateur planters. In flats of various sizes, the plants can be readily handled from the frame to the field. In commercial business, the young tomato plants are now rarely transplanted. They are thinned in the flats to stand two or three inches either way, or farther than this if the plants are started very early. Sometimes the plants are sheared if they become

146. Young seedlings of tomato (× nearly ½).

147. The round "smooth" tomato of the present day (× ½).

too tall and "leggy," although this is not the best practice. In the Middle and Southern States, cloth-covered frames are often used for starting tomato and other plants. The cloth is rolled up in the day.

In New York, seeds are sown in hotbed or house about the middle of March or first of April for field culture. If sown too early the plants become too tall and weak and they may be pot-bound. A good plant for transplanting should be short and stocky, and in vigorous growing condition. Very long-stemmed lopping plants are sometimes planted deep and the bare stem buried slanting; roots will form along the buried stem. In Figs. 145, 146, the seeds and seedlings are seen.

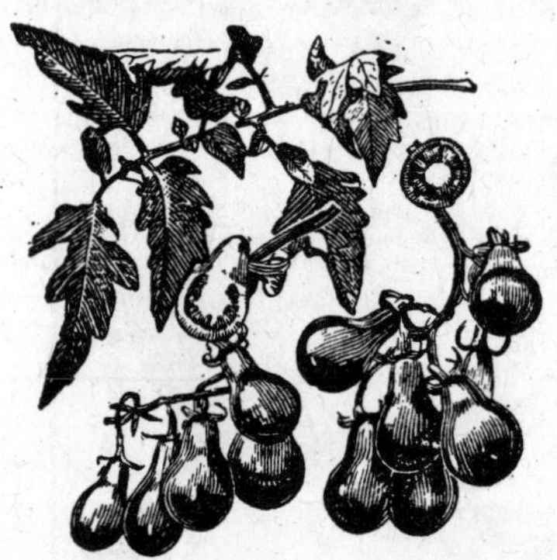

148. Pear tomato (× 1/5).

Training and pruning.

Tomatoes usually give earlier and better results when the vines are trained; but the expense of training precludes its use in large commercial plantations. The best mode of training for early results is to prune the plant to a single stem, tying it to a perpendicular cord. The cord is secured at top and bottom to horizontal strands stretched between strong stakes. When tomatoes are thus trained, they may be set as close as 18 inches in the row.

There are various styles of racks for supporting the tomato plants. The best are those that give the plants full exposure to sun and allow all the fruits to hang toward the outside of the trellis rather than to be covered by foliage. In commercial plantations, the plants are allowed to spread as they will, although the fruit-rot disease is usually more serious under such conditions, particularly if the surface soil contains much coarse manure. Pinching-in the shoots is thought to conduce to early bearing.

149. Plum tomato (× ¼).

Sometimes tomato plants are pruned. On this point, F. S. Earle writes as follows (Bull. 108, Ala. Exp. Sta.): "By pruning, commercial growers mean the pinching out of all lateral branches as soon as they appear, thus confining the growth strictly to one stem. When about three clusters of fruit are set the vines are topped, thus stopping all further growth of vine, and turning the energies of the plant entirely to the growth and maturing of the fruits that are already

150. The tomato of a past generation.

set. The advocates of this system claim that it greatly increases the size of the individual fruits and that the bulk of the crop ripens several days earlier than on unpruned plants. Of course each plant produces fewer fruits than when allowed to grow unchecked, but this is partly compensated for by increased size and by closer planting that is possible on this system, thus allowing a greater number of plants to the acre. In several of the more important tomato-growing regions this system is very widely followed."

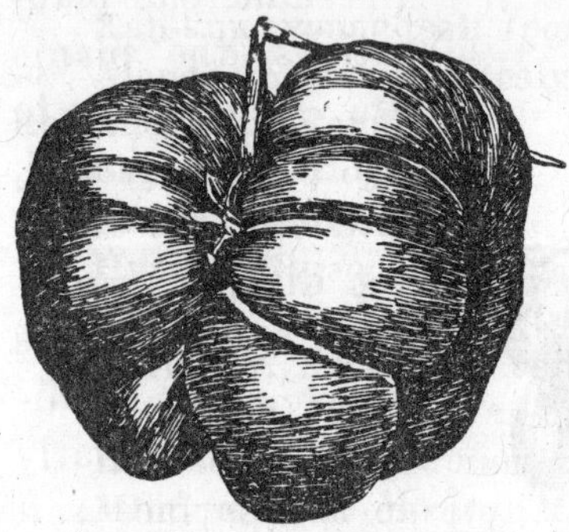

151. A stage in evolution.

Harvesting and marketing.

When frost threatens, the largest green tomatoes may be picked and allowed to ripen in drawers or in other dry and close places. Usually they color well and develop a good quality. If the fruits have not reached their full size, the whole plant may be pulled and hung in a barn or other dry place and the fruits will abstract nourishment from the vine and sometimes complete their ripening.

152. Foliage and flowers of the common tomato. Lycopersicon esculentum var. commune (× 2/5).

Tomatoes are now grown on a very large scale for canning factories. They are then a field crop, and are given no greater care than corn. A rather light warm soil is chosen. Frame-grown plants are used and they may be set with a transplanting machine. Thereafter no special treatment is given the crop except to keep the land well tilled. Plants are usually spaced 4 feet either way. The yield of the "can-house crop" varies greatly, from 3 tons to 12 and even 14 tons to the acre; 5 to 8 tons is a good crop. The legal weight of a bushel of tomatoes runs from 45 to 60 lbs. in different States; a yield of 8 tons is 320 bu. at 50 lbs.

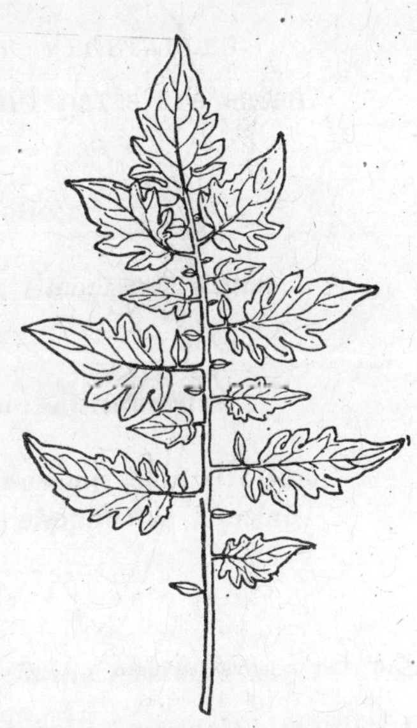

153. Detail of leaf of var. commune (× 1/5).

Harvesting is simple with tomatoes. They are hand-picked. For a near-by market and home use, they should be picked fully ripe, but for more distant shipment when they begin to color well. They are marketed in baskets or crates. If graded and of excellent quality and intended for the best market, the fruits should be wrapped. Early choice fruit is often sold in small splint baskets, like large berry boxes, about four or six baskets being contained in a carrier.

Kinds.

Varieties run out or vary, and fashions in tomatoes change frequently. Because the name of an old variety is still in the catalogues, it does not follow that the variety itself, as originally known, can still be identified. The round regular ("smooth") tomatoes are now almost everywhere grown (Fig. 147), in contrast to the angular wrinkled kinds of many years ago. Leading names at present are Stone, Ponderosa, Earliana, Acme, Crimson

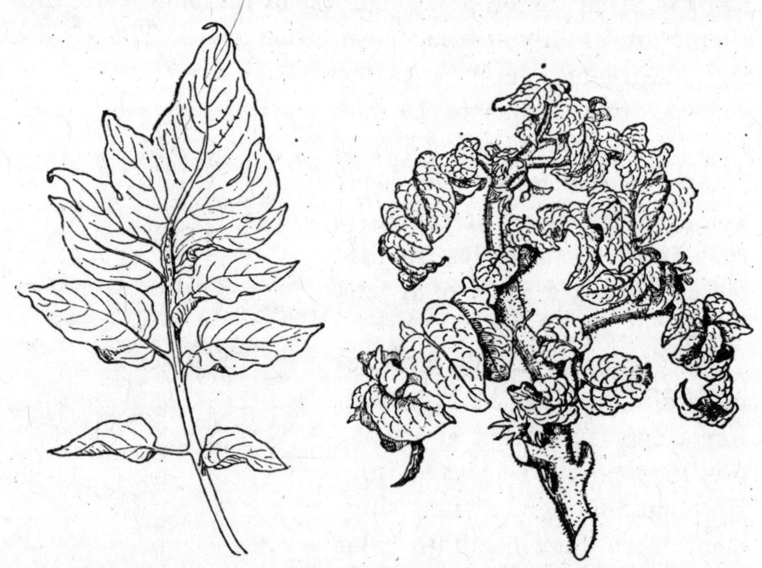

154. Large-leaved tomato (× 1/5). 155. Upright tomato (× 1/6).

Cushion, Beauty, Matchless, Dwarf Champion, Golden Queen. For preserving, the smaller kinds are grown, as Pear, Peach, Plum, Cherry tomatoes (Figs. 148, 149). Attention is now given to the breeding of disease-resistant varieties.

The Tomato Plant

Lycopersicon. *Solanaceæ.* A dozen or so weak branching herbs, perennial and perhaps some of them annual, of South

America, of which one or two are grown for food. The tomato was cultivated or utilized by American aborigines, but it is doubtful whether domestication was really ancient.

L. esculentum, Mill. Gard. Dict. No. 2. 1768. (*Solanum Lycopersicum,* Linn. Sp. Pl. 185. *L. esculentum* var. *vulgare,* Alef. Landw. Fl., 135. 1866.) TOMATO. Diffuse hairy-pubescent, grayish-green, the branches spreading but ascending, herbage strong-scented, perennial or at least plur-annual: leaves 6 to 18 in. long, odd-pinnate, leaflets stalked, with smaller nearly or quite entire ones interposed; main leaflets alternate or subopposite, 5 to 9, conduplicate or tending to curl or roll inward, ovate or oblong, acute or acuminate, bluntly toothed or jagged, the base unequal and sometimes with a supplementary secondary leaflet on one side: flowers nodding, 3 to 7 on forking and sometimes leaf-bearing peduncles borne near a leaf-insertion but on the opposite side of the stem, the yellow corolla ¾ in. and more across; calyx (much enlarging in fruit) green and hairy, cleft nearly to the base into 5 or 6 lance-linear acute lobes; corolla cleft into 5 or 6 long-pointed narrow lobes about as long as the calyx; stamens 5 or 6, with very short filaments, the long yellow green-pointed anthers connivent about the style: fruit a succulent red or yellow angled compressed berry subtended by the lengthened spreading calyx-lobes: seeds obovate, flat, densely hairy, 1/6 in. or more long, weighing 1 to 3 mg., and holding vitality 4 or 5 years.—Western South America. The plant here taken as the type of the variable species is the tomato of a hundred years and more ago which bore depressed (endwise flattened) fruits that were much furrowed or lobed on the sides, and presumably with the gray-green herbage, erect shoots and conduplicate leaflets that some of the last varieties of this old race bore when the writer began to study tomatoes now nearly forty years ago. The Large Red tomato, which was the prevailing variety 50 years ago, is shown in Fig. 150. Miller, in defining the species *L. esculentum,* described the fruit as "compressed at both ends, and deeply furrowed over all the sides." These lobes probably represent the additional carpels as the fruit began to enlarge and modify itself under the stimulus of cultivation. This angular type is passing out in the process of selection. The evolution is toward the round "smooth" (*i.e.*, not lobed or furrowed) tomato, as in Fig. 147. In the process, the intermediate forms, particularly common a generation ago, retained the lobing as they began to enlarge, resulting in the misshapen fruits shown in Fig. 151. Extra carpels are now thrust into the interior of the fruit, and the enlargement takes place on all sides, resulting in a globular tomato. The flower is modified so that the parts are more numerous and the pistil becomes broadened and many-celled. Originally the tomato fruit was probably 2-celled. The common tomatoes of the present day differ from these old ones in character of growth, leafage, and form of fruit, and they may be separated as:

Var. **commune** (Var. *vulgare*, Bailey, Bull. 19, Mich. Agr. Coll. 12., 1886, not Alef.). COMMON TOMATO. Plant green rather than gray-green due mostly to the leaflets being plane rather than "curled," and therefore not presenting the under-surface, the shoots and branches on mature plants usually not erect: fruit mostly globular or somewhat oblate, not distinctly furrowed or lobulate on the sides (Figs. 152, 153).

Var. **grandifolium**, Bailey, Bull. 19, Mich. Agr. Coll. 12. LARGE-LEAVED TOMATO. Leaves large and plane, the leaflets usually not more than 5, margins entire; secondary leaflets very few or none.—Here belong the marked varieties known as Potato-leaf, Mikado, Turner Hybrid, and others now apparently lost to cultivation (Fig. 154).

Var. **validum**, Bailey, Bull. 19, Mich. Agr. Coll. 12. UPRIGHT TOMATO. Plant short, compact, stiff and erect with small crowded curled leaves.—Probably not grown in this country except as a curiosity, although it has been a parent in breeding experiments when it was desired to obtain a tomato plant that might occupy less room and keep itself within bounds (Fig. 155).

Var. **cerasiforme**, Alef., Landw. Fl. 135. 1868. (*L. cerasiforme*, Dunal, Hist. Solan., 113. 1813.) CHERRY TOMATO.

Leaves thinner and smaller than in *L. esculentum*, the leaflets usually less acuminate, the shoots or branches rather more erect: flowers in longer racemiform clusters: fruit small, few-celled, globular.—Used for preserves, in the red and yellow kinds. Probably nearly or quite the original type of the tomato. Var. **pyriforme**, Alef. l. c. (*L. pyriforme*, Dunal, l. c., 112). PEAR TOMATO. Differs in bearing pear-shaped and usually somewhat larger fruits (Fig. 148).

L. pimpinellifolium, Mill. Gard. Dict. No. 4. 1768. (*Solanum pimpinellifolium*, Linn. Amœn. Acad. iv, 268. 1759.) CURRANT TOMATO. Weak and diffuse plant, very finely pubescent, not hairy, the herbage emitting only a mild odor: leaflets small, ovate, the margins obscurely toothed or entire, apex acute or obtuse, not acuminate: flowers in elongating 2-sided racemes: fruit small, 2-celled, like large red currants (about ½ in. diameter) 10 to 30 or more in the cluster, the acute sepal-lobes reflexed: seeds small, smooth.—Peru and probably elsewhere in South America. The botanical identity of the cultivated plant needs further consideration.

EGGPLANT

The essentials in eggplant culture are practically the same as in tomato culture, except that the plant requires a still longer season, and greater pains must be taken that the young plants are not checked but have a continuous rapid growth.

Eggplants are set in rows far enough apart to admit of horse tillage, usually 3 to 4 ft. for the large varieties. In the rows the plants are set from 18 in. to 3 or 4 ft. A common distance is 20 to 24 in., when the rows are spaced at 3 ft. The distance is determined largely by the variety. An ounce of eggplant seed should give 2,000 to 3,000 strong plants.

WILT (*Verticillium alboatrum*).—Affected plants make a stunted growth, and the lower leaves gradually yellow and wilt, causing defoliation. Many plants die prematurely. The wood of all parts of affected plants in the later stages of the disease shows a dark discoloration. *Control:* Care should be used to avoid introducing the fungus into new fields and crop rotation is desirable.

PHOMOPSIS LEAF, FRUIT and STEM DISEASE (*Phomopsis vexans*).—Irregular gray to brown spots on the leaves, and sunken spots on the fruit are covered with the tiny black fruiting bodies of the fungus. Slightly sunken cankers occur on the stem. *Control:* Disinfection of the seed with corrosive sublimate 1 to 1,000 has been recommended. Seed is soaked in the solution for ten minutes, rinsed at once in running water for fifteen minutes and planted immediately. Clean soil in the seed-bed and rotation of crops is important.

EGGPLANT TORTOISE BEETLE (*Cassida pallidula*).—A beautiful green or greenish yellow tortoise beetle about 1/5 in. long that feeds on the foliage. The eggs are laid in groups on the underside of the leaves. The larva is armed with branched spines and carries a mass of cast skins and excrement over its back borne on two long spines. Both larvæ and adults eat round holes in the leaves and sometimes attack the young fruit; restricted to the Southern States. *Control:* Spraying with arsenate of lead (paste), 2 or 3 lbs. in 50 gals. water often gives good results.

EGGPLANT LACE-BUG (*Gargaphia solani*).—A small lace-bug about 1/6 in. long, flat, with the prothorax expanded and covered with a lacework pattern. Both the young and the adults puncture the leaves and suck out the juices. *Control:* Spray with 7 or 8 lbs. whale oil soap in 50 gals. water.

COLORADO POTATO BEETLE (*Leptinotarsa decemlineata*).—See under Potato.

EGGPLANT FLEA-BEETLE (*Epitrix fuscula*).—This small black flea-beetle shows a preference for eggplant. The injury is most serious during the first three weeks after transplanting. *Control:* After the plants are taken from the seed-bed dip the foliage in a 2–3–50 bordeaux mixture. One week or ten days after transplanting spray thoroughly with 4–6–50 bordeaux mixture, to which has been added 4 lbs. of arsenate of lead (paste) to each 50 gallons.

POTATO APHIS (*Macrosiphum solanifolii*).—See under Potato.

SPINACH APHIS (*Myzus persicæ*).—See under Spinach.

RED-SPIDER (*Tetranychus telarius*).—Minute yellowish, green or reddish mites often attack the foliage both in the seed-beds and in the field, giving the leaves a whitish, blistered appearance. They sometimes kill the plants. *Control:* Many of the mites may be destroyed by washing the plants with a strong stream of water from a hose. Apply tobacco dust in the evening and drench the plants with water the next morning, then close the frames and allow them to remain closed for six or eight hours. Apply the treatment on alternate days until all the mites are killed. When they occur in the field, spray the fields thoroughly every few days with:

Nicotine sulfate	5 oz.
Fishoil soap	4 lbs.
Water	50 gals.

The eggplant, known also as Guinea squash in the Southern States, is emphatically a hot-climate crop. It is grown in the South to a large extent as a commercial crop and even as far north as New Jersey and Long Island. In the Northernmost States, it is grown mostly for home use. It demands a long season, a warm loose and fairly dry soil. It is not adapted to clay lands.

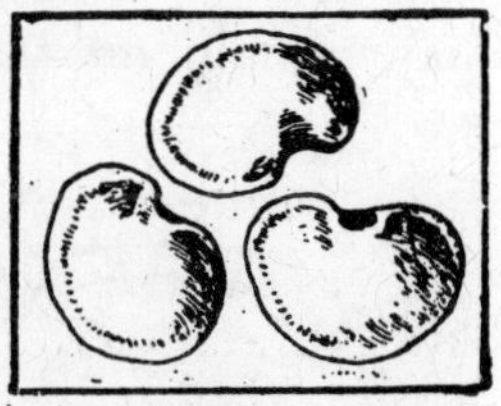
156. Seeds of eggplant (× 4).

The exposure should be warm and sunny. The land should not be as moist as that best adapted to early peas, beets and other cool-season things. The ground should be rich also, but whatever fertilizer is added should be quickly available so that the maturity of the crop may not be delayed. Take every precaution to forward the crop in order to secure it before the closing of the season, particularly in the Northern States. The ground should be kept in thorough tillage from first to last.

The plants are started under glass, and they should be 6 or 8 inches high and thrifty and stocky when placed in the field. In the Northern States the plants may be even larger than this when transplanted. It is important, however, that the plant receives no check from the germination of the seed to the setting of the fruit. If the plants in the forcing-house or hotbed become crowded and stunted, and the stems begin to harden, the crop will be much lessened. For home use, and sometimes for special market conditions, it is advisable to handle the young plants in two-inch or three-inch pots. They then suffer no check when taken to the field.

157. Young egg-plant (× ½).

The fruits are fit for eating from the time they are one-third grown until they are nearly or quite fully ripe. Even after the fruits have reached their full size and color, they may remain on the plant for a time without much de-

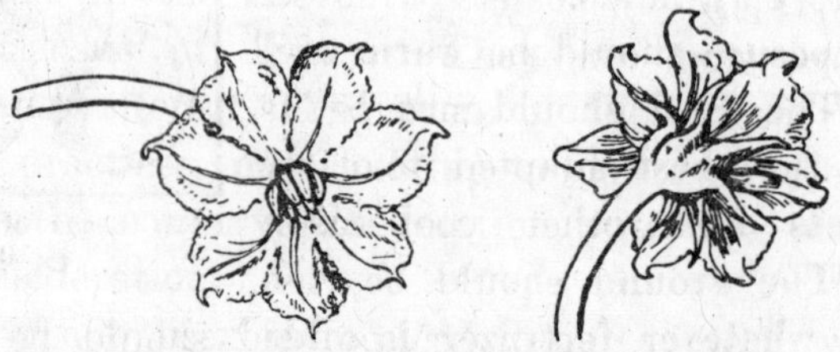

158. Flowers of eggplant, front and back (× ½).

terioration, although a very ripe fruit is worthless. A heavier crop may be secured by taking off the fruits before they reach their full size. It is necessary, however, that they be well colored in order to find sale in the market, and usually, also, the fruits of fair or rather large size sell best. In the Northernmost States the gardener is satisfied if he averages two or three good fruits to a plant of the large varieties.

The fruits are large and heavy, and they should be handled with care even though they are not perishable as are tomatoes. They may be cut from the plants with a knife, the large calyx being left on the fruit. They are usually handled in crates; in special cases, individual fruits are wrapped.

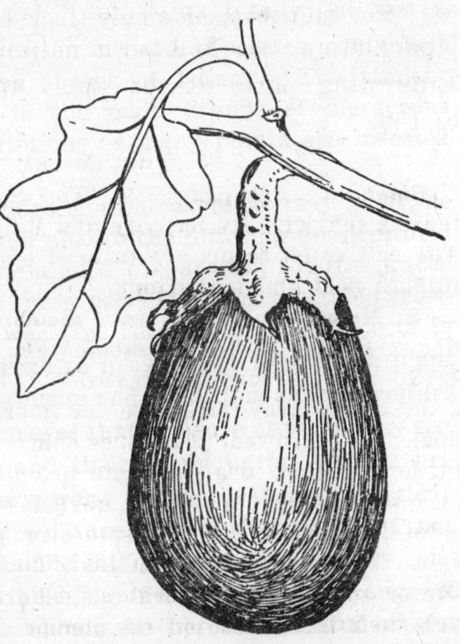

159. Eggplant of the improved purple type (Black Beauty) (× ¼).

The New York Improved, the Black Pekin and Black Beauty are leading commercial types. Good-sized marketable fruits of these varieties are 6 to 9 inches in diameter. Unless started very early and given a warm place and quick soil, however, these varieties are not likely to yield much before frost in the most northern States. In these short-season climates, some of the dwarf varieties, particularly the Early Dwarf Purple, are to be advised. The white eggplants are not popular, since the color is usually of a yellowish cast. There are varieties with striped fruits and others with long and coiling fruits, but these are known mostly as curiosities.

Eggplant has been grown from the earliest times. It is probably native to India. It is a low spreading bushy more or less hairy and spiny herb (or subshrub), with large blue flowers. It is known as aubergine by the French. Interesting parts of the plant are shown in Figs. 156 to 159.

The Eggplants

Solanum. See page 215. The eggplant is a puzzling botanical subject. Its nativity and its origin are both unknown. The cultivated forms are in need of thorough botanical study in both field and herbarium.

S. Melongena, Linn., var. **esculentum**, Nees. Trans. Linn. Soc. xvii, 48. 1832. (*S. esculentum*, Dunal, Hist. Solan. 208. 1813.) Cultivated Eggplant. A bushy leafy erect plant, 2 to 3 ft., gray-tomentose, or more or less scurfy: stem angled or furrowed, sometimes spiny: leaves alternate, 6 to 15 in. long, oblong, oval or ovate in outline, thick, stout-petioled, unequal at base, obtusely angled or lobed, apex acute or obtuse, commonly bearing spines on petiole, midrib and main veins, but sometimes spineless: flowers mostly single, opposite or subopposite the leaves, inclined or nodding, very large (1¾ to 2 in. across) in the commercial varieties, violet with lighter band in the center of the lobe, on a stout and lengthening jointed peduncle; calyx usually prickly, parted about half its length into 5 to 7 narrow pointed green lobes, the tube angled; corolla-limb rotate, with 5 to 7 acute lobes, the margins thin and more or less crinkled; stamens 5 to 8, attached to the corolla-tube, filaments very short, the long yellow anthers erect around the short style; ovary globular-conical, more or less hairy, particularly about the top, many-small-celled: fruit a large pendent berry, 2 or 3 in. to 1 ft. long, purple, yellowish, white or striped, smooth and shining, held by the greatly elongated hard peduncle and the immensely enlarging calyx: seeds numerous on many central placentæ imbedded in lines in the flesh of the fruit, round kidney-shaped, flat, smooth and shining, with many minute pits, 1/16 in. or more long, weighing 2 to 4 mg., retaining vitality 5 to 8 years. Var. **serpentinum,** Bailey, Bull. 26, Cornell Exp. Sta., 25. 1891. (*S. serpentinum*, Desf. Hort. Par. Ed. 3, 115, name only. 1829.) SNAKE EGGPLANT. Differs in the fruit, which is long and slender, 12 in. and more long and 1 in. or less in diameter, curling at the end.

Var. **depressum,** Bailey, Bull. 26, Cornell Exp. Sta. 25. DWARF EGGPLANT. Plant small and weak, spreading or even decumbent rather than erect, nearly or quite smooth and the growing parts often purplish, spineless except sometimes on peduncle and calyx: leaves small and thin, undulate and sinuate but scarcely lobed, nearly or quite smooth, the blade 2 to 6 in. long: flowers small, long-peduncled: fruit pear-shaped and perhaps oblong, 4 to 5 in. long, purple.—A race of good short-season small eggplants, very distinct in habit and foliage.

(The word "Melongena" is an old substantive, perhaps compounded with the Greek for apple.)

PEPPER OR CAPSICUM

Peppers require the treatment advised for tomatoes and eggplants, but they thrive in a rather cooler season and will endure some frost, although best results are obtained in a warm climate. Some of the varieties mature in a relatively short season.

Seeds are started indoors or in hotbed, and transplanted once (and preferably twice) before setting in the field. Sow seeds about ½ in. deep. Plants may stand 8 to 18 in. in the row, depending on variety; rows may be far enough asunder to allow of horse tillage (2 to 2½ ft.) or closer for hand tillage. One ounce of seed should produce 1,500 to 2,000 plants.

Plants should be in full crop in 120 to 150 days from seed. An acre may yield upwards of 30,000 good fruits.

CERCOSPORA LEAF-SPOT (*Cercospora capcici*).—Circular grayish brown spots with dark borders and gray centers are characteristic. Seriously affected leaves wilt and defoliation may occur. *Control:* Disinfection of the seed with corrosive sublimate 1 to 1,000 has been recommended. Seed should be immersed in the solution for ten minutes and then rinsed at once in running water for about fifteen minutes and planted immediately. Seed should be sown in disease-free soil; and plants in the seed-bed and soil may be sprayed with bordeaux mixture.

Peppers are attacked by the potato aphis, flea-beetle and potato beetle: see under Potato; also by the spinach aphis: see under Spinach.

The pepper (often called "red pepper," although there are yellow-, black- and white-fruited varieties) is a Cap-

160. Capsicum seeds, of the Bell pepper type (× 2¼).

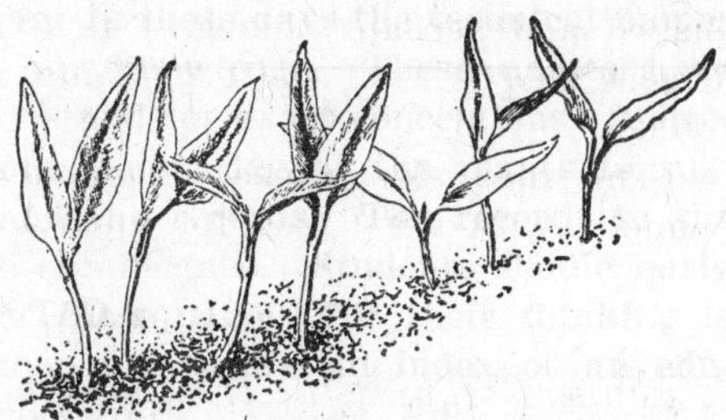

161. Young pepper plants (× ½).

sicum, the common garden forms now being referred to one species, *C. annuum.* It is very distinct from the pepper of commerce, which is the fruit of *Piper nigrum,* of another family. Part of the demand is for the making of mixed pickles and for seasoning, and for this purpose the Cayenne, Chilli, Tobasco and Cranberry varieties are grown. These are the "hot" or pungent varieties. The large "sweet peppers," of the Sweet Mountain, Chinese, Neapolitan, Pimiento and Ruby King type, are used for the dish known as "stuffed peppers," for sweet pickles and in other ways. Figs. 160 to 163 show the pepper from seed to fruit.

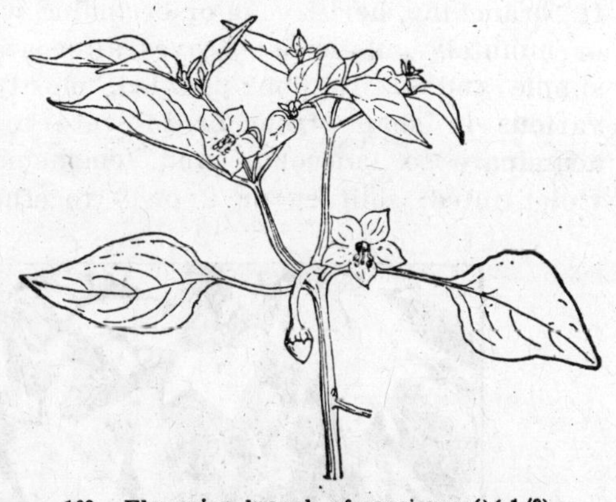

162. Flowering branch of capsicum (× 1/3).

The plants are started in frames, usually in boxes, as are tomatoes and eggplants. Care must be taken to keep the plants growing continuously, else they may not mature a full crop in short-season climates. Dealers usually sell pepper plants, ready for the garden.

For culinary purposes, the peppers are picked when about full grown but before they are colored. The fruits are often known in the market as "green peppers." Formerly little seen, they are now a common article in shops, and the culture of them, in a restricted way, is sometimes distinctly profitable. The hot kinds, or chillies, are usually allowed to color before being picked. Peppers are marketed in splint baskets, hampers and ventilated barrels.

The Pepper Plants

Capsicum. *Solanaceæ.* **Exceedingly variable plants, mostly of Central and South America, by some reviewers thought to be forms of one or two species and by others considered to be 30 or more; herbs and small shrubs.**

C. annuum, Linn. Sp. Pl. 188. RED PEPPER. Biennial or perennial, grown in gardens mostly as an annual, erect, 1 to 3 ft., branching, herbaceous or becoming woody at base, glabrous or minutely pubescent: leaves alternate and often clustered, simple and entire, long-petioled, mostly acuminate at apex, various in shape from broad-ovate to elliptical and short-acuminate to lanceolate and long-acuminate: fls. white or violet-tinted, solitary or 2 or 3 together, stout-peduncled in

163. Bell pepper (× 1/6).

the axils, usually decidedly inclined or declined but sometimes erect, ¼ to 1¼ in. across; calyx short, shallowly 5- or more-lobed; corolla much exceeding calyx, expanding or rotate, deeply and acutely 5-lobed (but with more lobes in some of the cultivated forms); stamens normally 5 (often 6 or 7), the bluish anthers separate and erect and surrounding the single straight central style, filaments equalling or exceeding the anthers and attached on the base of the corolla; ovary globular, about equalling the calyx, 2- to 3-celled: fruit a dryish or somewhat succulent many-seeded pod-like berry of many sizes and shapes, erect or declined, usually scarlet or yellow (sometimes black) when ripe: seeds flat and smooth, light colored, circular in outline with a hollowed projecting base on one side, those of the large bell peppers about 3/16 in. across and weighing 5 to 8 mg.; vitality 3 to 5 years.—Probably tropical American and now spontaneous in many countries, not certainly known as a wild plant; probably not of very ancient cultivation. Some of the many forms of the cultivated plant may be grouped briefly as follows: Var. **grossum,** Sendtner in Martius, Fl. Bras. x, 147. 1846-1856. (*C. grossum,* Linn. Manti. i, 47. 1767.) BELL PEPPER. Comprises the bell and bullnose sweet peppers: plants stout and tall, with large oblong-ovate leaves 4 or 5 in. long, flowers about 1 in. or more across, bearing very large usually puffy hanging fruits with a depression or basin at base and apex blunt, often furrowed at the sides. Var. **longum,** Sendt., Fl. Bras. x, 147. LONG PEPPER. Large-leaved stout plant, with peppers 3 to 12 in. long and narrow, sometimes 2 in. thick at base. Black Nubian, Long Red, Long Yellow, Elephant's Trunk, Ivory Tusk, are varieties in this group. Var. **acuminatum,** Fingerhut, Monogr. Caps., 13. 1832. CHILLI. Differs from var. *longum* in being a smaller and less coarse plant with smaller leaves, and the usually curved fruit very slender, 3/8 in. or less thick and 1/2 to 4 1/2 in. long, erect or declined. Long Cayenne and Chilli peppers belong here. Var. **fasciculatum,** Irish, 9th Rep. Mo. Bot. Gard. 68, 1898. (*C. fasciculatum,* Sturtevant, Bull. Torr. Club, xv, 133. 1888.) RED CLUSTER. Plant compact, with narrow clustered leaves and erect clustered slender very hot fruit which is about 3 in. long and 1/4 in. thick. Var. **conoides,** Irish, 9th Rep. Mo. 65. (*C. conoides,* Mill Gard. Dict. No. 1.

1768.) CAYENNE OR SHORT CAYENNE. Plant becoming woody, the leaves many and mostly small, 2–3 in. long: fruit erect, nearly conical to short-cylindrical, 1¼ in. or less long, very hot. Tobasco and Coral Gem are of this race. Var. **cerasiforme,** Irish, 9th Rept. Mo. 92. (*C. cerasiforme,* Mill. Dict. No. 5.) CHERRY PEPPER. Becoming woody, leaves of intermediate size, ovate or oblong, 2 or 3 in. long: fruit erect or declined, spherical or oblong-spherical, ⅛ in. or less thick, very hot.

HUSK TOMATO

Two or three species of Physalis are cultivated as husk tomato and strawberry tomato. They are very diffuse or even decumbent hairy herbs that produce a yellowish often glutinous berry inside a papery husk (enlarged calyx). There are several native species, some of which are known as "ground cherry." The soft sweetish fruits are sometimes used for preserves and pickles, or they may be eaten raw or cooked. It is a worthy plant.

The plants are of the easiest cultivation. In the North it is preferable to start seeds in frames and transfer to the open ground, in order to mature the largest crop of fruit. The plants spread widely and should be given abundant room, 2 or 3 feet apart being none too much if, in fact, sufficient for best results. The berries will keep all winter if put away dry in their husks.

164. Husk tomato (× 1/3).

The husk tomato is considered to be **Physalis pubescens**, Linn. Sp. Pl. 183, native in North America and other parts of the world. The cultivated plant is mostly a spreading annual but sometimes grows erect, pubescent, much branching: leaves ovate and mostly acuminate, more or less pubescent, oblique or semi-cordate at base, margins obtusely dentate or angled: flowers single, ⅜ in. or less long, yellowish with brown spots inside, the calyx much shorter than the corolla but enlarging in fruit and inclosing the globular yellow berry, which is ¾ in. across. The Cape gooseberry is **P. peruviana**, Linn. Sp. Pl. Ed. 2, 1670. 1763. It is a taller and later plant, not maturing well in the Northern States; leaves soft-pubescent, broad, not toothed, cordate at base; husk larger and somewhat hairy. Other species of Physalis are cultivated.

CHAPTER XI

THE CUCURBITS

Cucumber	**Pumpkin**
Muskmelon	**Squash**
Watermelon	

The cucurbits are annuals, grown for their fruits; they are tender to frost, and require a warm season and a full exposure to sun; they are long-season crops and with most of them a quick start is essential in order that they may mature the crop before frost; they are grown in hills, as a main crop, planted in the field or in frames, depending on the region and the period at which the crop is wanted; they transplant with difficulty when the roots are disturbed, and if the plants are started in advance of the season they are grown in pots, boxes or on sods.

The name "cucurbit" was employed by the writer many years ago for the horticultural designation of these plants of the Cucurbitaceæ, and the word has become current. Subsequently, the late Dr. B. D. Halsted proposed the shorter name "cucurb," but it appears not to have come into use. The cucurbits constitute a very natural group, both botanically and culturally.

Several other cucurbits aside from those listed at the head of this chapter are in cultivation in this country for food. The true gourds, *Lagenaria leucantha* (*Cucurbita*

Lagenaria, Linn.), are grown mostly as arbor covers, and for the great hard-shelled fruits from which dippers and other utensils are fashioned; the young fruits are sometimes eaten in other countries, but probably not here.

The dish-cloth gourds or vegetable sponges, two species of Luffa, are in cultivation as curiosities and for the fibrous interior, which is used, when dried and macerated, as a sponge. The young fruit may be eaten when cooked or dried, but it is scarcely known as a kitchen-garden product in this country. *Luffa acutangula,* with ridged fruits, is apparently more commonly cultivated in this country than *L. cylindrica.*

Of late years, *Benincasa hispida,* the wax gourd of the Orient, has been introduced as the Chinese preserving melon. It is used for the making of preserves and sweet pickles. The fruit is the size of a watermelon, hairy, and usually has a waxy covering. Cultivation is as for muskmelon.

The balsam apple (*Momordica Balsamina*) and balsam pear (*M. Charantia*) are very ornamental climbers with divided leaves and warty small fruits that split open and curl when ripe; they are common on porches and arbors far South, and the young fruits are edible. Their use in this country, except among the Chinese, seems to be for ornament only.

The chayote or christophine (*Sechium edule*) is grown in Florida and the tropics for its cucumber-flavored fruit. It is an odd plant, the fruit bearing only one seed, which is very large.

There are no fundamental differences in the cultivation of the various cucurbitous crops. They are all very tender to frost and they usually grow, at least in the North, till overtaken by frost or disease. They all demand light and very quick soil. Success lies in gaining an early start and in not allowing the plants to suffer a check. The one place at which most people fail in growing these crops is that the young plants do not secure a quick hold. This is usually due to the fact that the soil is not thoroughly well prepared or is not warm and well drained, and there is not sufficient available fertilizer within reach of the young plant. In the North, this quick start is exceedingly important, since the season is so short that every day must be made to count. In cucumbers, the quick start is not so important as in melons and squashes, since the plants come into bearing earlier. Many fields of squashes in the North are lost because the plants do not get to work before July or August, and then the dry weather comes and the blooming is delayed so long that the young fruits are caught by frost.

The land should be given the best of surface tillage. The plants and the fruits are succulent and need much moisture, and if this moisture is lost in the spring through lack of proper preparation of the land and neglect of surface tillage, a good crop may be impossible, even though the subsequent tillage is perfect. The land should also contain sufficient humus or vegetable matter to hold a good supply of moisture.

It is ordinarily best to have the plants so vigorous that several fruits set simultaneously. If one fruit sets two or three weeks in advance of the others, it is likely to consume so much of the energy of the vine that the subsequent fruits remain small. In fact, it may be well to pick off the first fruit if it sets much in advance of the main crop.

Although the land should be rich, the fertility should be available early in the season rather than late, else the growth may be delayed too long. Lands very rich in nitrogenous materials may cause the plants to grow to vine at the expense of fruit. If there seems to be a tendency to go to vine, it is good practice to pinch off the ends of the leading shoots. Usually, however, this practice is not necessary unless the season is very short.

All cucurbits are grown in hills. Each hill may be specially prepared, at least in the Northern States and on land that is rather hard and coarse. A space one or two feet across is spaded up loosely, and light loose earth and scrapings from the barnyard are mixed with it. A handful of fertilizer should be scattered in the soil. If the land is hard and late, it is well to remove the soil and to fill the space with fine earth and manure. In the warm and light melon lands of the South, where the seasons are longer, this precaution may not be necessary; nor is it practiced in the usual field culture of the crops.

Squashes, watermelons and cucumbers are usually planted in the field, although if early results are wanted and if the region is cold and the season short, it is well to start them in frames. Muskmelons are usually started in frames. It is advisable to plant the seeds on inverted sods, in small boxes or other receptacles; or in regular flower-pots, which are best. (See page 357.) It is imperative that the plants be stocky and firm when taken to the field, although they must not be stunted. If they have been grown too warm and are "soft," they will be injured by the sun and winds when transplanted, and will be later than plants started directly in the field.

The young plants are likely to be ruined by the attacks of the striped beetle and other enemies. It is important, therefore, that the seed be sown freely. If one-fourth or one-fifth of the plants escape their enemies, the grower may consider himself fortunate. In some cases growers plant pumpkin or squash seeds in the field very early to attract the striped beetle where they may be killed, and the later frame-grown melon or cucumber plants are then relatively safe.

CUCUMBER AND GHERKIN

Hills of cucumber are usually made about 4 x 4 or 4 x 5 feet; sometimes they are 4 x 6, for the large late varieties, or even 6 ft. either way in extensive field culture. At 4 x 4 feet, 2,722 hills are contained on an acre. Four or five plants are allowed to remain in each hill. About two pounds of seeds are calculated to plant an acre, or 1 ounce for 70 to 80 hills. Seed may be planted about ¾ in. deep. If the striped bugs are bad, plant heavily. An average acre should yield 100 bushels for pickling. Under the best conditions, 400 and 500 bushels of pickling cucumbers are raised to the acre.

BACTERIAL WILT (*Bacillus tracheiphilus*).—Affected plants droop and wilt within a short time. If a stem is cut across, sticky ooze will adhere to the finger and can be drawn out into thin threads. Certain biting insects of cucumbers are largely responsible for the spread of the bacteria causing this disease. *Control:* Spraying with bordeaux mixture and lead arsenate powder (4–5–50–2) to keep the plants free of the striped beetles is recommended. Applications should begin soon after the plants are started and should continue at about weekly intervals until insects are no longer present. Removal and destruction of affected plants is desirable. Dusting the plants with almost any dust mixture is also effective.

Mosaic.—Affected fruits may have raised areas and present a mottled appearance due to the presence of alternate green and yellow places. Leaves may exhibit a similar mottling. *Control:* The control measures recommended for bacterial wilt will aid in keeping this disease in check, as the virus is apparently distributed by insects.

Downy Mildew (*Plasmopora cubensis*).—Somewhat angular yellowish spots appear and under favorable conditions spread rapidly, resulting in the death and drying of affected leaves. *Control:* Spraying with bordeaux mixture, if begun as soon as runners begin to form and repeated every week or ten days throughout the season, will afford good control.

Angular leaf-spot (*Bacterium lachrymans*).—This disease is characterized by the production on the leaves of sharply angular spots bordered by the larger veinlets. The spots are at first water-soaked, later turning brown. Dead parts in old leaves may break away, causing a somewhat ragged appearance. The disease is evident on the stem as elongated lesions. Small water-soaked spots, circular and with white centers, may appear on the fruits. *Control:* Seed treatment with corrosive sublimate 1–1,000 for five minutes has been recommended. Spraying with bordeaux mixture as recommended for other cucumber diseases will hold this leaf-spot in check.

Anthracnose (*Colletotrichum lagenarium*).—See Watermelon, page 296.

Striped cucumber beetle (*Diabrotica vittata*).—A small yellow black-striped beetle, ¼ in. long, that attacks cucumber plants when they first come up, devouring the leaves and eating holes in the stem. The eggs are laid on the ground. The larva is a slender white grub, $\frac{3}{10}$ in. long, that burrows in the roots, sometimes causing the plants to wither and die. *Control:* In the home garden the plants may be protected by cheesecloth or mosquito-netting screens. In the field, spray the plants as soon as they come up with arsenate of lead (paste), 4 lbs. in 50 gals. of water. Although the beetles will not eat much of the poison, it serves to drive them away. It is better to use arsenate of lead for this purpose than bordeaux mixture, as the latter stunts the young plants. It is always well to plant an excess of seed and to use fish scrap fertilizer to make the plants grow rapidly. Tobacco dust or air-slaked lime and land plaster are of value as deterrents.

SQUASH BUG (*Anasa tristis*).—A dirty brownish black bug ⅝ in. long, with a highly offensive odor. The adults come out of hibernation and attack the plants as soon as they come up. They puncture the stems and petioles, sometimes killing the plants outright. The brownish eggs are deposited on the underside of the leaves and hatch in one or two weeks. The young bugs are nearly white and covered with a mealy substance. They puncture the leaves, causing them to wilt and turn brown. *Control:* The adult bugs are very resistant to contact sprays. Practice clean farming and thus reduce to a minimum hibernating shelter. After the crop is harvested, the vines should be raked up and burned. After the ground has been seeded, but before the plants are up, many of the over-wintered adults can be trapped under boards laid on the ground. Then all adults noticed on the young plants should be hand-picked, the eggs should be crushed or scraped off with a knife, and the young nymphs may be killed by spraying with "Black Leaf 40" tobacco extract, 1 part in 400 parts of water in which enough soap has been dissolved to give a good suds.

SQUASH-VINE BORER (*Melittia satyriniformis*).—A white brown-headed caterpillar, 1 in. long when mature, that bores in the stems of squash, pumpkin, cucumber, and melon. The adult is a moth having an expanse of about one inch. The front wings are nearly black and the hind wings are transparent. The eggs are deposited on the stems of the vine. On hatching the young borer burrows into the stem and then eats out a tunnel through the pith, often causing the death of the vine. *Control:* Practice a short rotation of crops. Do not grow susceptible crops year after year in the same field. Collect and destroy all vines after crop is harvested. Early squashes are sometimes planted as a trap crop around the field and between the rows of late varieties. The moths lay their eggs on the early plants, which should be pulled up and destroyed as soon as the early squashes are harvested. Some growers make a practice of covering the stem with earth two or three feet from the base in order that the vine may throw out a new root system which will sustain the plant in case the main stem is injured at the base.

PICKLE WORM (*Diaphania nitidalis*).—Restricted as a pest to the Southern States; the yellowish-white caterpillars, marked with numerous dark spots, are about ⅗ in. long when mature. In the last stage the spots are lost. The adult is a moth with yellowish brown wings marked with large irregular central spots of semi-transparent yellow. The eggs are laid on the flower buds and tender opening leaves. At first the larvæ feed on the blossoms or buds but later burrow into the fruit, often causing decay to result. *Control:* Spraying to control this pest has been unsuccessful. Clean farming and the destruction of the vines after the crop is harvested will help to decrease the number the following year. Waste fruits and fallen leaves should also be destroyed. A short rotation and the planting of a crop at a distance from fields infested the previous year has not been found of much value as the moths fly well from field to field. Much injury may be avoided by planting early so as to have the crop mature before the larvæ become abundant.

MELON APHIS (*Aphis gossypii*).—A yellowish or greenish plant-louse that often occurs in great numbers on the underside of the leaves. In the winged forms the head and most of the thorax is black. *Control:* Fumigation of the young plants with tobacco or carbon bisulfide may be practised in the garden. In large fields spraying is more practicable. The vines should be trained to run in rows. Spray with "Black Leaf 40" tobacco extract, ¾ pint in 100 gallons of water in which 5 or 6 pounds of soap have been dissolved. The first application should be made as soon as the lice appear. A fine nozzle with sufficient pressure should be used to give a fine mist. Use a short extension-rod and an upturned nozzle so as to hit the underside of the leaves.

GARDEN SPRINGTAIL (*Sminthurus hortensis*).—Adult cucumber plants are often badly injured, especially in the South, by a minute wingless jumping insect about 1/20 in. in length. It is dark purple, spotted with yellow. *Control:* Dust the plants with tobacco dust or air-slaked lime when they first come up and repeat the application in about a week if the insects are still present.

The cucumber is a staple garden and truck crop, of easy culture when the simple conditions are met and the diseases and pests are avoided or under control. In general prac-

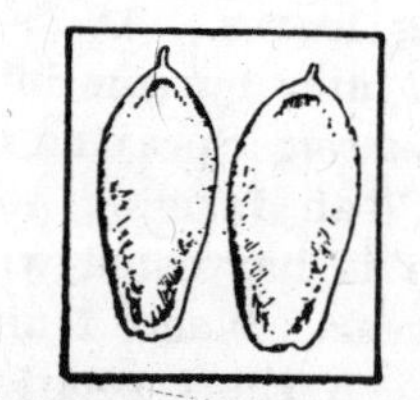

165. Seeds of cucumber (× 1 2/3).

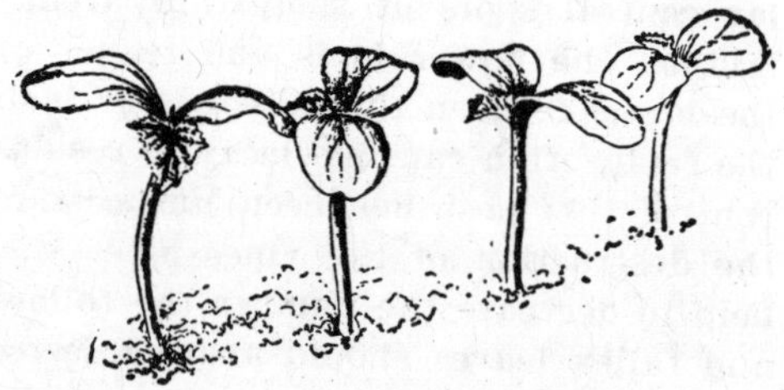

166. Young cucumber plants (× about ½).

tice, the seeds are planted directly where the crop is to mature, but early and choice crops are grown from plants started in frames or forcing-houses on inverted sods or turves, in berry boxes that soon decay, in pots or in knock-down boxes. Cucumber is a prime forcing crop for winter and spring; for amateur work, the English forcing varieties may be used, but the White Spine type is mostly grown under glass for market in this country.

Stages in the cucumber plant are shown in Figs. 165, 166, 167. In the last figure, a fruit is shown and also staminate (or male) flowers, which in the field are more numerous than the pistillate (or fertile) flowers.

The quality of cucumbers depends on the variety, vigor of the plant, when picked, and how kept and handled in transportation. The notion that cucumbers are spoiled by muskmelons planted near, and *vice versa,* is erroneous. Carefully selected seed should be obtained.

Since the fruits of cucumbers are used when young, the productivity of the plants may be greatly enhanced by picking the fruits as soon as they are fit. The patch

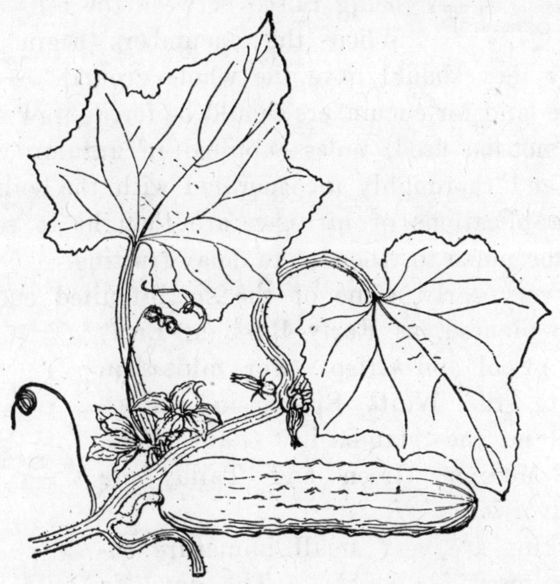

167. Cucumber of the White Spine type (× about 1/3).

should be gone over every two or three days at least, and if the area is large, it should be picked over every day. If one fruit is allowed to ripen it may prevent the setting of other fruits. If seeds of cucumbers are desired, it is best to reserve a few hills specially for that purpose.

Cucumbers for the main or pickling crop are usually grown from seeds planted directly in the fields when frost is past; sometimes they are in two or three plantings, up to even the first or middle of July. Cucumbers do not require as much heat as melons.

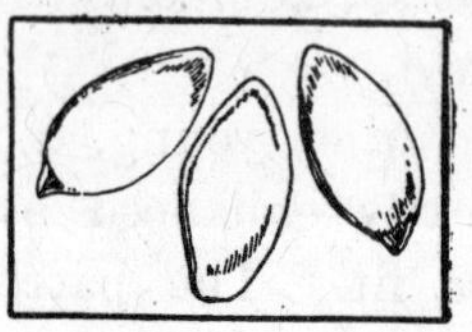

168. Seeds of Cucumis Anguria (× about 3).

Sometimes cucumbers are grown as a companion crop, beans or lettuce being raised between the hills or rows. When the cucumbers begin to run strongly they should have the whole ground.

While land for cucumbers should be fertile, raw manures should not be used, unless applied in autumn or early spring and thoroughly incorporated with the soil. Very heavy applications of nitrogen are thought to send the plants too much to vine and to delay fruiting.

For very early, some of the small-fruited cucumbers may be planted, as Early Russian, Early Cluster, Cool and Crisp. For midseason and late, the White Spine, in various strains, is the standard. Giant Pera, Nichol Medium Green and Tailby are older favorites.

169. Seedling of C. Anguria. (× 2/3).

Gherkins are very small immature cucumbers, used for pickles. The name is also applied to the small prickly fruits of *Cucumis Anguria,* a species known as the West Indian or burr cucumber. This is sometimes cultivated, and its fruits are used for pickles. It is grown in every way as is the ordinary cucumber. Seeds and seedling are seen in Figs. 168, 169.

For table use (slicing), cucumber fruits should be 6 inches or more long, green, fresh and plump. All cucumbers, whether for slicing or pickles, are picked before they begin to turn yellow. For good markets, the cucumbers should be graded to shape and size. They are marketed in baskets, crates, and hampers.

MELON

Four by six feet is a customary distance for the hills of muskmelons, making 1,185 hills to the acre. The quantity of seed required is about the same as for cucumber. Sometimes two crops are grown on the same land, a very early and a main-season crop. The early crop is planted 4 x 5 feet, and two or three weeks later the main crop is planted between. Three or four good fruits to the plant is a good yield. Seeds are covered about ¾ in., or somewhat deeper if planted directly in the field.

The melon is affected by wilt, mildew and mosaic, for which see the account under Cucumber; and by anthracnose, treated under Watermelon.

STRIPED CUCUMBER BEETLE (*Diabrotica vittata*).—See under Cucumber. The beetles not only attack the young plants but later in the season destroy the young blossoms. *Control:* In small gardens screen the young plants. Fish scrap fertilizer not only forces the growth of the plants but has a tendency to keep the beetles away. Air-slaked lime, tobacco and sulfur have a decidedly deterrent effect. Sow an excess of seed and thin the plants after the danger from the insects has passed.

GARDEN SPRINGTAIL (*Sminthurus hortensis*).—See under Cucumber. Apply tobacco dust, fish scrap or air-slaked lime just as the plants are coming above the ground. Repeat a week later. Sow an excess of seed, cultivate frequently and apply quick acting fertilizers to help the plant outgrow the injury.

SOUTHERN CORN ROOT-WORM (*Diabrotica duodecimpunctata*).—A yellowish green beetle, ¼ in. long, with twelve black spots on the wing-covers. The beetles are often destructive to cucumbers and melons which they attack in much the same way as the striped cucumber beetle. *Control:* Same as for the striped cucumber beetle (page 285).

MELON WORM (*Diaphania hyalinata*).—The adult is a moth with pearly white wings, marked with a shining iridescent brown band along the front and outer margins. The eggs are laid on the young buds, leaves and stems of the vines. When full grown the caterpillar is about 1 in. long and mottled greenish yellow. When partly grown it is yellowish or greenish with two white stripes on the back. The first brood of caterpillars feeds mostly on the foliage and does not cause much injury to the fruit. The larvæ of the later generations feed at first on the buds or foliage and then attack the fruit, feeding on the surface and burrowing through the rind, causing decay. *Control:* Plant summer squashes ahead of the main crop to serve as a trap. Spray with arsenate of lead (paste), 3 lbs. in 50 gals. water or bordeaux mixture. As soon as the crop is harvested, vines and waste fruits should be gathered up and destroyed.

The melon (or muskmelon) is a prevailing inhabitant of the vegetable-garden and is much grown by market-gardeners. It has a shorter season than most forms of watermelon, and lends itself to a wide variety of soils and conditions at the same time that it is rather exacting if the greatest success is to be attained. The commercial product is grown mostly in special and limited localities, and yet a given locality may not long hold its leadership. he melon is now a popular breakfast food, in its season taking the place of grapefruit. It forces well, and for this purpose some of the special English forcing varieties are most useful.

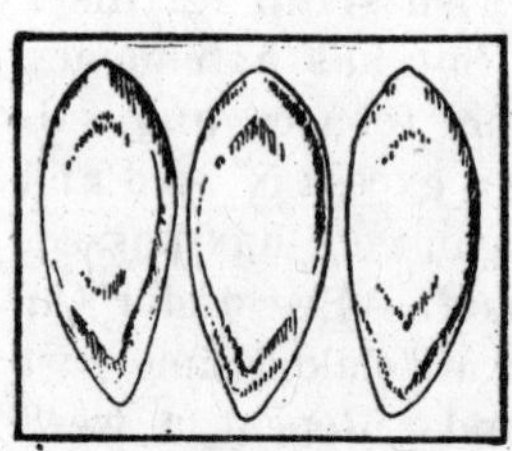

170. Melon seeds (× about 2).

The melon is displayed in Figs. 170 to 174. In Fig. 172 a young fruit is seen at P. Above at the left is a pistillate flower with the ovary beneath the corolla; above right is a staminate flower, lacking the ovary.

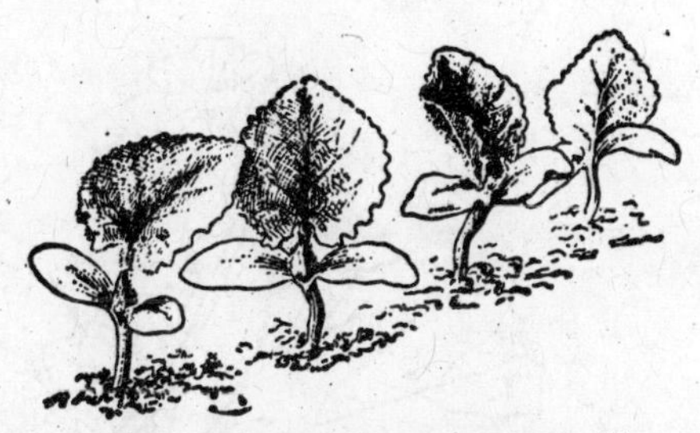

171. Melon seedlings (× about ⅓).

A light warm easily-tilled clean soil is usually chosen for the melon. Shallow tillage should be frequent, that the plant may grow strongly and continuously. It is useless to attempt to grow melons on cold, backward or hard heavy lands. The melon thrives particularly well in the irrigated regions; parts of the West and far Southwest are large producers.

The distance of planting depends somewhat on the variety and also on the room at the grower's disposal. If land is ample, 6 x 6 feet is a good distance, but 4 x 6 feet is commonly advised. If the soil is not deep and fertile, "hills" may be made by working well-rotted manure into the earth where the plants are to stand and perhaps by adding quickly available commercial fertilizer.

Seeds may be planted in the field as soon as the weather is finally settled and the soil warm. Twice as many seeds should be put in as are required to make the stand, to allow for insect injury and accidents. Two or three strong plants are finally left in each hill. For early melons, and also for late-maturing kinds, plants may be started in frames, on turves or in berry boxes, veneer boxes, or pots.

In the northern parts, the crop is often grown permanently in frames, the glass being removed entirely when the weather is fit and the plants established. Very choice

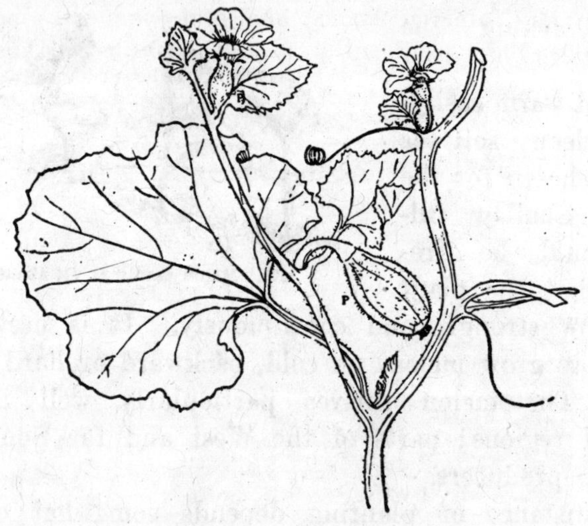

172. Flowers and foliage of melon (× ⅜).

melons may be grown in this way; much hand work is usually applied, and the fruits are sometimes lifted and a flat stone or shingle slipped under them; when the melons reach good size, the frames themselves, as well as the sash, may be removed. Every pains should be exercised to secure well-bred seed true to type, in any careful melon-growing.

Melons are picked when signs of ripeness appear. This period cannot be described, but must be learned by experience. The bright color begins to tone down to gray, signs of yellowness are apparent, the stem parts readily from the vine, and the fruit has attained the full size and development of the particular variety; the appearance of immaturity and "greenness" has passed.

The fruits are marketed in crates and open-topped baskets, the melons always as visible as possible through the openings. The soft-fleshed melons are often packed on straw in baskets and hampers. The melons should be graded to size, shape, markings and color.

173. Netted melon (× 1/5).

In this country, the various forms of reticulated melons are popular. It is customary to divide the varieties into (1) the green-fleshed kind and (2) yellow-fleshed and salmon-fleshed. Of the former are Netted Gem, Emerald Gem, Rocky

174. Cantaloupe, scarcely known in North America (× ¼).

Ford, Hackensack, Jenny Lind, Montreal. Of the latter are Emerald Gem, Osage, Banquet, Burrell, Tip Top. The Cassaba and Christmas melons are large types grown specially well in long-season irrigated regions, and thence shipped to the eastern markets. They are very unlike the common netted melons, lacking the odor and characteristic markings and keeping well.

WATERMELON

At 10 x 10 ft., 435 hills are contained in an acre; this is a common distance for planting the commercial crop, but smaller garden varieties may be set 8 x 8 ft. About 4 or 5 pounds of seed are used to the acre. Plant about 1 in. deep. A good commercial crop is about 12 tons to the acre. The watermelon is more tender to cold than the muskmelon.

WILT (*Fusarium niveum*).—Usually one branch after another of an affected plant wilts and dries up until the whole plant is dead. An examination of the stem shows the woody portion to be discolored. *Control:* Crop rotation, the control of drainage water to prevent overflowing uninfested soil, and the avoidance of contaminated stable manure on melon fields are important.

ANTHRACNOSE (*Colletotrichum lagenarium*).—This disease occurs also on cucumbers, muskmelons, and other plants of the cucurbit family. Irregular black dead spots appear on the affected leaves and die prematurely. Numerous blackened sunken spots appear on the fruits. *Control:* Thorough and timely spraying with bordeaux mixture is a preventive.

STEM-END ROT (*Dioplodia* sp.).—The first indication of the disease is a browning and shrivelling of the stem followed by a softening of the melon at the point of attachment. As the flesh softens, it becomes water-soaked in appearance. The disease causes severe loss in shipment. *Control:* Field sanitation is important in view of the fact that vegetation of nearly all kinds may harbor the causal organism. Thorough applications of bordeaux mixture are necessary, since the fungus developing on vines killed by other diseases will spread to the melons. The stems should be disinfected at the time of loading by painting them with a starch paste containing copper sulfate. To prepare the paste, eight ounces of copper sulfate are dissolved in three and one-half quarts of hot water and to this boiling solution are added four ounces of starch mixed with a pint of cold water.

MELON APHIS (*Aphis gossypii*).—See under Cucumber. Spray with "Black Leaf 40" tobacco extract, ¾ pint in 100 gals. water in which 4 or 5 lbs. soap have been added. Be careful to hit the underside of the leaves. When the first hills are infested, fumigate with tobacco papers under frames covered with oilcloth.

The watermelon is more popular in North America, probably, than elsewhere in the world. In fact, it is a feature of American living. The South Atlantic and Gulf States have occupied first place for size and quality of melons. More recently, the mid-continental States are coming to the front. The watermelon is a leading field crop in Georgia and elsewhere, great areas being devoted to it. The plant is little grown in market-gardens, for it requires too much space and the returns are not sufficient. It is primarily a truck crop or farm crop, on relatively low-priced land.

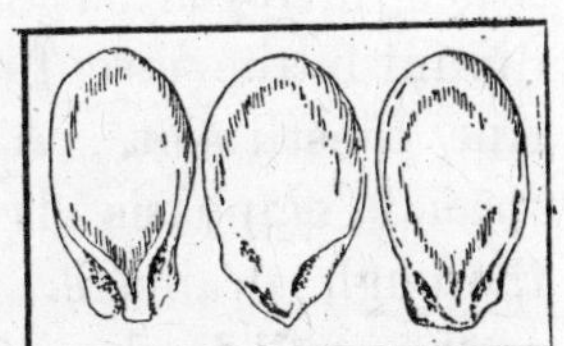

175. Watermelon seeds (× 1¼).

The reader will recognize the watermelon in Figs. 175 to 179. The outward distinctions between pistillate and staminate flowers are shown in Figs. 177 and 178, the presence or absence of the ovary (young fruit) being conspicuous.

The first requisite in watermelon culture is a location

with sufficient length of season and continuous warmth to insure maturity of crop. Many varieties of watermelons are catalogued by seedsmen. Only a few of them are commercial varieties, and the kinds that are popular in the South require a too long season for the North. Only in favored places are watermelons grown in the Northernmost States. They are more uncertain than muskmelons, because of the short and cool seasons. A number of varieties, however, ripen without difficulty in the Northern States and Ontario when a warm soil and exposure are at hand and where small boys are absent. The plants may be started under glass, as advised on page 283.

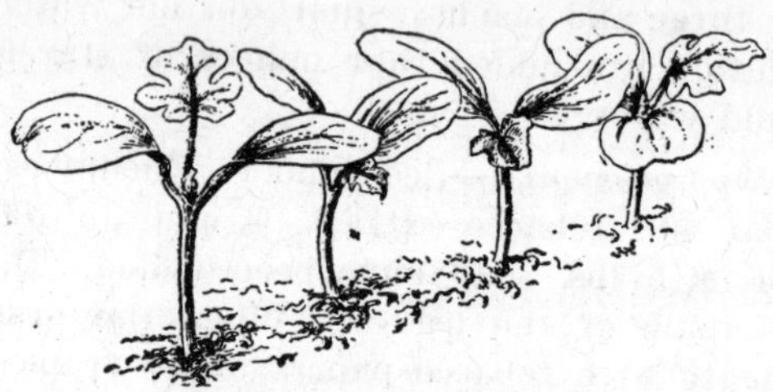

176. Watermelon seedlings (× 2/5).

The ideal soil is light sandy loam with only a medium or small amount of nitrogen. Much nitrogen is thought to diminish the essential saccharine constituent. A point of special emphasis is that of thorough drainage. Swampy or "soggy" land will not produce favorable results. In the South the field for melons is often plowed in the fall, to expose the soil to the pulverizing action of frost.

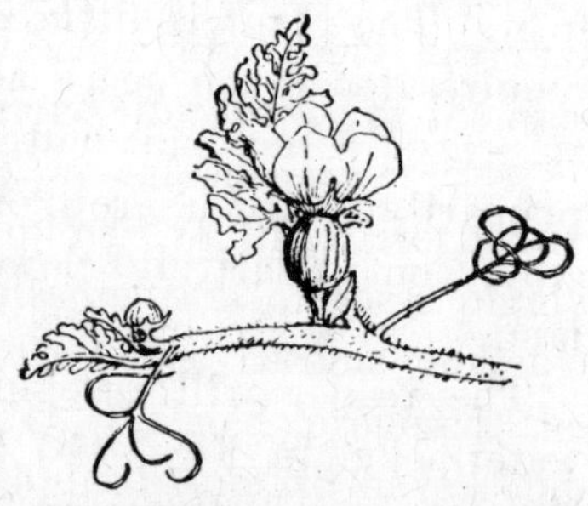

177. Pistillate (fertile) flower of watermelon (× 1/3).

Watermelons are planted in hills, which are usually 10

feet each way. The hills are made at the intersection of check-rows. This "checking" is usually accomplished with shovel- or turn-plow. The hills are made by mixing several shovelfuls of well-rotted manure with soil and then covering the whole with several inches of soft earth, into which the seeds are planted directly. All danger of frosts should be over before planting. Avoid baking or crusting of the earth on the hills, especially before germination of seeds. Only hand tools should be used in the cultivation of crop after the vines have begun to run, as lifting or turning the vines will injure quality and size of fruit.

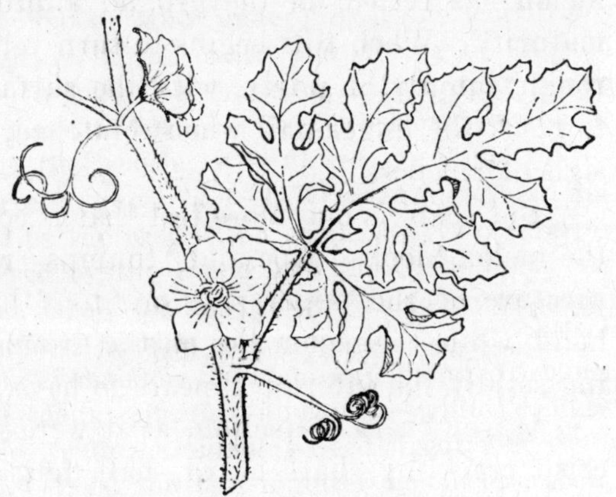

178. Staminate (sterile) flowers of watermelon (× 1/3).

"Rotation is all-important," as written by Starnes (Bull. 38, Ga. Exp. Sta.). "In no case should melons follow melons the next season, and at least four years should intervene before the land is again planted in this crop. By that time insect depredators, attracted by the first melon crop, will have probably become exterminated and the drain from the soil of specific plant-food (especially potash) will also have been, to a certain extent, at least, made good."

When is a watermelon ripe? According to Starnes,

"unquestionably the flat, dead sound emitted by a melon when 'thumped' is the readiest indication of ripeness, and the one most universally depended on. If the resonance is hollow, ringing or musical, it is a certain proof of immaturity.

"Frequently on turning the melon and exposing the under side, the irregular white blotch formed where the melon has rested on the ground affords an indication of maturity. When this begins to turn yellowish and becomes rough, pimply or warty, with the surface sufficiently hard to resist the finger-nail when scratched, it is usually a fair sign of ripeness.

"But there is one more test that is corroborative. After the melon 'looks' ripe and 'thumps' ripe, if, on a steady pressure of the upper side or 'top' by the palm of the hand, while the melon lies on the ground, instead of resisting solidly the interior appears to have a tendency to yield—a 'givey' sort of feeling, as it were—accompanied by a crisp crackling, half heard, half felt, as the flesh parts longitudinally in sections under the pressure, the melon may be pulled with absolute confidence. It is certainly ripe. This test should never be resorted to with melons intended for shipment, as their carrying quality is necessarily impaired thereby.

"Yet all this, as stated, comes largely by instinct to the expert, and it is rarely that one finds it necessary to 'thump,' much less to 'press,' a melon before deciding as to its maturity."

Many of the small early watermelons may be grown successfully in warm northern gardens. Fruits of superior quality, and picked when in perfection, may be had freely for home consumption; greater attention should be paid to the plant by the home gardener. Among the melons suitable for home gardens, not to mention others equally as good, are Peerless, Dark and Light Icing, Kleckley, Mc-Iver, Phinney, Halbert, Hungarian. The shipping watermelons, mostly requiring longer season, comprise such varieties as Kolb Gem, Rattlesnake, Dixie, Alabama Sweet,

179. The market watermelon (× 1/10).

Ironclad, Tom Watson. The oblong kinds (Fig. 179) may reach 2 feet in length. Fair-sized shipping watermelons weigh about 20 or 25 pounds, but they run to 30 pounds and more. Watermelons are shipped in bulk, by the carload.

PUMPKIN AND SQUASH

Seeds are planted 1 in. to 1½ in. deep. When grown by themselves, pumpkins and field squashes are planted in hills

8 to 10 feet apart. About 3 pounds of seed are required for an acre with the field or running varieties. Two or three mature fruits to a vine are a large crop.

The bush squashes are grown as close as 3 x 4 feet in gardens, but the hills should be 4 or 5 feet apart if possible. From 4 to 5 pounds of seed are required to the acre.

The pumpkins and squashes are affected by wilt disease and mosaic, for which see Cucumber.

The striped beetle and squash bug are also treated under Cucumber.

The pumpkins and squashes are of simple and easy culture. Warm well-drained lands are chosen. The seeds

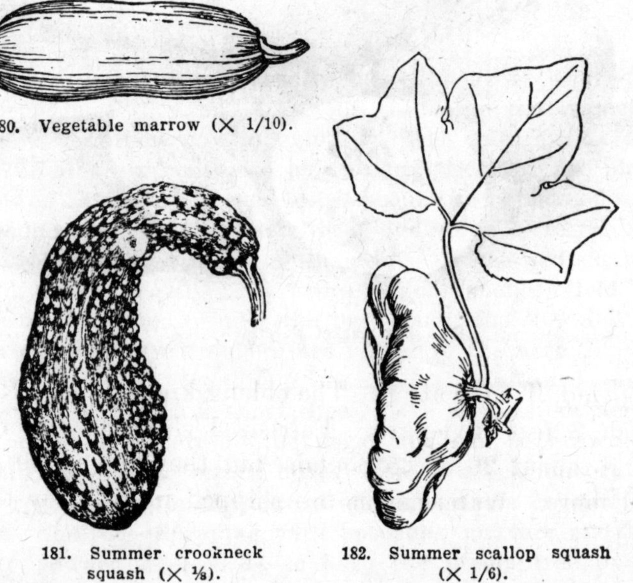

180. Vegetable marrow (× 1/10).

181. Summer crookneck squash (× ⅛).

182. Summer scallop squash (× 1/6).

are planted directly in the field, often in "hills" specially prepared by the incorporation of manure or fertilizer or both. Usually they thrive in good well-prepared corn land without special treatment. They must be got ahead early, in the Northern States, to yield the full crop before frost. Many or several seeds should be planted in the hill, and the plants thinned to two or three when the early dangers are passed.

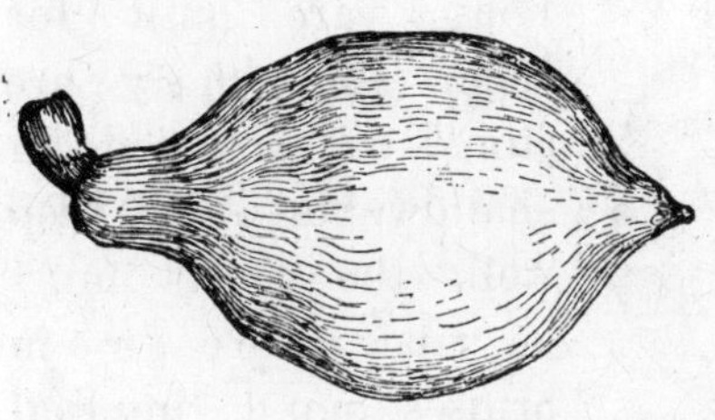

183. One of the marrow squashes (× ⅛).

In pumpkins, as the term is understood in this country, the standard variety is the Connecticut Field. It is a long-running plant. The large orange-colored sleek furrowed fruits are used for pies and to feed stock; and the small boy prizes them for "jack o' lanterns." It was formerly much grown in corn-fields. This plant is a form of *Cucurbita Pepo*. The summer squashes (Figs. 181, 182) are taken previous to full maturity before the shells harden. They are marketed in baskets and crates. They are interesting for the oddities in their shapes, as well as for their good comestible qualities.

184. One of the turban squashes; they are commonly less furrowed (× 1/6).

Of field or late squashes the leading types are the Hubbard, Marblehead, Boston Marrow, Turban (Figs. 183, 184). They are long-runners and sometimes are planted as much as 12 feet apart. The fruits have soft cylindrical stems. These squashes are kept for winter; they should have a dry and fairly warm place (temperature above 40°). When they are grown extensively, special stove-heated houses are built for them and they are stored on shelves or in shallow bins. To keep well, the fruits must be ripe, free from bruises and internal cracks, not frosted, and have the stem on. These squashes are *Cucurbita maxima.* They have a firm yellow flesh, and a richer quality than others. They lend themselves well to baking.

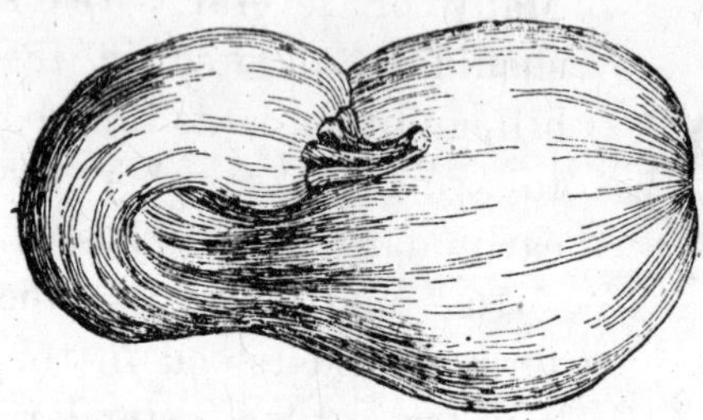

185. Winter crookneck (× ⅛).

A third specific type is *Cucurbita moschata,* to which belong the Cushaws, Winter Crookneck (Fig. 185), Dunkard, Tennessee Sweet Potato Pumpkin, and others. In the South the varieties of this species are common, but most of them are only indifferently successful in the North. They are famous pie pumpkins in the Southern States.

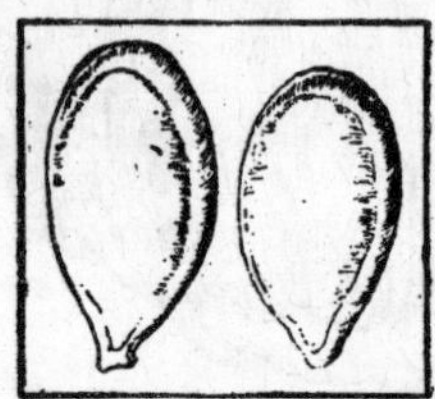

186. Seeds of squash (somewhat enlarged).

187. Seedlings of squash (× 2/5).

The illustrations will aid in distinguishing some of the classes; and Figs. 186 and 187 show seeds and seedlings of

C. maxima. The grower should familiarize himself with the interesting differences in foliage and flowers.

THE CUCURBITOUS PLANTS

For the purposes of this discussion, we may consider only three genera of the Cucurbitaceæ, comprising annual and perennial herbs of warm countries, those of the vegetable-garden being tender annuals: **Cucumis** with about 25 species, mostly African and Asian; **Citrullus,** 4 species in Africa; **Cucurbita,** about 10 species, perhaps American but the origin of the cultivated kinds unknown. All these garden species are monœcious,—the stamens and pistils being in separate flowers on the same plant; the staminate (male) flowers are more numerous than the pistillate, and soon perish. All are tendril-bearing, thereby grasping weeds and other supports and climbing over bushes and fences when allowed to do so; plants hirsute, pubescent or prickly-hairy; fruit a pepo (the word "pepo" is Latin for a pumpkin or related fruit) which is a normally 3-celled and mostly indehiscent more or less fleshy many-seeded pericarp, with the flower-parts at the apex.

Cucumis. Two species are in common cultivation; and the burr gherkin, *C. Anguria,* is sometimes grown for the making of pickles from its tuberculate fruit and also for ornament. They are all slender-running plants with simple (unbranched) tendrils. The cucumber has been cultivated from prehistoric times, but the melon appears to be of later domestication.

C. sativus, Linn. Sp. Pl. 1012. CUCUMBER. Trailing or climbing rough-hairy herbs with alternate long-petioled triangular-ovate angled or somewhat 3-lobed irregularly dentate leaves, the middle lobe usually pointed: flowers axillary, yellow, with hairy calyx; staminate 1 to several in the axil, 1 to 1½ in. across, very short-pedicelled, the calyx-tube campanulate and exceeding the 5 subulate spreading lobes, the corolla 3 or 4 times longer than calyx-lobes, the corolla-lobes acute and usually conduplicate, stamens 3 inserted on the corolla-tube and the bearded anthers produced into an erect appendage; pistillate flower mostly solitary, nearly or quite sessile, the long 3-celled ovary much constricted at its summit, the three 2-lobed stigmas very large, the staminodia usually not evident: fruit mostly oblong, sometimes nearly globular, prickly or tuberculate: seeds small (about ½ in. long and nearly half as wide), brownish-white, elliptic, flat, and apiculate or sharp-pointed at apex, smooth, 20 to 35 mg. in weight, keeping 8 to 10 years or even longer. Var. **anglicus,** Bailey, Cyclo. Amer. Hort., 408. 1900. ENGLISH OR FORCING CUCUMBER. Vine very strong and vigorous: leaves large and broad, short in proportion to breadth: flowers very large; ovaries and fruits very long and slender (fruit sometimes 3 ft. long), little furrowed, spineless or nearly so, ripening green or nearly so rather than yellow, the seeds few.

C. Melo, Linn. Sp. Pl. 1011. MELON. MUSKMELON. Trailing or climbing soft-pubescent or hairy herbs, with long-petioled reniform or round-ovate deeply cordate hairy angled but commonly not lobed apiculate-dentate leaves: flowers yellow, with hairy calyx, on short peduncles; staminate 1 or more in the axil, about 1 in. across, the 5 narrow calyx-lobes about as long as the tube, the 5 oblong nearly obtuse corolla-lobes 3 times as long as calyx-lobes, stamens 3 inserted on the corolla-tube and the anthers produced above into an erect appendage; pistillate flower single, with inferior 3-celled globular or oblong ovary, the 3 stigmas surrounded by 3 conspicuous staminodia (sterile anthers), which, however, are often polliniferous, making the flower perfect: fruit various, globular or cylindrical, more or less furrowed, pubescent but usually becoming glabrous: seeds elliptic or oblong, brownish-white, plump, about ½ in. long and $\frac{3}{16}$ in. broad, not apiculate, smooth, weighing 25 to 35 mg. and holding vitality 5 to 10 years.—Probably central Asian. (The word "Melo" is Latin for a form of melon.) Var. **reticulatus,** Naudin, Ann. des Sci. Nat. Bot. Ser. 4, ii, 50. 1859. RETICULATED OR NETTED MELONS. Small fruits with the surface net-ribbed, comprising the nutmeg melons. Var. **cantalupensis,** Naudin, l. c. 47. CANTALOUPE MELONS. (Fig. 174.) Fruits with hard rinds, often furrowed, warty, scaly or rough.

Practically unknown in this country, the name cantaloupe here being improperly applied to melons in general.

Var. **inodorus**, Naudin, l. c. 56. WINTER MELON. CASSABA MELON. Strong long-tendrilled plants with large less hairy leaves which often are lobed, sometimes round-ovate and deeply

188. Leaf and staminate flower of Cucurbita Pepo (× about ½).

cordate like leaves of *Cucurbita maxima:* flowers very large, often 2 in. across: fruit with little of the musky odor associated with the musk melon, ripening late and keeping into winter, often oblong and squash-like in shape and frequently striped and splashed.

Var. flexuosus, Naudin, l. c. 63. (*C. flexuosus*, Linn. Sp. Pl. Ed. 2, 1437. 1763.) SNAKE or SERPENT MELON. Plant slender: flowers large: fruit long and thin, 1 to 3 in. thick and frequently 18 to 36 in. long, often curiously curved and crooked.

189. Leaf and pistillate flower of C. Pepo; the ovary is at P (× about 2/5).

—Used sometimes for preserves, but grown mostly as a curiosity.

Var. **Dudaim**, Naudin, l. c. 69. (*C. Dudaim*, Linn. Sp. Pl. 1011. *C. odoratissimus*, Moench, Meth. 654. 1794.) DUDAIM MELON. Small and slender plant with more or less lengthened leaves: flowers relatively large: fruit size of an oblate orange, smooth, longitudinally marbled with rich brown, very fragrant.—Grown for ornament and for the strong scent of the fruit. ("Dudaim" is a Hebrew name, said to be scriptural.)

Var. **Chito,** Naudin, l. c. 67. (*C. Chito,* Morr. Ann. Soc. Gand. v, 341. 1849.) MANGO MELON. Slender plant with melon-like foliage but smaller: fruit size and shape of an orange or lemon, or sometimes oblong, not fragrant or variegated, yellow or greenish yellow, the flesh white and much like that of a cucumber, whence the name "Lemon Cucumber." —Used in the making of "mango" preserves and pickles; known also as Orange Melon, Melon Apple, Vine Peach, Vegetable Orange. (The word "Chito" is probably geographical.)

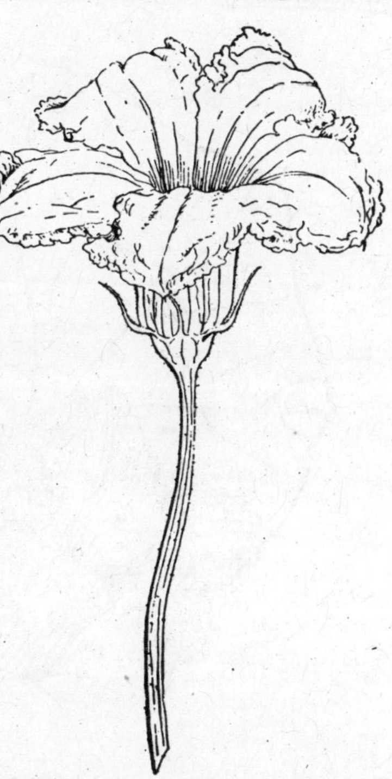

190. Staminate flower of Cucurbita maxima (× about 2/5).

C. Anguria, Linn. Sp. Pl. 1011. WEST INDIA OR BURR GHERKIN. BURR CUCUMBER. A very slender rough-hairy plant with angled stems and small leaves lobed or cut into usually 5 rounded lobes with open sinuses: flowers about ⅛ in. across, yellow, on slender peduncles: fruit oval or oblong, pale yellow, longitudinally furrowed and marked, prickly, about 2 in. long: seeds elliptic, whitish, about $\frac{3}{16}$ in. long, 6 to 8 mg. in weight.—Florida and Texas to South America. (The name "Anguria" is of Greek origin, applied to some kind of cucurbitous fruit.)

Citrullus. Aside from the watermelon, only the colocynth

(*C. Colocynthis*) is cultivated, as a curiosity for its small globular very bitter fruit which is also used in medicine.

C. vulgaris, Schrad. in Ecklon & Zeyher, Enum. Pl. Afr. Austr., 279. 1834. (*Cucurbita Citrullus,* Linn. Sp. Pl. 1010.) WATERMELON. Long-running hairy vine with branching ten-

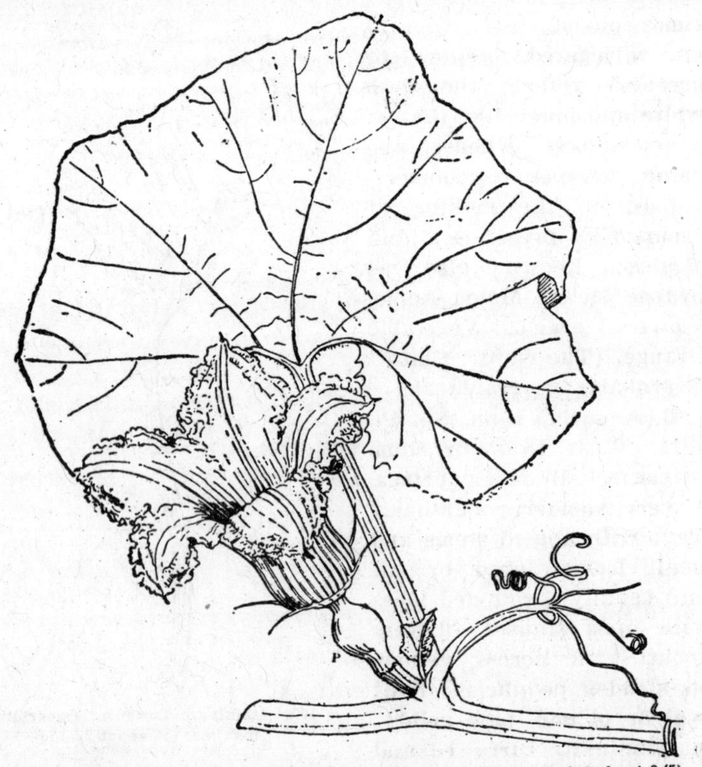

191. Leaf and pistillate flower of C. maxima; ovary at P (× about 2/5).

drils: leaves ovate to ovate-oblong in outline, short- or long-petioled, the blade pinnately divided into 3 or 4 pairs of lobes, the lowest one again lobed, and with small lobes and teeth variously placed, the base of the blade cordate: flowers axillary, light yellow, rather slender-peduncled, with hairy calyx;

staminate 1¼ to 1½ in. across, rotate, the shallow calyx-tube not equalling the 5 subulate spreading calyx-lobes, the broad obovate obtuse green-veiny corolla-lobes about 3 times exceeding the calyx-lobes, the 3 short stamens with very large curling anthers; pistillate flowers solitary, with 3 short very large 2-lobed stigmas and small not protruding staminodia, the ovary usually 3-celled: fruit globular or oblong, mostly glabrous, with a hard rind and sweet red or white flesh, on the outside green and commonly more or less marbled when ripe: seeds white or black, elliptic, flat, ridged on the edge, about ½ to ⅝ in. long and ¼ to ⅜ in. wide, with a characteristic prominence on either side at the point, weighing 90 to 120 mg., lasting 5 or 6 years.—Tropical and South Africa. The "citron" of housewives, used for making of a preserve, is a hard-fleshed watermelon. A special kind is grown in China for the seeds, which are eaten.

192. Staminate flower of Cucurbita moschata (× about ½).

Cucurbita. The three domesticated species of Cucurbita, comprising the squashes, pumpkins and the small yellow-flowered gourds, are readily distinguished in the field when the eye is trained to recognize the distinction, but may not be easily separated in herbarium specimens or by description. Following are visual features of separation:

A. Plant harsh and rough to the feel, due to the presence of many stiff sharp translucent hairs, the foliage stiff and more or less rigid, standing erect: leaves with a triangular or ovate-triangular outline, pointed, mostly distinctly lobed and the margins irregularly sharp-serrate, the

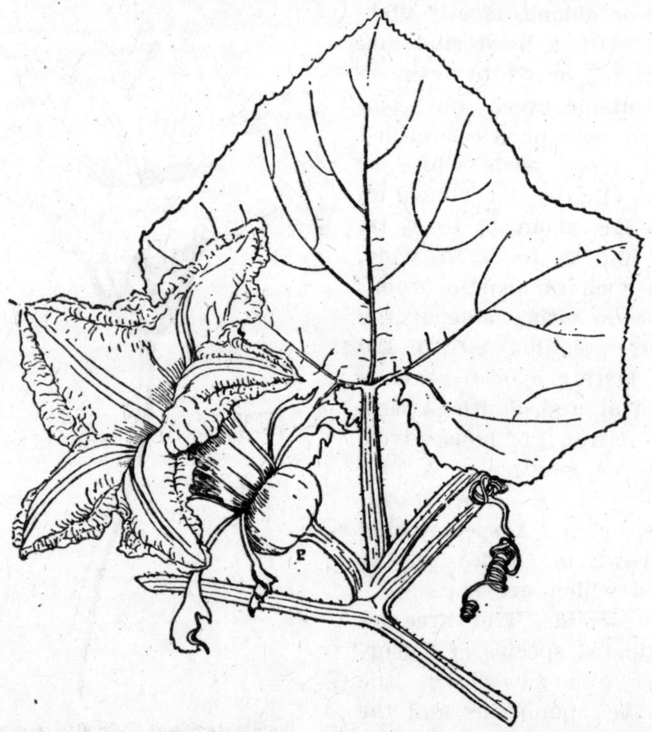

193. Leaf and pistillate flower of C. moschata; ovary at P (× about ½).

lobes and larger angles apiculate: flowers mostly with erect or spreading pointed lobes, the corolla-tube prevailingly flaring and narrowing toward the base; calyx-lobes short and narrow: peduncle strongly angled and expanding next the ovary and fruit: **C. Pepo**, Linn. Sp. Pl.

1010. FIELD PUMPKIN. Here belong the plants commonly known in North America as pumpkins, used for stock-feeding and for the making of pumpkin pie. There are several garden varieties, long-running coarse rough vines, the fruits ripening in autumn. The vegetable marrow is of this species. (Figs. 180, 188, 189.)

Var. **condensa,** Bailey, Cyclo. Amer. Hort. 409. 1900. BUSH PUMPKIN. SUMMER SQUASH. SIMLIN (Cymling). Not running or tendril-bearing, compact; fruits very various, ripening in summer and autumn. Here are included the Scallop or Pattypan squashes, and the common Summer Crookneck (Figs. 181, 182).

Var. **ovifera,** Bailey, l. c. (*C. ovifera*, Linn. Mant. i, 126. 1767.) YELLOW-FLOWERED GOURDS. Plants running, slender, the leaves small and commonly deeply lobed: fruits small, hard-shelled and keeping indefinitely, yellow or green or variously striped, apple-shaped, pear-shaped, oblate, sometimes warty. An interesting group of plants grown for the ornamental inedible fruits.

AA. Plants softer to the feel, the foliage less rigid and not so upright: leaves round or nearly so, not lobed, the cordate base with a very deep sinus, margins uniformly shallow-serrate with soft points to the serratures: flowers with broader lobes which are usually reflexed or revolute in full bloom, the corolla-tube with parallel sides or even bulging toward the base; calyx-lobes short and narrow: peduncle short and nearly cylindrical, not enlarging next the ovary and fruit, often developing its largest diameter at the middle: **C. maxima,** Duchesne in Lam. Encyc. ii, 151. 1786. AUTUMN and WINTER SQUASH. Here belong the Hubbard, Mammoth Chile, Lowe, Essex Hybrid, Boston Marrow, Marblehead, Turban and similar varieties. They are autumn-ripening fruits and keep well in winter. The flesh is firm and mostly golden yellow or orange-yellow. Some of the large or mammoth kinds are frequently called pumpkins; but they lack the light or bright yellow external color of the fruits of *C. Pepo* (Figs. 183, 184, 186, 187, 190, 191).

AAA. Plant soft to the feel, the foliage as if limp and velvety, not strongly upright: leaf-form and margins much as in AA, but sometimes distinctly lobed as in A, often with whitish marks or blotches: flowers with wide crinkly wide-spreading lobes, the tube broad at base but usually not bulging; calyx-lobes often long and expanding into a leaf-like structure at the end: peduncle much as in A, usually expanding more widely at its juncture with the mature fruit: **C. moschata,** Duchesne. Dict. Sci. Nat. xi, 234. 1818. CUSHAW and WINTER CROOKNECK SQUASHES. Many of the forms of this species appear to be oriental. The Canada Crookneck belongs here, as also the Yokohama, Quaker Pie, Japanese Pie, Jonathan (Figs. 185, 192, 193).

The three species of Cucurbita described above are coarse long-running plants (except that there are bush varieties of *C. Pepo*), with large alternate leaves on hollow petioles and forking tendrils arising from the side of the stem near the axils: stem angled, rooting at some of the joints: staminate flowers long-peduncled and therefore conspicuous; pistillate (female) flowers short-stalked and therefore lower down among the foliage; stamens 3, with very broad filaments separate near the base but upwardly joined and with the united anthers making a single central column in the flower; ovary inferior, 3-celled, the 3 large stigmas 2-lobed, the bottom of the pistillate flower provided with a prominent cup-like disc which leaves its scar on the "blossom end" of the fruit: the peduncle or stem of the fruit is characteristic of the species, as described above and shown in the illustrations (Figs. 180–2, 183–4, 185): seeds various in size between the species as also between varieties in the same species, elliptic-ovate in outline, flat, or somewhat plumper in *C. maxima* and *C. moschata*, those of the yellow-flowered gourds (*C. Pepo* var. *ovifera*) about ½ in. long and ⅜ in. broad and weighing 60 to

100 mg., those of the Summer Crookneck (*C. Pepo* var. *condensa*) about ⅝ in. long and ⅜ in. or more broad and weighing 90 to 100 mg., those of field pumpkin (*C. Pepo*) ¾ in. long and ½ in. broad and weighing 150 to more than 200 mg., of Mammoth Chile Squash (*C. maxima*) 1 in. by ⅝ in. and weighing nearly 500 mg.; vitality 5 to 7 years. The seeds of *C. Pepo* and *C. moschata* are much alike in form, dirty white color, and thin edge with raised border; those of *C. maxima* are whiter, round-edged, and without the same kind of elevated rim.

The nativity of these cucurbitas is not yet determined. Some authorities think them probably American and others ascribe them to Central Asia. It is not likely that the species intermix. *C. Pepo* and *C. maxima* apparently do not cross, and there are no known hybrids in cultivation between any of the species. It would be good to know whether a bee visits the three species indiscriminately in a single journey. The botany of the group is still imperfectly comprehended, and it is unsafe to make positive statements on these subjects; but for practical purposes it may be said that the species hold their identity.

CHAPTER XII

SWEET CORN. OKRA. MARTYNIA

The plants herein discussed are all warm-weather crops; they are annuals, or grown as such, cultivated for their immature fruits; they should have quick soil; usually they are not transplanted; other than good tillage, no special treatment is required.

Corn, okra and martynia are culturally somewhat related, but they have little else in common. They are placed together here because none of them fits well into the other groups.

SWEET CORN

Rows of corn are made at 3 to 4 ft. apart. In the row the hills (of 3 to 5 stalks each) are planted at about 2½ to 3 ft., or single kernels may be dropped every ten to twelve inches. At 2½ to 3 ft. apart, the crop may be tilled in both directions. Cover the seed about 1 in. deep, or somewhat deeper late in the season. When the corn is small, the ground may be harrowed without destroying the plant. In hills, one peck to the acre is required for planting; 8,000 to 10,000 ears should be secured from an acre.

CORN SMUT (*Ustilago zeæ*).—Enlarged galls or swellings that break open and expose a dark brown to black powdery mass of spores are formed on any actively growing part. *Control:* The practice of removing and destroying all smut boils while they are young is recommended as a means of reducing the smut developing in the field. Crop rotation is beneficial and it is desirable not to apply corn-fodder manure to a field on which corn is to be grown next season.

CORN EAR-WORM (*Heliothis obsoleta*).—A caterpillar, 1½ to 2 in. long, varying from light green to brown, highly variable in markings but usually with a longitudinal pale stripe along the side, edged above with blackish. The eggs are laid on the silk and the young caterpillars work their way down under the husk, where they feed on the green silk and unripe kernels. The broods coming late in the season are much more abundant and injury to late corn is therefore greater. *Control:* Experiments in New Jersey have shown that the injury to sweet corn may be greatly decreased by dusting the silk with a mixture of 50% arsenate of lead and 50% finely ground sulfur. The first application is made soon after the silk first appears, followed by one or two more applications before the corn is ready to pick. For regions where the pest is abundant corn for the cannery should be grown early in the season to avoid most of the injury.

EUROPEAN CORN BORER (*Pyrausta nubilalis*).—A yellowish gray brown-headed caterpillar, about ¾ in. long, minutely brown-spotted and indistinctly striped with reddish or dusky. The caterpillars bore in all parts of the plants except the roots. They are found in the stalks, ear, cob and in the tassel. There are one or two broods depending on the climate—two in the vicinity of Boston and one in the Mohawk River Valley, New York. This pest was recently introduced from Europe and as far as known is now restricted to central New England, New York and Northwestern Pennsylvania. *Control:* To prevent the spread of this pest, strict Federal and State quarantines have been established, governing the transportation of plants likely to contain the larvæ. No effective control measures applicable to ordinary farm conditions have been devised.

BROWN FRUIT CHAFER (*Euphoria inda*).—A thick-set yellowish brown beetle, ½ in. long, marked on the back with small irregular black dots. The beetles appear in the fall and attack the corn in the milk, often working down under the husk.

The larvæ feed on the ground in the vicinity of manure piles. Hand-picking is the only effective method of control known.

STINK-BUGS (*Euschistus variolarius* and *E. euschistoides*).— These two dull grayish brown stink-bugs, about ½ in. long, often attack corn by puncturing the kernels through the husk and sucking out the juices. The bugs are most abundant in waste land grown up to weeds. Clean farming will greatly reduce their numbers.

As a garden or horticultural crop, sweet corn or sugar corn is the only kind of maize that need be considered here. It is grown for the immature ears, which are eaten when the grains are yet soft. Although practically unknown in other parts of the world, it is a very important product in North America. Its importance has greatly increased in recent time because it is extensively canned. Sweet corn is not grown in the Southern States; or if it is, the seed is renewed every year. It holds its peculiar attributes in the short sharp seasons of the Northern States and parts of Canada. Eating corn from the cob seems to be an American enterprise. "Green corn" is a characteristic and highly desirable food product, and nothing seems to connect one closer with the soil and the open. Figs. 194, 195, 196 illustrate it.

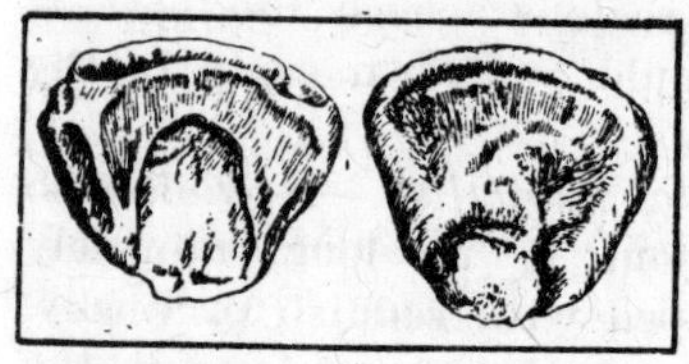

194. Kernels of a sweet corn (× about 2).

The cultivation of sweet corn is not unlike that of field corn, with the exception that greater attention is paid to earliness and to the development of each individual plant. It is therefore given, if possible, an earlier and warmer soil, with quickly available fertilizers, and it is usually grown in hills rather than in continuous drills. The idea is to secure as many ears as possible, and therefore each stalk should be given adequate room. In field corn, on the contrary, particularly since the advent of the silo, the fodder may be quite as important as the grain. If the season is short and the soil is hard and backward, it is well to add a little commercial fertilizer to each hill to start the plants off quickly. Maize does particularly well following sod.

195. Maize coming up (× 2/3).

The excellence of the crop depends to an important degree on the parentage of the seed. Seed-breeding plots should be maintained, or else extra discrimination should be exercised in the purchase of seed for planting.

Seed is planted for the early crop as soon as the ground is thoroughly warm. Since sweet-corn seed is particularly liable to rot in cold and damp ground, it is well to make the first planting rather heavy. It is possible to start in plots and transplant, but in practice it is planted directly in the field. The early plantings are usually made of the extra-early varieties, as Minnesota, Cory, Golden Bantam and others. The main crop is commonly secured from the later or main-season varieties, of which the Stowell Evergreen is a standard. Successional plantings may be made at intervals of one to two weeks, particularly for the home garden or for a continuous supply for the market-garden. In market-gardening, the value of the green-corn crop is often determined by its earliness. Two or three days in

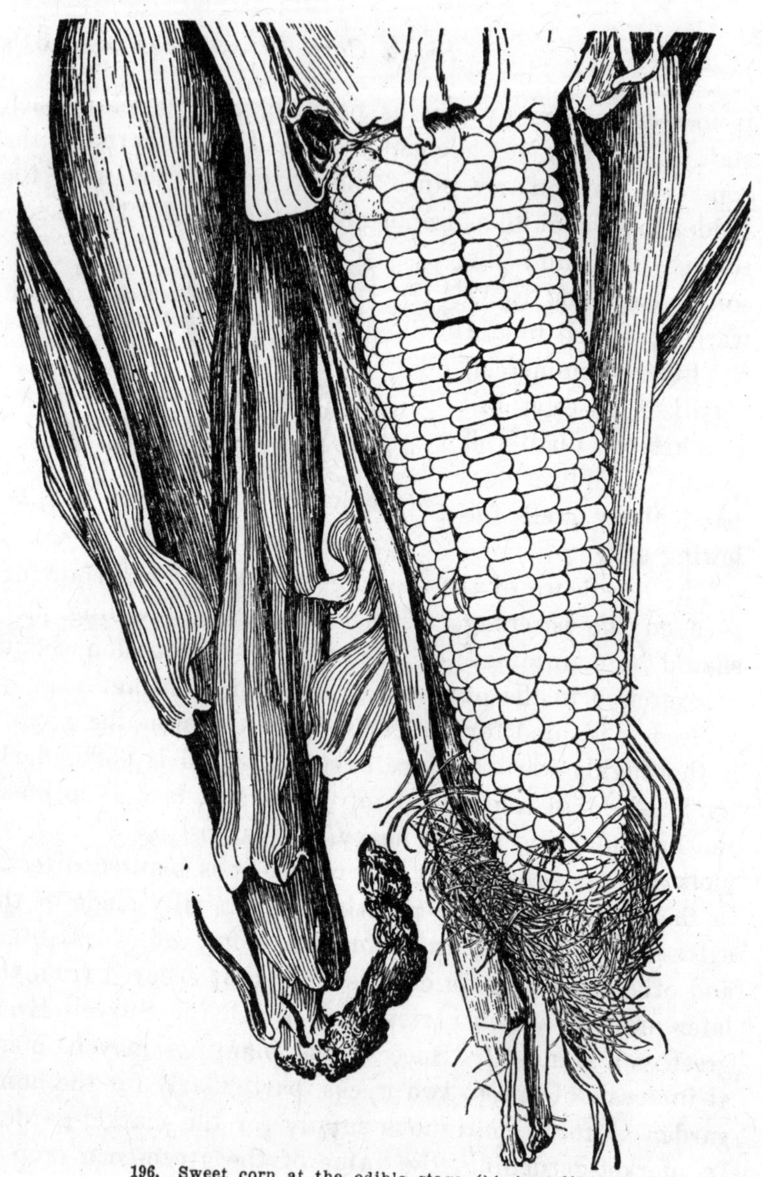

196. Sweet corn at the edible stage (× about ½).

time of ripening may make a difference between the profitable and unprofitable crop, particularly when one is under strong competition with neighboring gardeners. In such cases the grower secures the early crop by means of the very earliest varieties, carefully selected seed, and particularly by having quick and well-prepared land to which only readily available fertilizers have been added. If the land is inclined to be hard and rough, it is well to turn it up loose in the fall.

Tillage of sweet corn should be shallow and frequent until earing well begins; thereafter the tillage may be little or discontinued, but weeds should be kept down by hand or pulled as they appear.

Although corn is a hot-weather plant and thrives in the fullest exposure to sunlight, it nevertheless is not able to withstand drought as well as potatoes and many other crops. This is because it is relatively a surface feeder. Every effort should be made, therefore, to save the moisture in the soil. The moisture content is held by deep preparation of the land and by the incorporation of vegetable matter. Thereafter it is saved by surface tillage.

In the general market, corn is usually retailed by the dozen ears. As a field crop for the canning factories, the ears are ordinarily sold by the ton, after all small and imperfect ears are discarded. The ears of the second setting will develop better if those of the first setting are picked as soon as they are fit for use.

It is a frequent practice to pull the ears too soon, to get the benefit of early market. The kernels should be large and well formed when the corn is harvested, so that they make a continuous pavement-like surface on the ear, "well filled out." The ears are marketed in their husks, the outer loose leaves being pulled off, in baskets, hampers and barrels. Sweet corn makes a very attractive product if well graded, and sent to market in paper-lined hampers or in cartons (Fig. 235).

For home use, Golden Bantam is now the favorite because of its delicious sweet quality. At first objections were raised because of its yellow color but this is mere prejudice or lack of reason; there is no more reason why corn should be white than yellow. The demand for mere whiteness in food products is one of our precious absurdities. Golden Bantam yields small ears and therefore may not be wanted on certain markets. It is an 8- and 10-rowed variety. There is a long list of excellent varieties of sweet corn, of which Mayflower, Cory, Metropolitan, Perry, Minnesota, Crosby, Stabler, Champion, Country Gentleman, Black Mexican, Stowell Evergreen may be mentioned. Adams Early is not a true sweet corn, but is grown for the market because of its earliness and hardiness.

THE SWEET MAIZE PLANT

Zea. *Gramineæ.* Two dozen and more specific names have been given in the genus Zea, but the prevailing opinion reduces them all to forms of one polymorphous species, **Z. Mays,** Linn. Sp. Pl. 971. MAIZE. INDIAN CORN. The plant is unknown wild. Historical and other evidence indicates an American origin; probably Mexican. By some authors it is thought to have originated as a hybrid between other genera of grasses.

Z. Mays, Linn., var. **rugosa,** Bonaf., Mais, 39, fig. 19 pl. xi. 1836. (*Z. saccharata,* Sturtevant, 3rd Rep. N. Y. Exp. Sta. 156. 1884. *Z. Mays* var. *saccharata,* Bailey, Cyclo. Amer. Hort. 2006. 1902.) SWEET CORN. Plant of relatively low stature, 4 to 7 or 8 ft., strict, the culm smooth and glabrous, commonly with brace-roots from the lower exposed joints; stem or culm with prominent nodes or joints, above which extend the long tight often ciliate-edged leaf-sheath: leaves 1 at every joint, long linear-lanceolate, acuminate-pointed, 2 to 3 ft. long and 2 to 3 in. wide, with a short scarious ligule at top of sheath, the midrib prominent: flowers numerous, imperfect, the staminate (male) in the "tassel" or panicle terminating the culm, and the pistillate in "ears" or spikes from 1, 2 or 3 of the lower or mid-stem axils and facing a grooved internode, the ears covered with modified sheaths or husks; staminate spikelets 2 at the nodes of the rachises constituting the panicle (one of them pedicelled and one sessile), each spikelet 2-flowered and with 2 empty ciliate glumes and 2 thin palets and 2 lemmas, the stamens 3 in each flower and bearing large exserted dangling anthers; pistillate spikelets sessile, 8 to 24 rows on a long thick axis or cob, comprising a single pistil covered in the ciliate notched glumes but outgrowing the floral envelopes (which are 2 glumes, 2 palets and 2 lemmas) and leaving them as chaff on the cob, the single style arising from the apex and very much prolonged, the many protruding hanging styles constituting the "silk": fruit a hard dry angular kernel ("seed"), flattened on the sides, narrowed below to a point or in other kernels truncate at base, sulcate on one side, at maturity and when dry wrinkled on top and the outsides, a well-formed pointed kernel measuring at maturity ½ in. either way, weighing 200 to 300 mg. more or less; vitality 1 to 4 years. Sometimes pistillate flowers are borne in the tassel, producing kernels; and sometimes there is a staminate extension of the ear; these are unusual and abnormal states.

OKRA or GUMBO

Warm climate and soil, and the attention given to the growing of a good crop of corn or cotton, are the prime requirements for okra. It is usually planted directly in the field.

The large varieties of okra should go in rows 4 to 5 ft. apart, and the plants may stand 12 to 36 in. in the row; the dwarf varieties may go as close as 3 ft., and 10 to 15 in. in the row. Sometimes the crop is grown in hills, after the way of corn, 2, 3 or 4 plants standing together after the thinning. If land is abundant, the rows for large sorts may be as much as 5 ft. asunder. Seeds are covered 1 to 1½ or 2 in. deep.

THE OKRA CATERPILLAR (*Anomis erosa*).—A pale pea-green looking caterpillar, about 1¼ in. long, inconspicuously marked with five narrow broken yellowish lines above and with a broader yellowish white stripe on each side. The young caterpillars eat out small holes in the leaves and the older ones irregular areas in the side, often defoliating the plant. *Control:* When the caterpillars first appear, spray with arsenate of lead (paste), 2 lbs. in 50 gallons water or, to avoid the danger of using an arsenical, "Black Leaf 40" tobacco extract, 10 ounces to 100 gallons water, in which 5 or 6 lbs. soap have been dissolved, may be used.

CORN EAR-WORM (*Heliothis obsoleta*).—The corn ear-worm often attacks the pods. See under Sweet Corn. *Control:* Plant a row or two of corn near the okra to serve as trap crop. It should be cut before the caterpillars reach maturity.

GRAY HAIR-STREAK BUTTERFLY (*Uranotes melinus*).—The slug-like caterpillars of this dainty blackish blue-gray butterfly sometimes injure the buds and leaves. If necessary they may be controlled by spraying with arsenate of lead.

SPINACH APHIS (*Myzus persicæ*).—See under Spinach.

MELON APHIS (*Aphis gossypii*).—See under Cucumber.

Okra is a hot-weather plant, cultivated as an annual, the seeds being sown each spring. It is commonly grown in the Southern States, where its partially matured pods are in much demand for soups and stews, and salads are made from the boiled tender pods. These pods must be cut when still tender and pulpy, before they have developed strings or woody fiber. Pods are also canned (often with tomatoes), and dried for subsequent use. They are ready for picking a day or two after bloom.

197. Seeds of okra (× 2).

Okra is grown in essentially the same way as corn. The seeds are sown where the plants are to stand, as the young plants do not transplant with ease. In the Northern States, however, the plants are sometimes started in pots, boxes or on inverted sods in frames. Okra is a large-growing plant and the rows should be 3 to 4, or even 5, feet apart for the larger varieties. In the row the plants should stand 1 to 3 feet. In the North certain dwarf and early-maturing varieties are usually grown, and these may stand as close as 1 foot, or even less, in the row.

198. Seedlings of okra (× ½).

"As soon as the plants begin to set fruit," writes W. R. Beattie in Farmers' Bulletin No. 232, "the pods should be gathered each day, preferably in the evening. The flower opens during the night or early morning and fades after a few hours. The pollen must be transferred during the early morning, and the pod thus formed will usually be ready for gathering during the latter part of the following day, although the time required to produce a marketable pod varies according to the age of the plant and the conditions under which it is grown.

The pods should always be gathered, irrespective of size, while they are still soft and before the seeds are half grown."

Varieties are tall and dwarf; also long-podded and short-podded. Prominent names among the varieties, which are often more or less unstable and poorly defined, are Perkins, Tall Green, Long Green, Creole, and Velvet among the

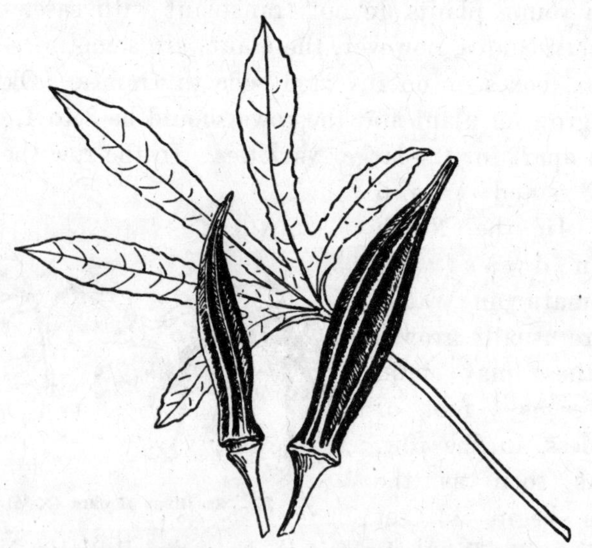

199. Leaf and pods of okra (× ¼).

long-pods, Little Gem and Dwarf Green among the short-pods. The pods should be picked every day, when the plant comes into good size, not only that the product may be tender but that the bearing season of the plant may be extended. The small green pods are marketed in berry boxes or other small packages. Figs. 197 to 199 show the okra plant.

The Okra Plant

Hibiscus. *Malvaceæ.* Nearly 200 species of herbs and small trees in many parts of the world.

H. esculentus, Linn. Sp. Pl. 696. Okra. Gumbo. (*Abelmoschus esculentus,* Mœnch, Meth. 617. 1794.) Stout erect branching nearly glabrous annual (biennial and perhaps perennial), 2 to 7 ft. tall: stems terete, pithy, more or less furrowed, often colored, glabrous or with few scattered hairs: leaves alternate, long-petioled, the blade various in shape from rounded and hollyhock-like to palmately 3- to 5-parted or compound, the margins coarsely and irregularly dentate, cordate at the base, with scattered hairs on the veins: flowers solitary in the upper axils, on stout erect furrowed or angled peduncles, large, yellow or straw-yellow with a red eye, subtended by very narrow bracteoles about 1 in. long; calyx-lobes large and broad, acute, about half the length of the bell-shaped corolla which is 2 to 3 in. long; petals large and showy, obovate; stamens united in a column in the center of the flower and surrounding the 5 styles: fruit a long straight or curved strongly ribbed pubescent or hairy 5-celled pod (4 to 12 in. long), which is expanded at the base and long-pointed at the apex: seeds gray or brown, skull-shaped or nearly globular, with fine concentric broken lines, and a whitish base-point, $\frac{3}{16}$ to ¼ in. diameter, about 45 to 60 mg. in weight, lasting about 5 years.—Supposedly native in Old World tropics, Africa or Asia or both; probably not anciently cultivated, although now widespread in warm countries.

MARTYNIA

Martynia is grown for the half-matured seed-pods, which are used for pickles. The plant requires a warm soil and exposure. Give much room, for a good plant will spread over an area of 3 or 4 feet across. It is a low-spreading plant of very rapid growth, with very large hairy leaves, odd showy flowers, and long-beaked hairy pods. It de-

200. Seeds of the garden martynia (× 1¼).

201. Seedling of the garden martynia (× ⅓).

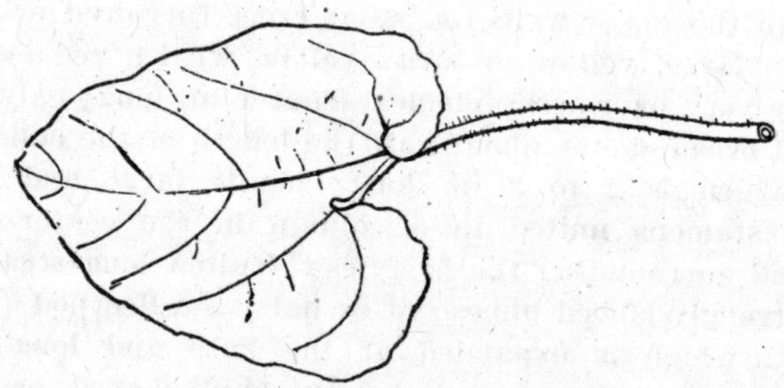

202. Leaf of martynia (× 1/5).

203. Flower of martynia (× ½).

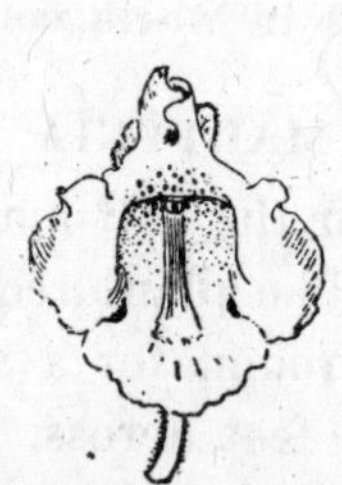

204. Flower, front view (× ½).

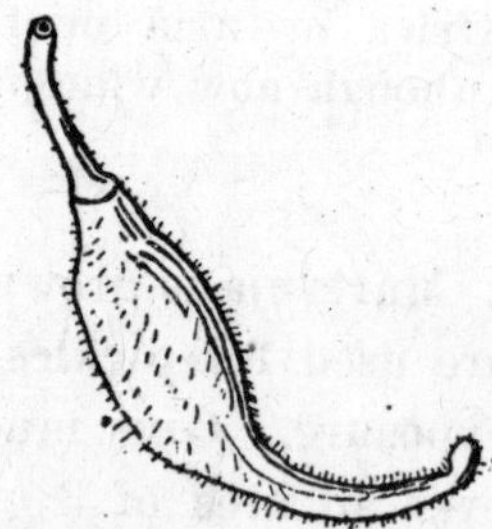

205. Young edible pod of martynia or unicorn plant (× ⅓ to ½).

mands no special treatment. Seeds may be started in frames or planted in the open as soon as warm weather comes. Its use is not extensive, but the seeds find a place in the standard catalogues. It often self-sows, coming up the following year; in this way it is reported as an introduced or escaped plant in regions far outside its native range. It is frequently grown for ornament or as a curiosity. Figs. 200 to 205 show the features of this odd plant.

The martynia or unicorn plant is one of the few species of the Martyniaceæ. As now treated, the family comprises three genera, and the genus Martynia proper has but a single species, while the common martynia of the gardens goes into the genus Proboscidea, characterized by a short corolla-tube and 4 fertile stamens.

The martynia mostly offered by American seedsmen amongst vegetable-garden seeds becomes **Proboscidea louisiana,** Wooton & Standley, Contr. U. S. Nat. Mus. xix, 602. 1915. (*Martynia louisiana,* Mill. Gard. Dict. Ed. 8. 1768. *M. proboscidea,* Gloxin, Obs. Bot. 14. 1785. *P. Jussicui,* Steud. Nomen. Ed. 2, ii, 397. 1841.) It is an odd densely clammy-pubescent low wide-spreading tender annual herb with thick opposite divaricate branches: leaves alternate or subopposite, soft and thick, mostly horizontal or nearly so, long-petioled, round-ovate to oblong-ovate, obtuse, wavy-margined but not lobed, strongly palmately ribbed, cordate and often unequal at the base, the basal auricles turned upwards: flowers large, in racemes that become central in the forks, square-ended in bud; calyx unequally 5-lobed, the upper lobes much longer, slit to the base on the lower side, subtended by 2 small pad-like or lanceolate deciduous bracts; corolla light violet to purple, 1¾ in. long, tube 1¼ in. long and ½ in. diameter crosswise at the mouth, hairy outside, limb unequally 5-lobed, oblique with the rounded middle lobe largest and projected forward and undu-

late, the side lobes spreading, the two upper lobes upright and the edges more or less rolled back, the floor of the tube marked with a broad straight yellow and sometimes striped band which enlarges and terminates with irregular end toward the center of the lower lobe, upper part of throat spotted; anthers 4, included in the roof of the tube, borne on the lower part of the corolla-tube, in two pairs joined by their 2-celled anthers, one pair ⅓ shorter than the other; pistil single, ovary oblong bearing a long upwardly expanding style, the stigma 2-lobed, the lobes closing to the touch: fruit hanging, with a thick body 3 in. long and a curved beak of equal or greater length, properly 1-celled but appearing 5-celled on cross-section, the fleshy pericarp finally rotting away and leaving the two bony horned valves with crests on the inner edge of the main part like the lower jaws of a tusked animal: seeds oblong or oblong-ovate, ¼ to ⅜ in. long, more or less angled and irregular, black, tuberculate and alveolate, weighing 30 to 50 mg., holding vitality a year or two.—Native from Indiana to New Mexico, sometimes escaped elsewhere. Many small insects become entangled in the sticky hairs of the stems, leaves, flowers and pods.

CHAPTER XIII

CULINARY HERBS

Although there is relatively small desire on the part of Americans for condimental and flavoring herbs, nevertheless every complete home garden should have a small area set aside for the cultivation of at least a half dozen of the leading kinds. They add a peculiar variety and charm to the kitchen-garden, and connect it with old rhymes and memories.

What are commonly known as "herbs" in the trade comprise a great variety of plants. Some of them are grown for medicinal purposes, some for flavoring, some for the decoration of culinary dishes and others for salads and minor home uses. What are commonly known as "the sweet herbs," however, are such plants as are used as an incident to cookery. Of these the most popular in America is sage.

Some of the culinary herbs are prized for foliage, and others for seeds or fruits. The species to which the name "sweet herb" should be restricted are those that have aromatic foliage. Of such are sage (Fig. 206), hyssop, thyme, mints, tansy, horehound, savory (Fig. 207). Most of these plants are members of the mint family, or Labiatæ, although some of them, as tansy and wormwood, are of the sunflower family. Those species of which the seeds are used are mostly of the parsley family, or Umbelliferæ. Of such are caraway, coriander and dill. The larger number of the seed-crop plants is annual.

The culinary herbs are of two classes as respects the general methods of cultivation: the annuals, or those that must be resown every year; and the perennials, or those that persist for a number of years. Even the perennial species, as sage and hyssop, should be resown or replanted frequently to keep the plants in vigorous condition, particularly if the climate is severe and if the plants are not given a little winter protection.

206. Sage (× ⅓).

207. Summer savory (× ¾).

It is well to grow all the kitchen herbs together on one side of the garden, whether they are annual or perennial, and to have a clump of a particular herb each year in its accustomed place. The "herb garden," in a place devoted to it, should oftener be part of the garden plan. A strip 3 or 4 feet wide can be made a collecting-place for the herbs; and the place will have more than a commercial or culinary interest.

Most of the culinary herbs are of the easiest cultivation. They thrive in any loose warm and open soil. Although the growth is usually most profuse in rather heavy and moist soils, it is thought that the aromatic qualities, for which they are particularly esteemed, are more pronounced in soils in which the plants do not make exuberant growth. The land should always be fertile enough, however, to produce a full development of the plant.

The strongest-growing perennial species may be propagated easily by division of the root. When the clump begins to fail, it is well to dig it up and discard all the older parts of the roots and to replant the younger and more vigorous parts. When such species are grown from seed, they are usually not strong enough to supply a heavy product until the second year, although some of them may give a cutting the first autumn if they are started early and if the land is good. Ordinarily a space 4 feet square will contain enough of any herb to supply a family, although twice that area may be desired for such popular species as sage, caraway and spearmint.

The plants grown for herbage are usually cut when they are in full growth and before they have become woody. The stems are cut off near the ground and are then tied together in bundles and hung in a dry cool place, as an attic. The dried herbage is then in condition for use in winter. Continual cuttings of the young herbage may also be made during the season for current uses. It is evident that if the plants are cut severely and continuously they will be weakened, and that it may be necessary to raise a fresh stock to take their places.

The species grown for seeds are allowed to ripen before the product is gathered. The plants are usually cut or pulled just before the seeds are ready to fall. The plants are then dried under cover and the seeds are threshed out.

Seeds of the seed-cropping herbs and dried herbage of the true sweet herbs are usually to be had at drug stores, but there is much satisfaction in growing one's own. Sometimes there is a fair market for home-grown herbs.

The following lists contain the leading species of sweet and culinary herbs cultivated in this country, arranged with reference to duration:

Annual or biennial, or grown as such

anise,	caraway (biennial),
sweet basil,	clary (biennial),
summer savory,	dill (biennial),
coriander,	sweet marjoram (biennial or perennial).

Perennial

sage,	rosemary,
lavender,	horehound,
peppermint,	fennel,
spearmint,	lovage,
hyssop,	winter savory,
thyme,	tansy,
marjoram,	wormwood,
balm,	costmary,
catnip,	tarragon.
pennyroyal,	

CHAPTER XIV

GLASS

To protect and forward plants, various covers are used; and these covers, of every kind and description, are usually spoken of as "glass," even though paper or cloth may sometimes be employed in place of glass. They comprise all the range of forcing-hills, coldframes, hotbeds and glasshouses.

Every vegetable-gardener, however small his area, needs glass. Thereby is he enabled to secure a crop in advance of its normal season. He becomes, in a measure, independent of season or even of climate. The vegetable-gardener is less subject to loss from vagaries of frost than is the fruit-grower. He can cover his plants. The plants are also more amenable to treatment: he can sometimes harden them off, so that they withstand frost. He can grow them at such times as to escape the dangerous season: the fruit-grower's plants must stand and take it.

The purpose of glass is to forward plants in advance of their season or beyond it. This result is obtained by protecting the plants from unpropitious weather or by actually forcing them. An example of the former object is the protection in winter of hardy plants started in the fall. The plants are kept alive in the cold weather by means of the covering, but they do not grow. There are two general types of forcing: the plants may be started under glass, and then transplanted into the open; they may be grown to maturity under glass.

1. QUANTITY OF GLASS REQUIRED

How much glass the vegetable-gardener needs depends (1) on how intensified are his operations, (2) in what season he wants the major part of his crops, (3) the region, (4) the kinds of crops. These factors are largely determined, in their turn, by the man's location with reference to market, and the price of labor and land. Very small areas sometimes have sufficient glass to cover them.

Glasshouses are increasing in number and popularity. They are driving out hotbeds for the forcing of winter stuff. But for general vegetable-gardening, the coldframe and hotbed remain, although their relative importance is likely to diminish. These humble structures are desirable because they are cheap, because they allow the person quickly to change or modify his business (a great advantage on rented land), and because they can be removed when the spring forcing is accomplished, allowing the land to be used for other purposes. The growing of winter vegetables in the North (under glass) is a special business, and is not discussed in this book.

Vegetable-gardening glass is usually computed in sashes. A normal sash is 3 x 6 feet in surface area. Sashes are combined into frames. A frame is a box covered by four sash,—that is, an area 6 x 12 feet. For general and mixed vegetable-gardening, about twenty-five sash are sufficient for an acre of garden, considering that the plants are to be transplanted to the field, not matured under the sash.

If one is growing particular crops, as tomatoes, fifteen sash may be sufficient. For the best kind of home gardening, when it is desired to mature spring lettuce and radishes under glass as well as to transplant stuff into the open, thirty-five to fifty sash may be needed to the acre.

In growing plants for transplanting, a sash may be estimated to accommodate 400 to 500 cabbage and cauliflower plants, 300 to 400 tomatoes and eggplants, 600 to 800 lettuces. When the plants are transplanted in the frames, only one-third to two-thirds these numbers can be accommodated. If the plants are started very late and are not transplanted, as many as 800 tomato or cabbage plants can be grown under one sash. In general, one may expect to gain three weeks to one month on the crop of hardy things like cabbages, and two to three weeks on tomatoes. To gain two weeks on the crop, however, it is necessary to gain three or four weeks on the sowing. In extra-good hotbeds, greater gain can be secured; but it is not common.

In calculating the amount of glass required, the gardener must remember that many of his plants may fail after they are set in the field. There are risks of frost, cold rains, droughts, worms, accidents. He may lose plants while they are still in the frames. The grower should start at least fifty per cent more plants than he expects to raise. The surplus may be left in the frames until the transplanted subjects are thoroughly established and safe.

The following sample estimate, by a gardener, illustrates the method of casting up one's outlay for the season's glass. It is an estimate for a market-garden of one acre, for a general line of vegetables. It supposes that half of the acre is to be set with plants from hotbeds:

One-eighth acre to early cauliflower and cabbage, about 2,000 plants; if transplanted would require two 6 x 12 frames, 200 to 250 plants being grown under each sash, or about 1,000 plants from each frame.

These frames may be used again for tomato plants for the same area, using about 450 plants. This will allow one sash for every 55 plants. Plants for this area may be grown in one frame, but would be crowded and not as stocky as if given more room.

One frame may be in use at the same time for eggplants and peppers, two sash of each, growing 50 transplanted plants under each sash.

Two frames will be required for cucumbers, melons and early squashes.

If one wishes to grow extra-early lettuce, an estimate of 60 to 70 heads may be made to a sash.

It is assumed that celery and late cabbages are to be started in seed-beds in the open.

If spinach is grown in frames, the sash used for one of the late crops above may be used through the following winter.

This makes a total of five frames; twenty sash and covers; manure, calculating at least three or four loads to a frame. This is a liberal estimate of space, and should allow for all ordinary loss of plants, and for discarding the weak and inferior ones. It supposes that most or all the plants are to be transplanted once or more in the frames. Many gardeners have less equipment of glass and do less transplanting.

2. THE MAKING OF FRAMES

In the planning of a coldframe or hotbed, the builder must have in mind the following objects to be attained: (1) sufficient and uniform supply of heat; (2) ample protection from cold; (3) means for ventilation; (4) facilities for obtaining water; (5) plants to be near the glass, and yet to have head-room for growth of tall kinds; (6)

ease and convenience of manipulation; (7) cheapness and durability.

Location and exposure.

Ideally, the place on which frames are set should slope gently to the south or southeast. The area should be well protected from the cold and prevailing winds. A windbreak is necessary. This may be a pronounced rise of land to the north or west, a building, a wall, or a hedge. If none of these shelters exists, a temporary one may be made. A board fence 5 to 8 feet high is the common resort; if it slants back somewhat, it provides a good support for mats and sash leaned against it. A screen of cornstalks, evergreen boughs, or other material may serve the purpose.

The frames should be near the buildings and easy of access. They need frequent attention, particularly in changeable weather. Frames far from the house, or which are cut off by snowdrifts or mud, are likely to suffer in critical times. Water supply should be at hand. If pipe-water cannot be had, a good well or cistern, with force-pump, should be provided. Some provision should also be made for warming the water in cold weather, for very cold water chills and delays the plants and wastes the heat of the bed.

If land is sufficient and the garden area remains year by year in approximately the same place, it is advisable to have a permanent frameyard. The windbreaks, water supply and other accessories can then be well provided.

Pits may be dug for the hotbeds and the sides stoned or bricked. These pits retain heat better than surface-built beds, are less exposed to winds, and are permanent; but they are more expensive in the beginning. The pits can also be filled in autumn with manure or litter, and if this is pitched out at any time in winter or spring, an unfrozen area is at once ready for the making of the hotbed. Pits should be tile-drained, unless the soil is very loose and the bottom is below the frost line of the surrounding unprotected land. If many frames are employed, they should extend in parallel rows, six or seven feet apart, so that a man walking between can water or tend two runs.

Building the frame.

The common type of frame is shown in Fig. 208. It is a little over 12 feet long, is 6 feet wide, and is covered with four 3 x 6 sash. It is sometimes made of ordinary

208. A frame. In this case the frame is mortised together, so that the material can be taken apart and stored.

boards loosely nailed together. If one expects to use coldframes or hotbeds every year, however, it is advisable to make the frames of heavier stuff, well painted, and to join the parts by bolts or tenons, so that they may be taken apart and stored. Fig. 209 suggests methods of making the frames so that they may be taken apart. The pieces for the sash to slide on are made of stuff three inches wide mortised into the frame. These pieces have a strip or mounting nailed along their middle to hold the sash to its place. Fig. 210 (from Cornell Reading-Course Lesson) shows the details of a two-sash coldframe before the parts are nailed together.

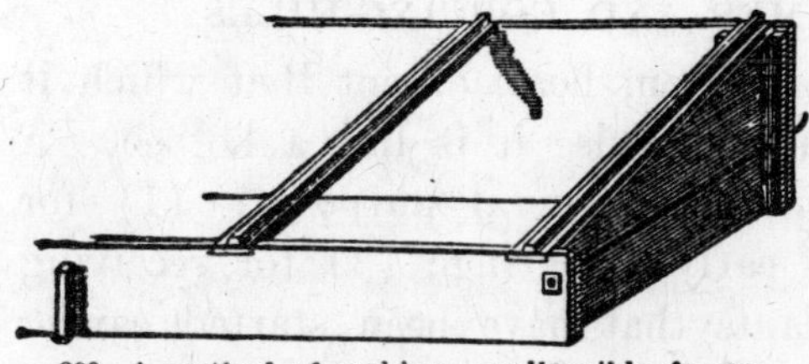

209. A method of making a collapsible frame.

The depth of the frame must be governed largely by the plants it is desired to grow, and by the length of time they are to remain in the bed. It is well to have the plants as near the glass as possible and yet give them room in which to grow. If the frame sets on top of the manure,

210. The five members of a two-sash frame.

the back side may be 12 to 15 inches high, and the front side 8 to 10 inches.

3. COLDFRAMES AND FORCING-HILLS

A coldframe has no bottom heat, except that which it receives from the sun; otherwise it is like a hotbed. A coldframe is used for three general purposes: (1) for the starting of plants early in spring; (2) for receiving partially hardened plants that have been started earlier in hotbeds and forcing-houses; (3) for wintering young cabbages, lettuce and other hardy plants sown in autumn.

Coldframes are ordinarily placed near the buildings, and the plants are transplanted into the field when settled weather comes. Sometimes, however, frames are made directly in the field where the plants are to remain, and the frames, and not the plants, are removed. When used for this latter purpose, the frames are made very cheap by running two rows of parallel planks through the field at a distance of six feet apart. The plank on the north is ordinarily 10 to 12 inches wide, and that on the south 8 to 10 inches. These planks are held in place by stakes, and the sash are laid across them. Seeds of radishes, beets, lettuce, and the like are then sown beneath the sash, and when settled weather arrives the sash and planks are removed and the plants are growing naturally in the field. Half-hardy plants, as those mentioned, may be started two or three weeks in advance of the normal season by this means.

When the heat is spent from hotbeds, they become coldframes. They can then be used, if empty, for the starting of late plants; or the plants may be hardened-off in them as they cool, thus, perhaps, obviating the necessity of transplanting to other frames.

Span-roof coldframes are useful, as they allow better and more uniform conditions for the growing of plants than the ordinary frame. They are covered with hotbed sash laid on a framework, as seen in Fig. 211, and the sashes pulled down from the top for ventilation. They are essentially forcing-houses, however, and the discussion of them is foreign to the purpose of this volume.

211. Span-roof coldframe.

Forcing-hills.

A forcing-hill is an arrangement by means of which a single plant or a single hill of plants may be forced where it permanently stands. It is a small or "individual" coldframe.

This type of forcing may be applied to perennial plants, as rhubarb and asparagus, or to annuals, as melons and cucumbers. Fig. 212 illustrates a common method of hastening the growth of rhubarb in the spring. A box with four removable sides, two of which are shown in end section in the figure, is placed about the plant in the fall. The inside of the box is filled with straw or litter, and the outside is banked thoroughly with any refuse, to prevent the ground from freezing. When it is desired to start the plants, the covering is removed from both the inside and outside of the box, and hot manure is piled around the box to its top. If the weather is still cold, dry light leaves or straw may be placed inside the box, or a pane or sash of glass may be placed on top of the box, to answer the purpose of a coldframe. Rhubarb, asparagus, sea-kale and similar plants may be advanced two to four weeks by this method of forcing. Some gardeners use old barrels or half-barrels in place of the box. The box, however, is better and handier, and the sides can be stored for future use.

212. Forwarding of rhubarb in the field.

Plants that require a long season and which do not transplant readily, as melons and cucumbers, may be planted in forcing-hills in the field. One of these hills is shown in Fig. 213. The frame or mold is shown at the top. This mold is a box with flaring sides and no top or bottom, and provided with a handle. This mold is placed with the small end down at the point where the seeds are to be planted, and the earth is hilled up about it and firmly packed with the feet. The mold is then withdrawn, and a pane of glass is laid on the top of the mound to concentrate the sun's rays, and to prevent the bank from washing down with the

213. The making of an elevated forcing-hill.

rains. A clod of earth or a stone may be placed on the pane to hold it down. This type of forcing-hill is not much made, because the bank of earth is likely to wash away, and heavy rain occurring when the glass is off will fill the hill with water and drown the plant. However, it can be used to very good advantage when the gardener can give it close attention.

A forcing-hill is sometimes made by digging a hole in the ground and planting the seeds in the bottom of it, placing the pane of glass on a slight ridge or mound made on the surface. This method is less desirable than the other, because the seeds are placed in the poorest and coldest soil, and the hole is very likely to fill with water in the early days of spring.

An excellent type of forcing-hill is made by the use of the hand-box, as shown in Fig. 214. This is a rectangular box, without top or bottom, and a pane of glass is slipped into a groove at the top. The earth is banked slightly about the box, to hold it against winds and to prevent the water from running into it. If these boxes are made of good lumber and painted, they will last for many years. Any size of glass may be used, but a 10 x 12 pane is as good as any for general purposes.

214. A hand-box.

After the plants are well established in these forcing-hills and the weather is settled, the protection is wholly removed, and the plants grow normally in the open. Forcing-hills are not well adapted to large-area work, as they require too much time in the tending. Neither do they have much advantage of protection from windbreaks, and, containing a less body of air, they do not give as early results as well-made coldframes.

For starting plants in a small way, a glass-covered box in the kitchen window may answer very well. An incubator is useful for the germinating of seeds.

4. HOTBEDS

A hotbed has artificial bottom heat. This heat is ordinarily supplied by means of fermenting manure, but it may be obtained from other fermenting material, as tanbark or leaves, or from heat in flues and pipes. The hotbed is used for the very early starting of plants, and when the plants have outgrown the bed, or have become too thick, they may be transplanted into cooler hotbeds or into coldframes. Some crops, however, may be carried to full maturity in the hotbed itself, as radishes and lettuce. The date at which the hotbed may be started with safety depends almost entirely upon the means at command of heating it and on the skill of the operator. In the Northern States, where outdoor gardening does not begin until the first or the last of May, hotbeds are sometimes started as early as January; but they are ordinarily delayed until early in March. In exposed places, it is well to have the glass as near the level of the ground as possible.

Handling the horse manure.

The heat for hotbeds is commonly supplied by the fermentation of horse manure. It is important that the manure be uniform in composition and texture, that it come from highly-fed horses, and is practically of the same age. As much as one-third or one-half of the whole material may be of litter or straw that has been used in the bedding. If the manure is very dense, it will not heat well, and it should have bedding, litter or well-decayed leaves mixed with it.

The manure is accumulated in a long and shallow square-topped pile, not more than four or five feet high as a rule, and is then allowed to ferment. Better results are generally obtained if the manure is piled under cover. The manure should be moist, but not wet. If it is dry when piled, moisten it throughout. If it is very wet, it will usually remain cold until it begins to dry out. Sometimes the addition of a little hen manure to one part of the pile will start the heating. If the weather is cold and fermentation does not begin, wetting a part of the pile with hot water may start it.

The first fermentation is usually irregular,—it begins unequally in several places in the pile. To make the fermentation uniform, the pile may be turned, taking care to break up all hard lumps and to distribute the hot manure throughout the mass. It is sometimes necessary to turn the pile five or six times before it is finally used, although half this number of turnings is ordinarily sufficient. When the pile is steaming uniformly throughout, it is fit to be placed in the hotbed. From the first piling of the manure until it is fit to put in the bed will be a period, ordinarily, of two weeks.

In some cases the material will not need to be turned to induce fermentation, particularly when the manure is from grain-fed horses. Sometimes the manure heats so quickly and so violently that it has to be wet to prevent it from burning, although the admixture of straw or litter with the manure will remedy the trouble. Each case is a law unto itself.

Making the manure bed.

Hotbed frames are sometimes set on top of the pile of fermenting manure, as shown in Fig. 215. The manure should extend some distance beyond the edges of the frame; otherwise the frame will become too cold about the outside, and the plants will suffer. It is preferable to have a pit beneath the frame in which the manure is placed. The pit should be a foot wider on either side than the width of the frame, and should be about two feet deep. Fig. 216 is a cross-section of a standard pit hotbed (H. Ness, Circ. 3, n. s., Tex. Exp. Sta.), showing the position and proportion of the manure. On the ground under a bed an inch or two of any coarse material is laid to keep the manure from the cold earth. On this, twelve to thirty inches of manure are placed. Above the manure is a thin layer of leaf-mold or some porous material, that will serve as a distributor of the heat, and above this are four or five inches of soft garden loam, in which the plants are to be grown.

215. Hotbed with manure on top the ground.

It is advisable to place the manure in the pit in layers, each stratum to be packed or settled down before another one is put in. These layers should be four to eight inches in thickness. By this means the mass is easily made uniform.

Only by experience can one learn what is the proper "body" or texture of good hotbed manure. That with too much straw, and which therefore soon parts with its heat, springs up quickly when the pressure of the feet is removed. Manure with too little straw, and which therefore does not heat well or spends its heat quickly, packs down into a soggy mass underneath the feet. When

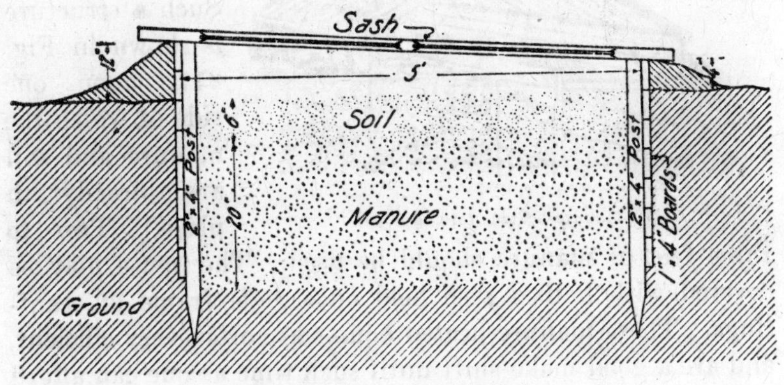

216. A manure-heated hotbed.

the manure has sufficient litter, it gives a springy feeling to the feet as a person walks over it, but does not fluff up when the pressure is removed.

The quantity of manure to be used depends (1) on its quality; (2) the season in which the hotbed is made; (3) the kind of plants; (4) the skill of the operator in managing the bed. Careless watering, by means of which the manure is kept soaked, will stop the heat in any hotbed. The earlier the bed is made, the larger should be the quantity of manure. Hotbeds that are supposed to hold two months should have about two and one-half feet of manure, as a rule. This is the maximum. For a light hotbed to be used late in the season, six or eight inches may be sufficient.

Various modifications of the common type of hotbed will suggest themselves. If the hotbed were high enough and broad enough to allow a man to work inside, we would have a forcing-house. Such a structure is shown in Fig. 217, upon one side of which the manure and soil are already in place. Two to three feet of manure should be used. The house may be covered with hotbed sash held on a rude frame of scantlings. These manure-heated houses are often very efficient, and are a good make-shift until such time as one can afford to put in flue or pipe heat.

217. Manure-heated forcing-house.

Pipe-heated hotbeds.

Hotbeds may be heated by means of steam or hot water. They can be piped from the heater in a dwelling-house or greenhouse. Exhaust steam from a factory can often be used with very good results. Fig. 218 shows a hotbed with two pipes, in the positions 7, 7, below the bed. The soil is shown at 4. Doors in the end of the house, shown at 2, 2, may be used for ventilation or for admitting air underneath the beds. The pipes should not be surrounded by earth, but should run through a free air space.

A flue-heated or pipe-heated hotbed may be likened to a greenhouse bench, and the arrangement of piping for the two should be similar. Two to four steam- or water-pipes are carried underneath the bed. If, however, one has plenty of exhaust steam, which is usually under considerable pressure, it may be carried directly through the soil in ordinary drain pipes. It will rarely pay to put in a hot water or steam heater for the express purpose of heating hotbeds, for if such an expense is incurred, it will be better to make a forcing-house.

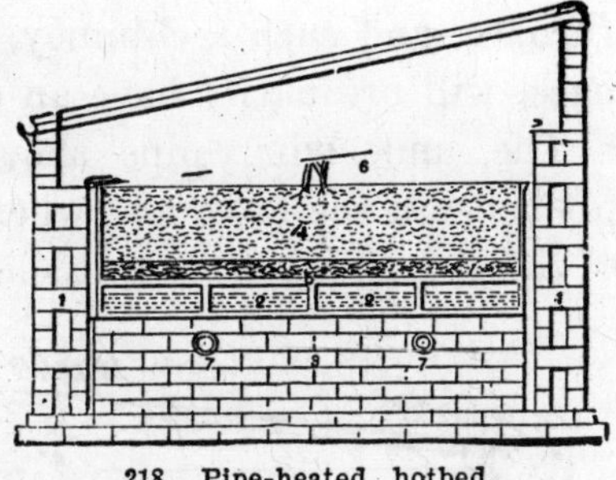

218. Pipe-heated hotbed.

Flue-heated beds.

Hotbeds may be heated with hot-air flues with very good results. A home-made brick furnace may be constructed in a pit at one end of the run and underneath a shed, and the smoke and hot air, instead of being carried directly upwards, are carried through a slightly rising horizontal pipe which runs underneath the beds. For some distance from the furnace, this flue may be made of brick or unvitrified sewer pipe, but stovepipe may be used for the greater part of the run. The chimney is ordinarily at the farther end of the run. It should be high, to provide a good draft. If the run of beds is long, there should be a rise in the underlying pipe of at least one foot in twenty-five. The greater the rise in this pipe, the more perfect will be the draft. If the runs are not too long, the underlying pipe may return beneath the beds and enter a chimney directly over the back end of the furnace, and such a chimney, being warmed from the furnace, will ordinarily have an excellent draft.

The underlying pipe should occupy a free space or pit beneath the beds, and whenever it lies near to the floor of the bed or is very hot, it should be covered with asbestos.

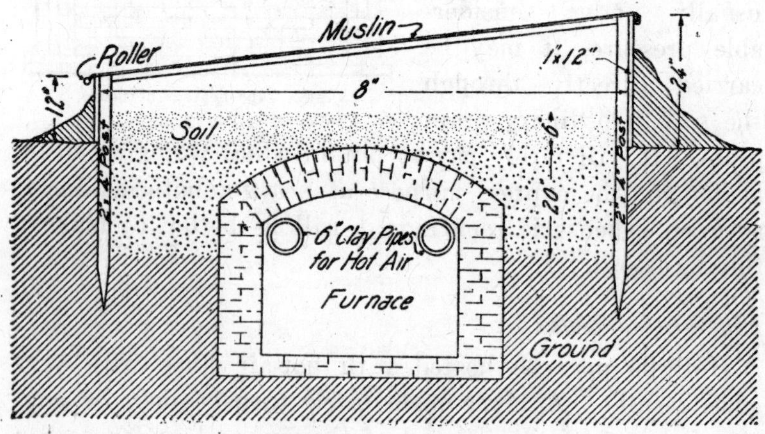

219. Cross-section of a flue-heated hotbed.

The construction of a flue-heated hotbed is thus described by Ness (Tex. Exp. Sta.), and shown in Fig. 219: "A furnace is constructed outside the frame and sunk about two feet below the level of the surface. From this furnace two lines of 4- to 6-inch vitrified sewer pipe are placed obliquely through the soil to the opposite end of the frame, where vent to the smoke from the furnace is given on the outside of the frame. The pipes should run parallel at equal distances from the walls of the bed and each other, with a slant upward from the furnace to the chimney at the other end of the frame. They should be covered with a layer of earth sufficiently deep to secure as nearly as possible an equal distribution of the heat. The heat secured from this apparatus is much more difficult to control than that from fermenting manure, but the construction may be made permanent with only the removal of the layer of earth, in which the seeds are sown for each new crop."

Substitutes for glass.

It will be noted that the bed in Fig. 219 is covered with muslin. On this point Ness writes: "Instead of glass, the sashes may be covered with cotton cloth, saturated with pure raw linseed oil. Before using such cloth, care must be taken that the linseed oil is thoroughly dried, as the fumes given off by the too fresh oils are liable to kill or severely injure the plants, when enclosed in such an atmosphere."

Various prepared papers and fabrics have been advised from time to time to substitute for glass on late-started hotbeds or in the Southern States, and on coldframes. Some of them may give much satisfaction, reducing cost, breakage, and labor of handling. For late work or warm climates it may not be necessary to oil the cloth.

"Most commercial growers in the South [writes F. S. Earle, Bull. 108, Ala. Exp. Sta.] use cotton cloth for covering coldframes, as it is much cheaper than glass, and is much easier to handle in opening and closing the beds (Fig. 220). Ordinary unbleached, double-width or ten-fourths wide sheeting is used. One side is nailed fast to the back side of the bed or in double beds to the ridge-pole, and the other is nailed between two 1 x 2-inch strips, thus making a square roller on which the curtain is rolled up when it is wished to open the bed. By starting with one short and one long piece, so as to break joints, such a roller can be made any desired length. It will be necessary to provide some extra cover for each coldframe to use on very cold nights, for the single thickness of cloth will not turn more than a slight frost. The beds should always be well banked at the ends and sides with earth."

220. Cloth-covered beds, showing the cloth rolled up in the day.

Hotbed covers.

Some protection, other than the glass or muslin, must be given to early hotbeds. They need covering on every cold night, and sometimes the entire day in very severe weather. Very good material for covering the sash is matting, such as is used for carpeting floors. Old pieces of carpet may also be used. Burlap makes excellent cover; it may be doubled; and it may have straw, shavings or wool quilted in it. Various hotbed mattings are sold by dealers in gardeners' supplies.

In addition to the coverings of straw or matting, it is sometimes necessary to provide board shutters to protect the beds, particularly if the plants are started very early. These shutters are made of half-inch or five-eighths-inch lumber, and are the same size as the sash—3 x 6 feet. They are used above the matting to keep it dry and to prevent it from blowing off. In some cases they are used without matting.

In very cold weather, it is sometimes necessary to keep the mats and shutters on the hotbeds for two or three days at a time. During this time, when the plants are in comparative darkness, they are likely to become somewhat soft and tender, and great care must be taken that they are not scalded when the covers are removed and the sun comes out. The stockier and the tougher the plants are grown, the less is the danger of sun-scalding; but after a long period of cloudy weather, this danger is greater and the operator must watch his beds closely.

Hotbeds are usually more difficult to manage than forcing-houses, since the operator can be inside the forcing-house whatever the weather. In very cold and windy weather, hotbeds cannot be opened. The operator works from the outside. In many of the Plains regions, the strong winds make it difficult to handle the hotbed sash. In such case, the cheap forcing-house structure made of frames and heated either with fermenting manure or with pipes is more advantageous.

Beginners are likely to start the hotbed too soon. The age of the plant does not count for so much as its stockiness and vigor. If, therefore, the hotbed is started so early that the plants have to be "slowed up" and stunted in order to hold them until the field is ready, very little is gained. In the Northern States, cabbages and cauliflower may be started with profit about six weeks before the field is expected to be ready; tomatoes, six to seven weeks; onions and beets, four to six weeks.

In summer, after the frames are stripped, the old beds may be used for the growing of various delicate crops, as melons or half-hardy flowers. In this position, the plants can be protected in autumn. As already suggested, the pits should be cleaned in the fall and filled with litter, to facilitate the work of making the new bed in the winter or spring.

Sowing seeds in the hotbed.

Ordinarily the manure will heat very vigorously for a few days after it is placed in the bed. A soil thermometer should be thrust through the earth to the manure, and the frame kept tightly closed with sash and covers. When the temperature is passing below 90°, seeds of the warm plants, like tomatoes, may be sown, and when it passes below 80° or 70°, the seeds of cooler plants may be sown. By the time the beds are ready for planting, the weed seeds probably will have germinated. Loosen and aërate the soil before sowing. Sow in rows four to six inches apart.

More and more, gardeners are coming to start all plants in boxes or flats, for the plants can then be carted to the field or put on the market with ease and with little loss. The flats can also be shifted from one part of the frame to another, or from bed to bed, as conditions may require.

Vegetables that do not transplant well, as melons and cucumbers, may be grown in pots, old berry boxes, or on inverted sods, rather than directly in the hotbed earth. Pots are best.

The following practice in the handling of muskmelon plants from hotbed to field is the experience of a commercial grower of the crop:

Sow seed in flats in greenhouse about four weeks before plants should be set out, say April 20th for region of New York, in flats 2½ in. deep and with 2 in. of soil; use sifted soil, which should be rather sandy or leaf-loam, and cover with burlap (if fertilizer bag is used, be sure to wash before using). Keep the burlap damp but be careful not to overwater. As soon as the melons are up, say 1 in., take up very carefully by using a little stick or an old table fork. Sow plenty of seed and always throw away all poor plants and have the plants uniform. Use 2½-in. flower-pots. If they are dry, soak them in a tub of hot water to kill the germs. Place the pots on the greenhouse bench or in a coldframe with just a little hot manure sprinkled on the bottom and a little soil or sand or sifted coal ashes. Level up and place the pots level; sift on some good earth, one-third soil, one-third sand, one-third well-rotted manure; sift this carefully over the pots and be careful not to pack the earth in the pots; take a piece of board, say ½ in. thick, 2 in. wide, and 12 in. long, and stroke off until you see the top of the pots. Now begin to transplant. Put one good plant in a pot. If it is cloudy you will not need to shade the plants, but if the sun shines they should be shaded by day and the shading taken off at night. Keep a temperature from 75 to 90 degrees to start with; then as the plants start nicely, give more ventilation. As soon as the pot is full of roots, plants should be transplanted, usually about the fourth week. Sometimes they will have to be held back if it threatens frost; if so, give plenty of ventilation.

Now we are ready for the field, which should be a good warm well-prepared soil with a good sod, or if not some manure plowed under. Now drag or roll down. We use a one-horse moldboard plow to open the furrow, going both ways in the furrow. Apply any good manure, loaded in a manure-spreader; use some canvas on each side so that the manure is thrown in the furrow. Level off, then put on fertilizer, 50 to 100 lbs. 16% phosphate and potash. Close up the furrow with a plow. Level the ridge with a plank or pole; ridge when finished should be 3 or 4 inches high.

Previous to these operations, someone should have been taking the plants out of the pots, which ought to have been well watered the night before. Baskets are best in hauling the plants to the field. Set the basket at an angle of about 45 degrees; take the plants out of the pots, handling carefully. Then carry the basket to the field. Straddle one row and plant two on each side of wagon. This is done by a boy who is careful to carry his basket on one arm, using the other hand to pick out the plant. Take hold of the ball of dirt, hand to the planter who walks straddle of the ridge, using a garden trowel. Press plants in with the hands or with the feet slightly. Before planting, plants should be sprayed with bordeaux and arsenate of lead.

Plant in rows 7 ft. apart and 20 to 24 in. in the row. Cultivate with a spring-tooth cultivator which does not tear out the vines as does the five-tooth cultivator. Never cultivate deep for melons.

5. THE MANAGEMENT OF FRAMES

Only by experience can one learn how to manage a hotbed. There are a few principles and cautions, however, which will help.

The objects to be sought, so far as the plants are concerned, are specimens (1) ready at the required season; (2) stocky; and (3) that have made a continuous healthy growth.

The dangers to avoid are (1) the chilling of the plants; (2) too hot and close atmosphere, which tends to make the plants soft; (3) crowding, which tends to make the plants weak and spindling; (4) growing plants too far from the light, which also tends to make them soft and weak; (5) the scalding of the plants by the sun, an injury very likely to occur when the sun comes out after a long "spell" of dark or cold weather; (6) the wilting of the plants, due to too great heat and too little moisture.

Translated into the actual management of a hotbed, these objects may be grouped as follows: (1) maintaining the heat; (2) watering; (3) ventilating; (4) hardening-off.

Above all things, the plant should be stocky (and healthy) when put in the field. A stocky plant is comparatively short and thick, able to stand alone, and has a normal bright green color throughout. Plants not stocky are said to be "leggy" or "drawn," since their general tendency is to grow too long and weak for their bulk. A stocky plant, however, may be stunted. The perfect plant is both stocky and freshly vigorous.

The maintenance of the heat in the ordinary hotbed depends primarily on the quality and the amount of manure; but one can do something by subsequent management to maintain it. Heat will ordinarily fail sooner if the hotbed is above the ground and much exposed to winds. It may also be lessened by careless watering, particularly by soaking the manure. Manure that is too heavy and concentrated may heat violently, and wetting it may tend to cool it to the point at which plants can grow; but a better way is to mix leaves or other litter with the manure, thereby preventing too rapid fermentation. Not only should the heat from the fermenting manure be maintained, but care should be taken to prevent too much of it from escaping. This is an important caution in very cold nights and windy weather, at which time the frame should be protected by mats or other covering. A cold and wet soil also tends to lessen the heat in the hotbed. For this reason, hotbeds should be placed in a sandy or gravelly place, if possible; or if not, the greatest precaution should be taken to insure perfect drainage.

Watering should be performed with caution and care. Careless watering tends (1) to pack or to puddle the soil, (2) to chill the plants, and (3) to soak the manure and to check its fermentation. If watering is from a hose, the danger of packing the soil is greater than with a watering-pot, since the water is applied with greater force. Hotbed soils should be rather loose and fibrous to prevent the puddling. As compared with outdoor or field conditions, the amount of water applied to a hotbed is usually excessive, and the physical texture of the soil is likely to be injured unless one exercises considerable care.

It is better, as a rule, not to water hotbeds toward night or when the temperature is falling, for the application of water and the subsequent evaporation tend still further to cool the bed. It is particularly inadvisable to allow the plants to go into the night with wet foliage. This caution applies with special force to cucumbers, melons and other "warm" plants; and also to the early season, when it is necessary to keep the frame close. It is commonly better to water in the morning, or at least when there is still enough sun heat left to warm the soil before nightfall. It is well also to avoid ice-cold water. The water should have a temperature of 60° to 65°, if possible, particularly for warm-growing plants and early in the season. Avoid dribbling or merely wetting the surface. The soil should be wet thoroughly at each watering, and not wet again until the plants need it; but on the other hand, one should avoid drenching.

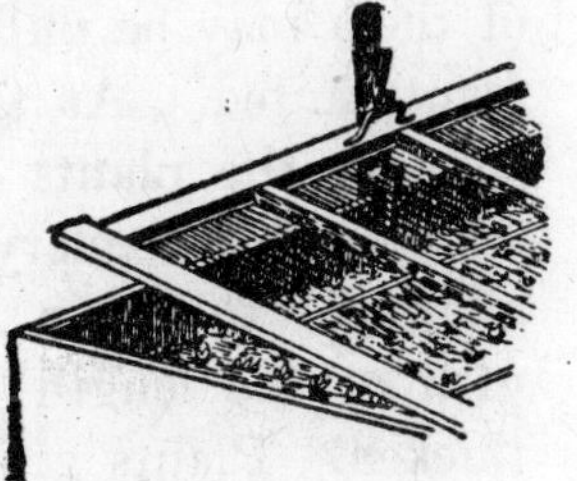

221. The sash raised for ventilation.

Ventilation is important, (1) to dry the air, (2) to aid in controlling the temperature.

Plants kept close and wet tend to grow too tall and soft, and to lack in stockiness. On pleasant and sunny days, ventilation should be given by raising the sash, resting it on a notched block (Fig. 221), or by sliding down the sash. The general tendency with beginners is to ventilate too little rather than too much. One is likely to judge the temperature by the wind and air about his face and ears, whereas the hotbed, being on the surface of the ground, is considerably warmer and more protected.

Whenever the air in the bed is so moist that drops of water collect on the panes, ventilation should be given if the temperature will permit. In fact, it is the aim of good gardeners not to have the atmosphere very moist when the temperature is low and ventilation cannot be given. As the plants grow, more and more ventilation should be given until finally in sunny days the sash may be stripped from the frames. In this way the plants become accustomed to the lower temperature and to normal conditions of the atmosphere; they become "hardened." Careful attention to ventilation is one of the important means of making plants stocky.

Hardening-off is also promoted by giving the plants plenty of room. As soon as they begin to crowd, some of them may be pulled out, or better, all of them may be transplanted. At the transplanting, it may be well to transfer the plants to a somewhat cooler and more airy frame. With celery and some other plants, it is often allowable to shear the tops, cutting off a fourth or a fifth of the length of the plant to make it branch and thicken. Plants grown in pots, berry boxes, oyster buckets are likely to be more stocky than those grown directly in the soil of the hotbed, since they have more room; and such plants may not need transplanting. If it is found that the heat is failing, it will be necessary to harden-off the plants more rapidly.

Certain plants, of which lettuce, cabbage and cauliflower are common examples, can be so completely hardened-off as to withstand considerable frost; and in this toughened condition they may be carried over two or three weeks of cold weather before it is safe to transplant them into the open. The general tendency is to do little transplanting in the frames because of the high price of labor, but transplanting is always advantageous to the plants, particularly if they are started very early.

Wintering autumn-sown plants.

It has been said (page 342) that one use of coldframes is to carry fall-sown plants over winter and to have them ready for transplanting into the field very early in the spring. The plants are not to grow in winter: they are only protected. Hardy plants (lettuce, cabbage, kale, cauliflower) are used for this purpose.

Seeds are sown in autumn, and when the plants have grown four or five weeks they are ready to be transplanted into frames. It is not well that they make much growth in bulk after transplanting to the frames; but they should secure a good root-hold before freezing weather comes. Some persons sow the seeds directly in the frames, but better results are usually obtained if the plants are made extra stocky by transplanting. All soft, weak and imperfect plants are likely to be destroyed by the cold. Very young and flabby plants usually perish. Those too old tend to run to seed in spring. Only by experience can one determine the proper age at which the plants should go into the winter; and this experience is likely to vary with different varieties of the same vegetable. A plant which has begun to thicken up and to show signs of a tendency to form a head usually runs to seed in spring. Cabbage plants with three or four true leaves should be able to pass the winter and to give satisfactory results the following year. The novice should undertake these experiments in a small way, particularly at the North, where the practice is not common and the results are precarious.

Keep the frames uncovered until stiff freezing weather comes. Then use sash and covers. Gradually the plants and soil may freeze. Give freely of ventilation. Strip the frames on all fine days. If the ground is frozen, the plants may stand several days under a cover of snow on the sash; but if the ground is soft, so that root action goes on, the plants should not be kept close and dark for more than a day or two.

In the middle States, the plants are sometimes carried over winter without frames, if they are in a protected place. As far south as Norfolk, cabbages are planted in the field in late fall. Check all winter growth, prevent sudden thawing if possible, try to avoid sun-scalding.

CHAPTER XV

THE LAND AND ITS TREATMENT

Soil and climate are determining factors in the locating of a vegetable-garden; and if the garden is a commercial enterprise, the market adds a third factor. The gardener may control the climate for the plants, at certain stages, and this subject we have considered in Chapter XIV under the name of "Glass."

Good market-gardening land is "quick." It warms up early in spring; it comes speedily into workable condition after a rain; it is easy to keep in good tilth; it responds quickly to fertilizing materials. Its physical condition is more important than its original richness in plant-food: the latter can be added. In the determination of a soil for vegetable-gardening purposes, two coordinate factors are to be considered—the structure or physical make-up, and the content of plant-food.

Most general market-gardens are on sandy loams. A few crops, of which onions and celery are examples, demand particular types of soils for best results; but if one has a deep and uniform sandy soil, one can make an ideal garden of it, other things being equal. If the land is well drained and if rainfall is sufficient, this sandy land can be made immensely productive by a combination of three things,—good tillage, the incorporation of plant-fiber or humus, the direct addition of plant-food. When thus ameliorated, it becomes a sandy loam.

Muckland gardening has come to be almost a special department of the business in recent years, but this enterprise grows special or particular crops and can hardly be called general market-gardening. Reclaimed swamps usually afford excellent land for vegetables, if the area can be thoroughly well drained, so that it is "early," and if the vegetable-matter or peat is well decomposed and comminuted. Soils that are nearly all muck have little body, and suffer from drought; these are mostly the deposit of peat and moss bogs. The fine loams that have accumulated in beds of shallow ponds or lakes are usually ideal vegetable-garden lands, provided the area is not too frosty.

Mucklands differ in productiveness, as do other lands. A good muck naturally grows a heavy stand of trees or other herbage. The mucks that dry out in midsummer are to be avoided unless a stream can be diverted through them. It is necessary to drain mucks so that there is no standing water, and the water-table should be two to three feet deep; but the ditches or a brook should hold enough water to prevent drought. It is the common opinion that mucks are very rich, but this is usually true only as respects organic matter and nitrogen; much of the heavy yield is due to the physical condition and the constant supply of moisture. Muckland gardeners use fertilizers and manures freely. The potash addition usually should be liberal and probably also the phosphoric acid. Poorly handled and unfertilized muck areas are likely to fail rapidly. Onions, celery, lettuce are the main crops, but other crops thrive, particularly on old and well-consolidated mucklands, as spinach, carrot, horse-radish, parsnip and asparagus when the water-table is low enough.

The land to avoid for vegetable-gardening is hard clay. It is cold and late. Plants start slowly in it. It cannot be worked when either wet or dry; and the period in which it can be tilled is so short that much labor and equipment are required to enable one to handle it quickly and efficiently. Clay is excellent for some fruits (particularly pears and plums) and for some general farm crops; but it is not the land for vegetable-growing. However, a friable clay loam may be excellent: this loamy condition may be obtained from hard clay soil by judicious tillage, the incorporation of humus, the addition of amendments in special cases, and by underdraining. Clay loams are good lands for main-season crops of many kinds, as cabbage, pea, bean.

Vegetable-gardening land should be fertile. It should contain much plant-food material; and this material should be quickly available, for on its availability depends the earliness or "quickness" to a great extent. The plant should grow quickly and continuously. Slow-growing and intermittent-growing vegetables may not only fail to reach the market or the table at the desired time, but they are usually poor in quality. To secure this quick growth, the land should be thoroughly prepared before the plants are put on it; and in most cases, an application of concentrated (or commercial) fertilizer will help.

It is usually more profitable to secure land already in productive condition than to take that of inferior quality and to improve it. This is true of all intensive farming;

for such farming demands rapid, positive, and large results. The closer one is to his market, the smaller his area, and the greater the variety of crops he is to grow, the greater is the necessity of securing land in prime condition. Yet, so important is the market for these products that the wise gardener sometimes buys an accessible piece of land even though it is not the best and puts extra effort into the improvement of it.

If one has small capital, he may not be able to obtain highly productive land. In such case, one takes land either naturally inferior or run down. It is said to be impracticable to attempt to reclaim run-down lands. This kind of advice has been over-emphasized. If run-out land lies right and is naturally well drained, it can be brought into profitable condition, in most cases, with comparatively little trouble and expense, if only the person goes at it right. It requires time and patience. The first thing is to till well and to add fiber (preferably by means of clover at the North). The common notion that commercial fertilizer is the first resort in such cases is usually an error. The fertilizer is for the purpose of adding plant-food, not of ameliorating the soil. If market-gardening is attempted on run-down land, the operator should choose the best part of the area for his more intensive efforts, giving it what manure he has and bestowing on it his best efforts in tillage. The remainder of the place can then be slowly brought into condition by cover-cropping, rotation, liming, and other cheaper means. Four or five years should suffice to bring the average worn-out land into good condition, without great expenditure of capital. The "run-down" character of a farm is usually more a matter of dilapidated fences and buildings, poor drainage, weedy fields and slovenly appearance, than of exhaustion of plant-food.

1. THE AMELIORATION OF THE LAND

Land that is " quick " is in good physical condition. It is finely pulverized, " mealy," mellow, deep. It is unprofitable to apply expensive plant-food to poorly tilled and intractable land. The first efforts, therefore, must be given to drainage, tillage, the addition of fiber, rotation.

Drainage.

The best drainage is provided by nature; that is, land naturally well drained comes into condition more quickly, as a rule, and is in more continuous good tilth than that which it is necessary to drain artificially. However, the very best results may be secured by a good system of tile-drainage.

Underdraining is practiced for two purposes—to carry off the superfluous water, and to improve the physical structure of the soil. All low and boggy lands need to be drained for the first purpose. Very stiff clay lands, which are normally dry and hard, usually can be much improved by a good system of underdrains. The reason of this is simple. If water stands long in clay lands, it tends to cement or to puddle the soil. If the superfluous water is quickly taken off, however, this cementing or puddling does not take place. The soil is thereby looser or more friable. This friable condition enables the soil to hold more moisture than when it is hard and brick-like. It therefore results that draining to remove the superfluous water puts the land in condition to hold more capillary moisture in its own tissues, and improves it for agricultural purposes.

For vegetable-gardening purposes, particularly if quickest results are desired, it is necessary to underdrain hard clay lands, even if they are not wet. It makes them workable early in the spring after rains, and enables the plants to obtain a quicker foothold. These same lands might be used for orchards, however, without underdraining, and they might also be very productive of general farm crops; but in such cases the crops may occupy the land for a term of years, and very early results are not essential.

For temporary purposes, surface drains may be employed, or the land may be ridged so that the surface water is taken off in the dead-furrows. This surface drainage, however, results only in carrying off superfluous water and does not have the effect of ameliorating the land. Surface drains are temporary creeks.

In most cases, it is better and cheaper in the end to use tile underdrains. Board drains were formerly sometimes used, but they are not so efficient nor so permanent. In stony countries, excellent drains may be made by partially filling the ditch with stones, particularly if flat stones are to be had so that a conduit can be laid in the bottom. Such drains not only provide the advantages of underdrainage, but also afford a means of disposing of superfluous stone. If they have a good fall, and care is exercised not to fill the spaces between the stones with earth, they may be nearly or quite as efficient as tile drains.

The deeper the drains, the deeper will be the ameliorating effect on the soil and the greater the area they drain.

As a matter of practice, however, it is found that 4 feet is usually the maximum depth, and about 3 feet the minimum. Wet lands, or very hard clay, should have drains at a distance of not more than 2 or 3 rods, if the lands are to be put in the very best condition for market-gardening purposes. It may be advisable, however, to use such lands for the later, cheaper and general-purpose crops rather than for the very early ones if the gardener has other land that can be tilled for the crops desired for the early market.

Tillage.

At present great emphasis in agricultural practice is placed on tillage. We have passed through that era in which we have looked to recipes and special practices for the improving of the land. The fundamental consideration is to till: the later and incidental thing is to fertilize the land.

We till (1) to prepare the land to receive the crop; and (2) to maintain the soil in good condition for the growth of the crop.

To prepare for the crop, the land should be loosened and pulverized as deep as ordinary roots go. To maintain the soil in ideal condition, the surface should be tilled or stirred as often as it becomes crusted or compacted. It is essential that every farmer keep in mind the differences between preparation-tillage and maintenance-tillage, for these ideas are associated with two classes of effort. Cultivating should be thought of as maintenance-tillage, not as preparation tillage.

1. *The tillage of preparation* insists that the land be

broken and pulverized. The depth to which this pulverization or plowing shall extend must be determined for each particular case: it depends on the character of the land and the crop. Land that is very hard, or in which there is a high sub-soil, usually needs to be plowed deep; the effort must be to deepen the soil. Sandy or leachy lands may need to be plowed shallow and approximately the same depth every year: the effort is to compact the under soil and thereby to prevent the leaching. The root-crops demand deep soil, that the roots may grow long and symmetrical. This is emphatically true with such long-growing roots as parsnips, late beets, carrots and horse-radish. Once it was the general advice that land be plowed deep. But neither deep plowing nor shallow plowing is the unit. The depth of plowing is a question of conditions.

It is a favorite practice with gardeners to plow in the fall. There are three objects of fall plowing: (1) To render the land earlier in the spring; (2) to be forehanded with the work; (3) to improve the physical character of the soil. Land plowed in fall usually can be worked several days earlier than that plowed in spring. It dries out sooner. Especially is this true of stiff and loamy lands. Clay lands may be much improved by being plowed in the fall, so that the weather may break down and slack the lumps. It is important, however, that such land should contain more or less vegetable matter; otherwise it may run together and puddle during the winter season and be difficult to manage in spring. If the land bears stubble of grain or grass, or if it has a covering of manure, such danger is averted. If land is clean and in good condition, it will not need to be plowed again in the spring, but can be worked down with heavy tools, like the spading harrows; but this spring working must not be delayed. Whenever land is needed very early in the spring, it is advisable to plow it in fall. This remark applies with little force to light and sandy lands, for they can ordinarily be plowed very early.

Lands may be made earlier to work if they are thrown into beds or ridges by the fall plowing, so that the dead-furrows lie every eight or ten feet. The surface water is then carried off and the ridges stand so high that they dry out quickly. This operation is sometimes spoken of as trenching, but it is more properly ridging. The term "trenching" should be reserved for its legitimate use to designate the spading up or loosening up of the land deeper than the original furrow. Ridging and trenching are only special practices.

Sub-soiling is a frequent practice in market-gardening lands. It is advisable in lands that are hard or that have a high sub-soil, and also for the long root-crops, which demand a deep soil in which to perfect their growth. Sub-soiling is not a permanent corrective, for the soil settles back into its original and hard condition, and the operation must be repeated. The fundamental corrective for such lands is underdraining and incorporation of humus. The growing of clover or alfalfa, which sends its roots deep into the soil, is also a great aid. But even with all these aids, sub-soiling may be very useful in certain cases. The sub-soil plow does not turn a furrow; it merely breaks the bottom of the original furrow. It is drawn by a separate team and follows in the furrow immediately behind the first plowman.

2. *The tillage of maintenance* should occur at least as frequently as once in ten days for the best market-garden conditions. Surface-tillage enables the land to drink in the water of rainfall. It also saves the soil water by hindering evaporation: it maintains a loose and dry layer that acts as a mulch to the moister soil beneath. The depth of this mulch must be determined by the character of the soil, kind of crop, frequency of tillage, and character of tools; but, as a rule, from three to four inches of loosely stirred earth is sufficient. It also solves the difficulty of weeds.

The soil in the surface-mulch is relatively dry, and it is moved so often that roots do not secure a foot-hold in it. It is therefore out of use for the time being as a source of plant-food; but it is more useful as a conservator of moisture than as plant-food. Its nutriment comes into use when it is turned under the following season, and it is also carried down by the rains, particularly by those of spring and fall.

All tillage of preparation—all fitting of the land—should be completed before the crop is put in: thereafter, only the surface-mulch is to be kept in repair. But many times the preparation-tillage is not completed in its season, and the land must be fitted after the crop is sown by means of deep and heavy cultivating; it is usually a loss of effort and efficiency when preparation-tillage and maintenance-tillage must be performed at the same time.

The rainfall of the growing season is often insufficient for the crop. The plants draw on the moisture stored in the soil by the winter rains and snows. Therefore, it is exceedingly important to save this winter rainfall, and this is accomplished by fitting the soil and making the surface-mulch the moment the land is dry enough to work in spring. Even if the land is not to be used until June, it should be fitted early, and lightly harrowed at frequent intervals before the crop is planted. The principles and practices of dry-farming, which is the recourse in non-irrigable semi-arid regions, should be understood by all farmers.

Addition of humus.

Land is rapidly improved by the incorporation of fiber. This fiber is obtained by plowing under any kind of vegetation or organic matter, as rye, clover, manure or the refuse of the garden. When this fiber decays it becomes humus. The humus improves the physical condition of the soil by making it loose, open and mellow; by enabling it to hold moisture; by preventing the puddling or cementing of clay soils; by decreasing the heat of the surface in summer; and by improving the chemical character. Humus itself contains plant-food. It also affords solvent acids. If it is derived from leguminous plants, it also adds nitrogen. The chief reason for the almost extravagant use of stable manures by market-gardeners is the addition of humus. Lands thus manured year after year become quick and amenable to treatment. Fertilizers work speedily in them. The lands can be tilled at almost any time in the growing season, and when one crop is off another can be put in quickly.

In the addition of plant-fiber much will be gained if it is thoroughly decomposed. It thereby becomes quickly incorporated with the soil, and its plant-food soon becomes available. This is the explanation of the general desire of market-gardeners to have what they call "short" or well-rotted manure, and also the common practice of composting manures and refuse.

Composting consists in piling the various materials together in long, low, flat-topped piles, which may catch and retain the rainfall, and then forking over two or more times in the season. If the materials are well disintegrated and mixed, they are in fit condition to be put on the land. Tomato vines, potato vines and even corn stalks, which are too raw and coarse to apply directly, may be made into useful and valuable material when they have been composted for several months or a year; although if serious diseases infest the refuse, the material would better be burned. The addition of quick-lime hastens the decomposition of raw materials. The florist, who must have his soils in perfect condition, is familiar with methods of composting, for he usually provides his soils a year in advance, rotting his sod for this purpose.

Rotation.

One great value of the rotation of crops is that it adds fiber and humus. It is probable that there is a tendency to use stable manure in excess in garden lands; that is, the same results in the incorporation of humus can be had in many cases more cheaply by the growing of catch-crops. Particularly is this true of those areas some distance from the market and in which it is not necessary to practice rapid succession of market crops. With the passing of the great city stables, in recent days, substitutes must be found for the barn manures in the market-garden.

Land that receives identical treatment year by year

tends to depreciate. A rotation is useful because (1) it provides different treatments for the land, the fault of one year tending to be corrected by the management in another year; (2) no one element of plant-food is exhausted, the rotation tending to even up the demands; (3) one crop leaves the land in good physical condition for another; (4) it incorporates humus; (5) it destroys pests and weeds; (6) it economizes labor; (7) when green crops are turned under, available or digested plant-food is incorporated with the soil, and nitrogen may be supplied. The rotation of crops means, also, rotation in tillage, manuring and other treatment; and one of these may be quite as important as the other.

The reason for the "resting" of land is hereby explained. It is not due to any need of recuperation in the soil; but the good effects are the compound results of the various benefits derived from tilling and rotation. Gardeners find that when soil becomes unproductive for a particular crop, a change to another crop may result in profit. Soils that have been long kept in market-gardens may be benefited by seeding down for two or three years. Whenever possible, attempts should be made to practice some kind of a rotation in the market-garden area. Now and then, a part of the land may be laid down to clover for a year or two until it recovers; this provides a form of rotation and destroys insects and other organisms.

2. THE DIRECT FERTILIZING OF THE LAND

When the soil has been thoroughly fitted and improved by all the foregoing means, a gardener may think of adding plant-food. This plant-food may be supplied in a concentrated fertilizer; it is also added when green-crops are plowed under, or when manure or compost of garden refuse is applied. It will now be seen that the best results are usually to be expected when there is something like a rotation in the fertilizing of the land, stable manures being used alternately with concentrated or commercial fertilizers. If such manures are not to be had for the entire plantation, they should be applied to the hardest spots or be reserved for the more exacting crops.

The kind and quantity of fertilizers are to be determined by several circumstances: (1) the earliness or quickness with which the crop is to be obtained; (2) the intensity of the operations to which the man is committed; (3) the nature of the land as regards tilth and texture; (4) the character of the land as regards richness in plant-food; (5) the kind or species of crops.

There is no infallible means by which one can determine what fertilizers to apply. The grower must study his conditions and judge as best he can. A little experiment with different kinds of fertilizer on two or three of the leading crops at one side of the plantation, is the readiest means of answering the question. If one is in doubt, it is well to seek advice how to lay out and conduct a demonstration plat; this advice will be given by the college of agriculture or farm-bureau agent. There are standard forms for such plats.

The chemical analysis of the plant, while of the greatest use to the chemist in giving him suggestions, is of no practical use to the farmer in determining the kind of fertilizers or what amount shall be applied, notwithstanding a still more or less prevalent notion to the contrary.

The chemical contents vary in the different seasons and in the different parts of the plant, and also with the soil in which the plant grows; the plant may take up more than it needs when some element is abundant. Even the widest variation in any one ingredient will be amply covered by the large quantity of fertilizer ordinarily applied. Consider, for example, that the fruit of a tomato comprises .05 per cent of phosphoric acid and .27 per cent of potash. If the crop is ten tons of fruit to the acre, more than the average quantity of required phosphoric acid is ten pounds and of potash fifty-four pounds. It is safe to assume that the land itself would supply at least three-fourths of these amounts. We will assume that one-fourth is to be supplied by the addition of fertilizer. We should then apply to the acre two and one-half pounds of phosphoric acid and about fourteen pounds of potash. As a matter of fact, however, the smallest quantities ever applied are many times in excess of these figures. Fertilizers must necessarily be applied in excess of theoretical needs. It is impossible to distribute a very small quantity; roots do not occupy every part of the ground. Much is risked in the chance that some of the material may be used.

Another difficulty in the giving of advice is the variable nature of the soil. This is particularly the case in the Northern States, in which the soil is largely drift and is therefore very uneven in kind and depth. In the long stretches of sand on the Atlantic coastal plain or in the red clays of the South, and in nearly all alluvial soils, the problem of choosing a fertilizer is less complex. The sandier and more uniform the land, the more marked, as a rule, the effects of commercial fertilizers. The harder the clay, the less marked, in general, is the result, although amendments (as lime) may have great effect in making such soils granular.

Again, the state of tillage has much to do with the efficacy of a fertilizer. The element the plant needs may be afforded more cheaply by giving better tillage than by adding fertilizers; for tillage sets at work forces that unlock plant-food. On the other hand, fertilizer is more usable by the plant on well-tilled soils: the plant can get hold of it because the material is more evenly distributed; there is more moisture to dissolve it; the plant is more comfortable and vigorous and thereby better able to appropriate it. The good gardener is the one who gets the most out of his land by means of tillage and then adds fertilizer to get more out of it. He uses fertilizer for the purpose of securing an extra yield, not to prevent the soil from becoming exhausted. As a rule, the men who till best buy most plant-food. Fertilizer is usually a losing investment for a poor farmer.

When to apply a fertilizer depends on (1) when it is needed by the plant; (2) the kind of fertilizer; (3) the soil; (4) the kind of plant; (5) the season of normal rainfall of the district.

The more soluble the fertilizer, the looser the soil, the shallower the roots, the later the material may be applied. With trees, it matters little whether fertilizer is applied in fall or spring, for it is usually one or two years before it affects the plant. With the general run of vegetable crops and on soils in good tilth, it is usually best to apply fertilizer in spring, sowing it on the surface and harrowing it in. On ordinary soils, very little of it is lost by leaching. Nitrates are most likely to leach. They are soluble and pass down quickly. Therefore, nitrate of soda and sulfate of ammonia should not be applied much in advance of the planting. With annual crops, fertilizer should not be applied much, if any, in advance: the fertilizer is needed near the surface, and it should be quickly available.

There is discussion whether fertilizer should be applied broadcast or in the hill, which proves that both methods give good results. If one wants to enrich the land, or to afford sustenance to the plant throughout its growing season, apply broadcast. If one wants to use fertilizer to start the plant off and to maintain it until it gets a firm hold on the soil, apply in the hill. A most important use of commercial fertilizer in vegetable-gardening is to hasten the plant in the beginning. It has been likened to kindling-wood to start the blaze.

It is not sufficient, however, that the plant be well started. Continuous growth of radishes, lettuce, spinach, turnips and many other crops means a tender and palatable product, with the minimum of fiber and stringiness; the fertilizing should be liberal enough to maintain this growth.

While the gardener must regulate his fertilizer practice by his own experiments and experience, he is not wholly dependent on his own resources. Investigation and general agricultural experience indicate what probably will take place in a given case. The general advice, for example, is to apply a complete fertilizer—that is, one containing nitrogen, potash and phosphoric acid in about the proportions that experience has found to be useful. This advice is particularly good when the person does not wish to experiment or to give the subject careful study. It is less useful, perhaps, when one does not wish to enrich the land as much as to give a stimulus to the young plant.

It is generally considered that nitrogen promotes rapid vegetative growth. It therefore may be used most freely on plants desired for their foliage parts. If it promotes growth, it also delays maturity. Therefore it should be used sparingly, or only early in the season, on fruit-bearing plants that tend to mature too late, as tomatoes and eggplants. Experiments at Cornell years ago showed that a little nitrate of soda is better than much for tomatoes; also, that a given quantity applied all at once early in the season is better than the same quantity applied at intervals, for in the latter case it promoted growth too late and the fruits did not ripen (page 257).

For the person who has studied the subject and his soil, it is preferable to buy the elements in the form of high-grade chemicals and to apply each by itself. He can then apply little or much of any element to this place or to that, as he thinks best. Good commercial sources of nitrogen are nitrate of soda, sulfate of ammonia and dried blood; of potash, muriate of potash and unleached wood ashes; of phosphoric acid, bone meal and acid phosphate made from the rock phosphates of Tennessee, South Carolina and Florida. Of nitrate of soda, 150 to 300 pounds to the acre is a good application; of muriate of potash, 200 to 400 pounds; of treated rock, 200 to 400 pounds.

The grower should conceive of a basic formula, and then add or subtract to meet special needs. Voorhees

("Fertilizers," 2nd edition, 286) defines a basic formula as "one containing large quantities of all the best forms of plant-food to be used as a base for supplying market-garden crops with their general needs, with the idea that amendments may be made of nitrogen, or of other constituents, as the conditions seem to require"; and he continued: "apply a reasonable excess of all the essential fertilizer constituents to all of the crops." He recommends "a good basic fertilizer for market-garden crops" as follows:

Nitrate of soda	250 lb.
Ammonium sulfate	100 lb.
Dried blood	150 lb.
Acid phosphate, 16% A. P. A. . . .	1000 lb.
Sulfate of potash	400 lb.

"A mixture of these materials of standard quality would show an average composition of 4 per cent nitrogen, 8 per cent phosphoric acid and 10 per cent potash. Such a mixture is an excellent basic formula for such crops as asparagus, cucumbers, onions, cabbage, cauliflower, celery, eggplant, melons, peppers, squashes and the like, but any mixture of the composition 4–8–10 which supplies the plant-food constituents in good forms may be used as a basic formula for all market-garden crops, leaving the specific needs of the different plants to be met by top-dressings, or applications of the other constitutents. The fertilizer ingredients, nitrogen and phosphoric acid, should preferably consist of the different forms, rather than to be all of one form, though the cost of the element will naturally regulate this point to some extent. That is, a part of the nitrogen should be nitrate or ammonia, and a part organic; a part of the phosphoric acid should be soluble (from superphosphates) and a part insoluble (from ground bone, tankage or natural phosphates). The soluble portions of both nitrogen and phosphoric acid contribute to the immediate needs of the plant, and the less soluble to its continuous and steady growth, and to the potential fertility of the soil."

All of this formula, or part of it, may be used on an acre. Commonly, 1,000 to 1,500 pounds are recommended. More specific advice may be found, as recommended by Voorhees or others, under the different vegetables. It is to be said, however, that the figures given for any vegetable are only by way of suggestion, for there is no invariable rule to follow. The grower learns by experience how to vary the indications for his land and for his method of handling the crop.

3. THE IRRIGATION OF THE LAND

In many regions the crop is determined by the amount of rainfall rather than by the plant-food. The crop often requires more water than is supplied by the normal rainfall of the growing season. Tillage can save much of the water that fell in the early rains and the winter snows, but there may still be insufficient moisture for a good crop. Irrigation may be necessary to supply the deficiency.

In the arid parts, irrigation is a necessity. It is a general practice. In the humid parts of the country—east of the plains—irrigation is often helpful and it reduces the risk of a poor crop. It is an exceptional or special practice. Evidently, in all regions in which crops yield abundantly without irrigation, the main reliance is to be placed on good tillage.

Irrigation is an economic question. If, by irrigation, one can produce enough better crop more than to pay the cost, the practice is to be advised. Too often the farmer thinks of irrigation as he thinks of fertilizer—as a means of giving him crops when he does not work for them. It is only the well-tilled and well-handled lands that pay for either irrigating or fertilizing. The intenser the cropping, the more the capital invested, the better the market, the more likely is irrigation to pay. Ordinary crops will not pay the cost and risk of irrigation in the East. The feasibility of it depends, also, on the lay of the land, the availability of water, the price and supply of labor, the character of the given climate.

Most vegetable-gardeners in the East do not find it profitable to irrigate. Now and then a man who has push and the ability to handle a fine crop finds it a profitable undertaking. If the grower contemplates putting in an irrigating plant, he should visit a garden in which one is in operation, if possible. He should take advice and buy a special book on the subject.

In general garden operations, the water may be applied on the surface, in the furrows between the rows. The main conduits—which may be ordinary wrought-iron water pipes—are carried along the highest land. The pipes may be laid in ditches or on the surface. At intervals, hose-bibs are provided, so that a rubber hose can be attached and the water conveyed into the furrows. When box sluices are provided, there may be openings or water-gates opposite the furrows. If iron pipes are used, faucets must be provided at the lowest point of the run and in the sags for the purpose of emptying the pipe of water in the fall. The water supply must be ample, for when irrigation is most needed, the air is dry and hot and evaporation is rapid. It should be the aim to convey the water in narrow streams or furrows close to the plants, rather than to cover the entire space between the rows. The farther end of the rows should be supplied quickly (by providing sufficient fall, head and quantity), otherwise most of the water will be taken up at the near end of the row.

Imitation of rainfall is now employed in many high-class gardens. This means an overhead installation, with the water forcibly thrown from small openings in lines of pipe. There should be a good head, for pressure is essential. The mains (consisting probably of 1½-inch iron pipe) are laid either on the surface or beneath it. From these mains, smaller pipes are carried overhead; they should be about 7 feet above ground to allow of easy working beneath them and to give sufficient "throw" to the jets of water. The pipes are punctured or bored at intervals of about 3 feet, special plug-nozzles being employed. The stream is small and solid as it leaves the nozzle, but soon breaks into a rain-like spray. With a head of 30 to 40 pounds, the spray should reach 20 to 30 feet, and this, therefore, determines the distance apart of the runs. Gates are provided in the pipes, which, when turned, throw the water in one course or another.

The overhead irrigation has given good commercial results in certain types of intensive gardening, greatly reducing the risks. The system should be installed by an experienced man, for the weak points in the operation have now been well worked out.

The other extreme from overhead watering is sub-irrigation. This method has long been employed in greenhouses. The water is there conducted underneath the soil in drain-tiles, and it distributes itself from the uncemented joints. The bed or bench is provided with a hardpan in the form of cement, so that the water does not leach away. Rarely do comparable conditions occur in nature; yet sometimes in reclaimed bogs and swamps a hardpan lies a foot or two beneath the surface and tiles may be laid on it and receive water from the higher end. These tiles may also serve the ordinary purposes of drainage. Sometimes the outlets of ditches and drains in mucklands are closed in dry weather and the water is held or even backed up, affording a kind of irrigation.

CHAPTER XVI

VEGETABLE-GARDENING TOOLS AND IMPLEMENTS

The tool multiplies the power of the man. Relative to the price of land, labor is expensive in America. It must be economized. Tools and implements are a necessity.

To an important degree it is true that the successful American farmer is known by the number and variety of his tools. The man who has many useful implements emphasizes brain above brawn. He is tactful and resourceful. He means to be master of the situation. He is to accomplish the given result with the least expenditure of mere muscular energy. He will do his work better and more expeditiously than the man who depends on his hands and his strength. Good tools educate the man. Their use cultivates ingenuity. They teach him to think.

On the other hand, the man who is rich in large agricultural implements has less intimate contact with his plants than has the hand-worker. The machine is between him and the plant. He depreciates the value of painstaking human care in the growing and the training of the plant. If he becomes machine-minded rather than plant-minded, he ceases to be a gardener.

In American conditions, a large equipment of tools is necessary to an abundant and cheap crop. The nicest judgment is required to make a proper choice; for the kinds should be determined by (1) the character of the soil; (2) the size of the plantation; (3) the comparative earliness of the required product; (4) the kinds of plants to be grown; (5) the personal ideas of the farmer.

Tools adapted to the working of clay soils may not be adapted to sand. There should be a tool for each diverse type of labor. An advantage of the variety in tools offered by American dealers is the fact that a tool may be found for each particular purpose. Some farms, however, are overstocked with tools. Too much capital is locked up in them. This fault is usually the result of duplication,—the various tools are too similar, they do not perform different kinds or types of labor.

It requires nearly as many tools to equip one acre of market-garden as to equip five acres. Consequently, it is relatively cheaper to till a fairly large area, so long as it can be tilled well. The greater the capital invested in an acre of land, the more intensive should be the cropping and cultivation.

In choosing a tool, the buyer should know (*a*) what labor is to be performed; (*b*) what implement will best perform it. Many persons buy a tool because it is perfect as a mechanism or merely because it is an improvement on what they already have. This is well; but it should be borne in mind, after all, that the tool is not the first consideration,—it is not the unit. The unit is the work to be performed or the condition to be attained. A farmer may not ask, therefore, whether he shall buy a spading-harrow: he should consider his soil and what he wants to do with it, and then search for the tool that will best meet the work.

In general, it is well to avoid combination tools which, by means of various attachments, are designed to perform very unlike kinds of labor. They are likely to be less efficient than tools made directly for the given labor, and are also more liable to get out of repair. They are usually cheaper than separate tools, however, and some of them are very satisfactory.

Market-gardening tools, implements, carriages and machines may be roughly classified as follows:

I. *For tillage.*

- *a.* Tools to prepare the land for planting:
  - Plows,
  - Harrows,
  - Cultivators,
  - Rollers,
  - Hand-tools of various kinds, as spades, wheel-hoes.
- *b.* Tools for subsequent use,—to maintain the condition of the land:
  - Cultivators,
  - Weeders,
  - Hand-tools, as wheel-hoes, hoes, rakes, scarifiers, finger-weeders.

II. *To facilitate hand-work.*

- In distributing manure and fertilizer,
- In marking the land,
- In sowing,
- In planting,
- In spraying,
- In harvesting,

In threshing,
In grading and packing,
In preparing the product for market or sale.

III. *For transportation.*
Carts and barrows,
Stone-boats and sledges,
Wagons,
Motor trucks.

IV. *For power.*
Water motors,
Wind mills,
Steam engines,
Gas engines. tractors, trucks, and other motors,
Electric motors.

For a market-garden large enough to be worked by horses or mechanical power, the following general-purpose tools and implements, at the least, will be needed:

1 2-horse plow,
1 1-horse plow,
1 furrowing or single shovel plow,
1 spading- or cutaway-harrow, if the land is heavy,
1 spring-tooth harrow,
1 roller or slicker,
1 smoothing harrow,
1 spike-tooth cultivator,
1 wide-tooth or shovel-blade cultivator,
1 or more hand cultivators,
1 marker,
1 seed-sower or drill,

1 or more hand wheel-hoes,
1 or more wagons,
1 stone-boat,
1 wheelbarrow,
1 spraying outfit,
Spades, shovels, hoes, rakes, forks, hand-weeders, trowels and dibbers, hose, watering-cans, carpenters' tools.

Implements of secondary importance, but which the well-equipped market-garden must possess, are:

Gang-plow, if the area is large,
Sub-soil plow,
Swivel plow,
2 or more types of spading, cutaway, or disk harrows, if the land is heavy,
Acme and other harrows,
Wire-tooth weeder,
Various patterns of cultivators for special work,
Plant-setter,
Fertilizer distributor,
Trucks and wagons.

Aside from these various devices, there are special implements for special crops, as celery-hillers, asparagus-bunchers, potato-diggers, potato-sorters, graders, and the like.

The implements and their work.

The plow is the primary or fundamental general-purpose farm implement. Its office is to prepare the land, not to maintain it in condition. As a class, stiff and heavy soils require heavy plows and deep plowing. Sandy soils may be the better for shallow plowing, for it is often desirable to compact the sub-soil rather than to loosen it. There are conditions and conditions.

Plowing has three general offices: (*a*) to break and pulverize the soil to fit it for the growth of the crop; (*b*) to turn under and cover the surface herbage, or the manure; (*c*) to begin the preparation of a seed-bed in which the plant may get a start. In the plowing of the sandy soils, the second office may be sought; only a good seed-bed is desired, for the land is loose enough without the plowing. In the clay field, all offices are sought. Not deep plowing nor shallow plowing is a principle: it is only a means of accomplishing a desired result.

The seed-bed of the general field is finished by the harrow. The soil is maintained in tilth by the harrow. The harrow, therefore, is an implement both for preparing and maintaining the soil condition.

If the land is light, loose or sandy, tillage presents few difficulties and relatively little expense. If it is hard clay, tillage must be nicely managed for best results. Many persons expend more time and muscle on clay lands than are required. The one important item is timeliness. When the soil is betwixt wet and dry, it breaks as it turns from the plow. Turn it up loose and open. Then let it lie for a few hours or a day. As the clods begin to dry, work roughly with a strong harrow, as a spading-harrow, spring-tooth, or acme. Do not try to work it down fine. As the lumps begin to dry after the next rain, hit them with the boot. If they break and crumble, work the land again, this time with a lighter harrow. A few timely workings when the soil is just right will accomplish more than thrice the labor at other times. Persons often make the mistake of tilling their clay lands until they become too fine. Then a rain packs and cements them, and the trouble begins all over again. The addition of humus enables one to make a clay soil mealy.

Gradually, as the texture improves, lighter tools may be used to maintain the surface mulch,—for the tillage of maintenance really has no other primary office than to keep the surface loose. When finally the wire-tooth weeder can be used, the gardener may know that his surface soil is in perfect condition. To most general farmers the weeder is a useless tool, but market-gardeners prize it,—which illustrates the differences in tillage between the common farm and the market-garden.

A one-horse harrow is usually known as a cultivator. But there are two types of cultivators,—those that only stir the soil and repair the surface mulch, as the spike-tooth cultivators; and those that move the soil or even invert it, as the shovel-tooth cultivators. Perhaps shovel-tooth cultivators are too common and spike-toothed cultivators too rare.

222. Leveling device attached to a cultivator frame.

Rollers have two uses: (a) to break clods and level the ground; (b) to provide moisture for seeds or newly set plants. Rolling establishes capillary connection with the under soil, and brings the particles into contact with the seeds. It destroys the surface mulch. The water rises and passes off into the air: in its passage, it moistens the seeds. As soon as the seedlings or transplanted plants are established, therefore, restore the surface mulch. The farmer patted his hill of corn with the hoe, in the former days, thereby accomplishing the result which he secures on the wheat field with his roller. The gardener walks over his row of seeds.

223.
Three kinds of wheel-hoe.

If the roller is employed only to break the clods, the land should be tilled again to restore the surface mulch. The roller is a poor tool in the hands of a thoughtless man. For the leveling of land, a home-made planker or slicker is a useful tool. A similar device may be attached to a cultivator frame (Fig. 222).

In the garden, the wheel-hoe is important (Fig. 223). It saves immensely of hand labor and usually leaves the soil in better condition than does hand-work. There are a number of patterns, large and small. Choose a large wheel with a broad tire, that it may ride over lumps and travel on soft ground. Soil must be in good condition to be worked with wheel-hoes; therefore, they should be introduced for their educational effect. Aim at the onion-bed condition of tilth.

A hand-hoe is a clumsy and inefficient tool. Its one merit in this regard is the fact that it can be used between the plants, where many other tools cannot enter; but it leaves no efficient surface mulch and does not often improve soil-texture. The common hoe has two types of legitimate uses on the farm,—to aid in planting, to kill weeds. As a tillage-tool, the rake is far superior. Most persons use the hoe as they would a pick,—to chop the earth. Much hoeing usually wastes soil moisture.

The gardener should secure a spraying outfit of large capacity. It is more efficient and more economical of labor. Be sure that the pump is strong, carefully made, well lined, and has much power. Clean it thoroughly inside before putting it away for winter. Get it out a month before it is wanted in spring; it will probably need tinkering. Year by year, spraying machinery is improving. The gardener is practically powerless before the multitude of bugs and fungi unless he has good spraying and dusting devices and a proper stock of insecticides and fungicides.

For the home garden one needs many small hand tools and helps, some of which can be made on the premises. These aids include hand-weeders, light hoes, sprinklers, watering-cans, garden line and reel, labels, stakes, and others. A tool shed or stall is one of the most interesting adjuncts to a garden, expressing the gardener's interest in deft and neat handicraft.

CHAPTER XVII

SEEDS AND SEEDAGE

Most vegetable-gardening crops are grown directly from seeds. Therefore, the character of the seed is of vital importance to the vegetable-grower, whether he is an amateur or a commercial man. The grower is interested in seeds from these points of view: (1) whether they are viable; (2) whether the sample is unadulterated, carrying no seeds of weeds and no foreign matter; (3) whether the seeds are true to name; (4) whether they represent an improvement on the variety or strain, or at least maintain the merits of it. The quality of the seeds may determine both the quality and the yield of the crops. Land, fertilizing, seeds,—these are the essential considerations at the beginning in the growing of vegetables.

1. THE TESTING OF SEEDS

Seed tests are of three leading kinds: (1) to determine the purity or content of the sample as respects admixture of foreign matter; (2) to determine viability; (3) to determine whether the variety is true to name or kind.

Testing for purity or content, and for viability or germinating power, are relatively simple.

But the determination of the nature of the sample as concerns its trueness to name and its peculiarities attained through heredity and environment is more difficult, as it must be made from the product of the plants, often requiring special and expert training on the part of the investigator. Such determinations have apparently not received the attention they deserve, largely from the prevalent opinion that these matters lie beyond the control or check afforded by the tests of impartial investigators, an opinion no doubt strengthened by the so-called contract often printed on seed-packets to the effect that the seller assumes no responsibility for the contents of the packet. The seed dealer certainly cannot be held responsible for failures that may be fairly associated with conditions of weather, soil, or method of growing; but the disclaimer cannot shield him if he is negligent or remiss, or if he fails to exercise reasonable caution in the care and selection of his stock. Undoubtedly the quality of seeds is improving, as seed-control laws become more exact, as the good practice of plant-breeding becomes better understood, and with the increasing care on the part of seedsmen.

Testing for impurities.

Testing samples to determine the foreign matter (as sand, stones, sticks, chaff, empty seeds), or the presence of seeds of other species is performed by carefully examining small lots under a lens. The operator should have at hand for comparison reliable samples of the seeds of weeds and other plants likely to occur in any sample.

In the vegetable-garden seeds there need be little fear that many weeds will be introduced. Such seeds are sold in small quantities and they are most carefully cleaned. Even if weeds were to be introduced, the thinning and tillage of a vegetable-garden would eradicate them. The greatest risk in the buying of seeds is the chance that they may not be true to name or that, if true to name, the particular strain may not be the best. There are differences within varieties which may make all the difference between profit and loss. If the grower wants to be very sure of his product, it is not enough that he buys seeds of Early Snowball cauliflower: he should know what kind of Snowball he is buying. There is no way of testing the seed except to raise the crop. One must rely on his seedsman. This he can do with safety if he chooses a reliable seedsman and if he is willing to pay a good price for his seeds. The cheapest seeds may be the dearest.

Testing for viability.

The testing of seeds for viability, or for the ability to grow, is preferably made in the soil under uniform conditions, for then they can be carried completely through the process of germination rather than merely through the sprouting stage. The best place for the test is in a greenhouse, but the living-room of a dwelling house may answer very well. Use a "flat" (Fig. 230) or other shallow box or earthenware pan. As a rule, the best results are to be obtained by planting in the soil in conditions as nearly as possible approaching the normal requirements of the particular species or variety. A light loose loam with a good mixture of sand is the best soil for this purpose. A good method is to place two or three inches of loam in a flat, wetting thoroughly without puddling it; then cover the soil with an inch or less of sterilized (baked) sand, in which to sow the seeds. The loam keeps the sand supplied with moisture.

The inexperienced operator usually applies too much

water. Gardeners are well aware that very conflicting results may be secured from the same lot of seed by different degrees of watering. The same remark applies to variations in temperature. Celery, for example, gives very poor tests in widely fluctuating temperatures; it is also injured by being kept at a uniformly high temperature, whereas melons and beans give the best tests in a high temperature.

The seeds should be sown carefully at uniform depths and at equal distances apart. To gauge the depth, nail a cleat of the required thickness on a thin block and press this cleat or tongue into the soil to its full extent: the furrow is then of uniform depth. The seedlings should be allowed to remain until large enough to show whether they are likely to make strong or weak plants. Not every seed that germinates is worth the planting.

If one desires to know what percentage of any sample of seeds still retains life, he should resort to a sprouting test. This test is made in an apparatus in which all agencies are under perfect control, and the seeds are counted and discarded as soon as they have sprouted. There are various patterns of germinating apparatus. An incubator may be made to answer the requirements. Samples of seeds which give the highest sprouting tests are not necessarily the most reliable, for it is probable that the percentage of vegetation, or subsequent growth, does not always bear a direct ratio to percentage of latent vitality.

Sprouting tests may be made in dinner-plates, on blotting-paper. The paper is kept moist, the seeds are placed on it, and another plate is inverted over it to hold the moisture intact.

Perhaps a better device is the sawdust box. Two or three inches of clean sawdust that has been soaked with warm water is placed in a box. In the smoothed packed sawdust is spread a stout wet cloth, on which the seeds may be placed or scattered. Cover with another warm wet cloth, over which place a thick cloth sawdust pad, well pressed down. Keep the box at a living-room temperature. When the time has come for examination (six to nine days for corn, less for radishes and some other things, more for carrots, parsnips, and celery), the pad and upper cloth are removed and the seeds exposed. Determine the percentage of seed that has germinated, and what proportion is most vigorous and apparently strong enough to make good plants. If just one hundred seeds were placed on the cloth, the calculation will be easier. Sometimes the under cloth is ruled off into squares, by pencil, and the seeds from each ear or fruit placed together. Any ear showing a poor or weak kernel should be discarded for seed.

The "rag-doll" tester is now popular. It is merely a canton flannel roll of seeds. A strip of the cloth about 6 inches wide and 30 inches long is laid on the table and the seeds are spread on it. It is then rolled up and tied loosely, and placed in a pail of lukewarm water for about 12 hours. The water is then poured off, and the doll is kept in the covered moist pail until the seeds sprout.

Percentages and longevity.

Seeds should hardly be expected to give 100 per cent of sprouting. Some species are habitually lower than others. Perhaps 85 to 90 per cent may be considered a good expectation, although it runs in honest samples from 75 (or even less) to 95 per cent. In the case of beet and sea-kale, fruits, not seeds, are sown, and each fruit contains one or more seeds: therefore the figures are often above 100 per cent. In some years all seeds are much better than in others. In many cases the percentages of germination are increased by cleaning the sample, thereby eliminating the weak and light seeds. Varieties of the same species may differ in germinating qualities.

The longevity of seeds is determined (*a*) by the species; (*b*) by the season in which they are grown; (*c*) by the way in which they are grown and harvested; (*d*) by the conditions in which they are kept or stored. The umbelliferous seeds (parsnip, celery, carrot) are usually good for only one or two or three years, whereas the cucurbits (pumpkin and squash, melon, cucumber, watermelon), may hold five to ten years. The gardener soon learns by experience what seeds he may safely hold over. In the botanical accounts of the various species, in this book, the usual expectations of longevity are stated.

2. THE GROWING OF SEEDS

The growing of seeds has come to be a business by itself, requiring expert knowledge of soils and climate, and of methods of handling every kind of crop. The demand for seeds is large. Competition is great. The quality constantly improves. Plant-breeding has come to be an important factor. Under the present-day conditions, it is only the exception that a man can afford to grow his own seeds. With the development of intensive market-gardening interest, seed-buyers are becoming more cautious and discriminating. Seeds are now wanted for their inherent quality rather than merely to represent a varietal name.

The breeding of seeds.

This means that plants, as well as animals, must be "bred"; that is, they should have a known history, coming from parents of accepted quality and attributes. The breeding of seeds has come to be an extensive business. The discriminating farmer makes sure that his oats represent a carefully chosen parentage and that the "seed" has been produced under accepted safeguards. He is willing to pay the extra cost of producing such seeds. Crops of the grains, cotton and vegetables, as well as florists' flowers, have been much improved in quality and yield by the work of plant-breeders, and greater gains are yet to come. The gardener may not desire to enter the larger fields of plant-breeding, but he should at least be aware of the importance of the subject and he should be able to practice intelligent selection.

The usual means at the disposal of the grower is to "select" his seed plants. He must understand that the quality is usually an attribute of the plant as a whole and not of a single fruit or branch; he therefore looks for plants that bear the produce he wants and does not take seeds from miscellaneous good fruits or pods he finds in the market.

Finding a plant in his field that has strong and useful variation, he marks it and saves seed from it. The plant may be a tomato; perhaps he finds two or more plants.

He saves seed from each fruit separately, recording the parentage; he raises the plants in separate rows, a row from a single fruit, or at least from a given plant; some rows show the characters persisting or even improving and other rows do not; again he selects seeds from the best plant, and repeats the operation until the desired attribute or product is reproduced with fair constancy from seed: then he uses his selected seed for the raising of his crop.

He must not suppose that the developed strain or variety is permanent. He must constantly select from the plants nearest his ideal or pattern, to keep the stock up to grade. He will do well to have a breeding-plot in which the seed-stock may be grown, if he is raising a specialty of a particular kind.

It is seen, therefore, that it is a particular business to grow good seeds. The seed-grower must have an idea or type and work to it. His plantations must be "rogued." That is, all plants that do not meet the breeder's type are pulled up and discarded, and the true or typical stock is left to produce the seed. The truer and higher the man's idea, the better his stock should be. It requires experience to enable one to make for himself a true and practical ideal of any variety of plant. He must know what the market wants. He must know what his customers want. He must know what will be good and useful under the greatest number of conditions. He must know what will be likely to be most stable and invariable. The type once apprehended, the seed-breeder must thereafter discard every plant that does not closely approach it; his stock must be uniform. As soon as the "roguing" or selection is neglected, or when new notions are introduced, the varietal characteristics tend to disappear or to change.

Seed-growing.

Experience has demonstrated that certain soils and climates produce the best seeds of certain species. No longer are all kinds of seeds grown indiscriminately in one place or merely where they will mature. The price of labor is an important factor. Seeds that require much care and trouble in the growing are raised, if possible, where labor is most abundant and cheap. It is no accident that radish seeds are grown in France, and lima beans in California.

Only when a man is making a specialty of some vegetable, and lives in the place in which the seeds can be produced most advantageously, or is under the necessity of developing a kind or strain of his own, can he afford to grow his seeds; and even then it is a question whether it would not be better and cheaper to delegate the business. The man who desires to secure the very best results in the growing of some specialty should know where his seeds are grown, particularly if his business success depends on the crop in question.

When one is engaged in a high-class vegetable-growing business one should not buy seed indiscriminately in the general market. There are particular strains of leading varieties of vegetables which are better for certain markets and conditions. These strains are likely to be most useful in the geographical area in which they are bred. Seeds of these strains are often sold as "market-gardeners' private stock." Under general conditions and in other geographical regions, these private stocks may be of no advantage, but in special places and for particular purposes they may make all the difference between success and failure; and yet the differences in the resulting crop might be of such a character that they could not be definitely described in a seed catalogue or in an experiment station bulletin.

The yield of seeds (in lbs.) that may be expected from an acre, under good conditions, is approximately as follows:

	When crop is as near maximum as 20 bus. of wheat would be, or average of "good crop"	A maximum crop corresponding to 50 bushels wheat	Yield seedsmen would figure on in making contracts for large quantities
Bean	600	1,500	500
Cabbage	250 (two years)	800	200
Cucumber	150	700	100
Muskmelon . . .	125	600	100
Pea	900	2,500	800
Squash, Winter . .	100	400	100
Squash, Summer .	100	700	100
Sweet corn . . .	1,000 to 2,500 (according to var.)	2,500 to 4,000	800 to 2,000
Tomato	100	400	100
Watermelon . . .	150	1,000	100

How the seeds are grown is told briefly in the chapters that deal with these different vegetables; but this book does not purport to discuss seed-growing, and the person who desires detailed information should go to special literature.

3. THE SOWING OF SEEDS

The gardener should buy his seeds in bulk and in advance, if possible, if he is growing large areas and for a critical market. He can then demand the best. He will also secure a cheaper rate. It may even be well to engage them of the seed dealer a season in advance, to be sure that he has the kind and quantity he desires. Since seeds are poor in some seasons, it is well for him to keep at least a partial stock on hand from year to year, particularly of those kinds that retain their vitality for several years. He is then relatively independent. The gardener who grows largely for a special market of such important crops as beet, carrot, cabbage, cauliflower, cucumber, melon, lettuce, radish and tomato, will do well to purchase double the quantity of seed he requires for the one season, in order that he may preserve stock of the strains that prove to be particularly desirable. The capital thus locked up in seeds is small, as compared with the risk of being unable to secure a desirable strain.

Congenial temperature and a continuous supply of moisture are the two requisites of germination to be provided by the gardener. He provides these agents by placing the seeds in a loose, moist, granular medium, as mealy and friable soil. If this soil lies on other soil, the moisture is drawn up by capillary attraction and as it passes off it moistens the seeds and promotes germination. If the soil is very loose, open or lumpy, the capillary attraction is broken and the moisture does not rise to the seeds. Or, if it does rise, the seeds are not in intimate contact with the particles of earth and do not receive much of the soil moisture; moreover, the air held in the large interstices tends to dry out the seed. To a large extent, a continuous and uniform supply of moisture is a regulator of temperature. It is therefore apparent why a finely divided and compact soil is the proper medium in which to sow seeds.

Whenever the soil is likely to become drier rather than moister, as at the usual germinating season, it is important to firm the earth over the seeds. In large field operations, as in the sowing of the cereal grains, the roller is ordinarily used. Under market-gardening conditions, the soil is usually compacted by a roller which is a part of the seed-drill and which follows just behind the delivery spout. When seeds are sown from the hand, the soil is compacted with a hoe or by walking over the row. Since this compacting of the surface establishes capillary connection with the under soil, thereby drawing up the water and passing it into the atmosphere, it is important that this condition be allowed to remain only until the seeds have germinated and are able to shift for themselves. Therefore, as soon as possible restore the surface mulch by rake or smoothing-harrow (pages 394–5).

Particulars in seed-sowing.

Seeds that are planted very deep, as peas, may have the earth compacted about them, and the surface layer may be loosened immediately thereafter, thereby preventing, to some extent, the escape of the soil moisture. The space between the rows should be kept well tilled, even before the seeds germinate, thereby saving the moisture in that area.

Seeds that germinate very slowly, as parsnips and celery, should be sown thick in order that the combined forces of the germinating plantlets may break the crust on the soil. This caution is always necessary on soils that tend to bake, whatever the kind of seed. It may be well to sow a few strong and quick-germinating seeds with those of slow-germinating species to break the soil, and also to mark the row so that tillage may be begun before the main-crop seeds are up and before the weeds have taken possession of the land. Seeds of radish, cabbage or turnip may be sown in the row with celery, parsnips, carrots and the like. In some cases, a crop of radish may be obtained in this way before the main crop occupies the land, but this is only an accidental gain and there is danger that the major crop may be injured.

The cost of seed is ordinarily a small matter in comparison with the expense of the season's labor and the value of the crop. Therefore, seeds should be sown freely to avoid the risk of failure. Even if five or ten times more seeds are sown than plants are required, the extra expenditure may be justified. Another great value of thick seeding is that it allows of more extensive thinning of the plants; and thinning is a process of selection, and the best are allowed to remain. It is evident that the chances of securing the best are greater when the gardener leaves one plant out of ten rather than one plant out of three. The selection in the seed-bed or the seed-row is undoubtedly one of the means by which cultivated plants have been so greatly ameliorated or improved.

Most of the recommendations of writers on the quantity of seed for a given length of row are in excess of the number of plants actually required. It may be that some of these recommendations are higher than even the risks will warrant; but it is much safer to sow even the most excessive amounts than to sow just as many seeds as are theoretically needed on a basis of the number of mature plants to the row or the acre.

Seeds ordinarily germinate best in freshly turned or freshly worked soil. This is because there is more moisture in the fresh soil than in that which has been exposed to the weather. We shall find in the succeeding chapter that gardeners expect to secure better success in transplanting when they can set plants on freshly plowed land.

The depth at which seeds should be sown depends (1) on the soil, as to whether it is moist or dry, well tilled or poorly tilled; (2) on the species and size of the seed; (3) on the season. The finer and moister the soil, the shallower the sowing may be. The larger the seeds, the deeper they may be sown. Seeds may be sown shallower in spring than in summer, for at the latter season the surface soil is dry. An old gardener's rule is to cover the seeds to a depth equal to twice their diameter. This applies well to greenhouse conditions, in which the soil is finely prepared and kept continuously moist but in the open ground, the seeds are usually planted deeper than this.

Horticultural plants are ordinarily divided into three classes in respect to hardiness: (1) hardy, or those able to withstand the vicissitudes of climate in a given place; (2) half-hardy, or able to withstand light frosts or other uncongenial conditions; (3) tender, or wholly unable to withstand frost. Seeds of the hardy plants may be sown in spring as early as the land can be made fit, or even in autumn. Examples of such seeds are sweet pea, onion, leek. In the Northern States, however, few seeds are sown in the autumn; but the land is often prepared in autumn, and the seeds are sown as soon as the earth is dry enough in spring. The seeds of half-hardy plants, as beets and lettuce, may be sown two or three weeks before settled weather is expected to come—that is, when it is still expected that there will be hard frosts. Tender seeds, as beans, tomatoes, eggplants, cucumbers, melons, are sown only after last frost has passed and when the ground is thoroughly settled and warm.

The seed-bed.

Of plants normally transplanted, it is better to start the seeds in a seed-bed. These beds may be in the forcing-house, hotbed or coldframe; or, if it is not desired to force the plants beyond the normal season, it may be made in the open. There are three chief advantages in sowing in a seed-bed, rather than where the plants are to grow: (1) it insures better germination, since the conditions are more uniform and congenial; (2) it saves time and labor; (3) it enables the gardener to guard against insects, fungi and accidents, since plants in a compact body can be sprayed, fumigated, covered or otherwise treated to advantage. In forcing-houses and frames, it is now a common practice to start seeds in flats or boxes.

The seed-bed should be a small area on land that is in the best of tilth. It should be near the buildings and the water supply. If the season is hot and dry, it may be well to shade the bed until the seedlings appear. The best shading ordinarily is a lath screen laid on a frame standing two to three feet above the ground. Such a screen gives a partial shade and also allows of a free circulation of air; and the screens may be removed and the bed weeded at any time. A covering of brush is sometimes used, but it is less handy than the lath screen; if it is laid directly on ground, the bed cannot be weeded and it is likely to become foul. Sometimes boards, matting or other dense covers are laid directly on the bed. This may do very well for a few days, until the seeds begin to break the ground, but thereafter the covering should be removed, else the young seedlings will be injured. The seedlings should be given sufficient head-room and light and air to enable them to develop to their normal condition. If the seed-bed is kept too wet and the seedlings are too soft, the damping-off fungi are likely to work havoc. Sometimes the seed-bed is made underneath a tree, but this is rarely advisable, since the earth usually requires too much watering and the shade may be too dense.

If it is desired to secure a quick germination of seeds in a summer seed-bed, it is well to prepare the bed the fall before, or at least very early in the spring, and to keep it covered with several inches or a foot of well-rotted manure until needed. When the bed is needed, the manure is removed; the soil is then full of moisture and the seeds germinate quickly. The fertility leached from the manure also enables the plantlets to secure an early foothold. This method is practiced in some of the market-gardening centers, particularly those in which late cabbages and cauliflower are grown.

When sowing in the open field, the use of a seed-drill should be encouraged, not only because it saves time and labor, but also because it enforces good preparation of the land. A drill cannot be worked in hard, dense and lumpy soil. Seed-drills, wheel-hoes and smoothing-harrows make better gardeners. If a seed-drill is not used, the seed-furrows for ordinary use may be made by drawing the end of a hoe handle or rake forcibly through the soil. A garden line should be used to keep the rows straight.

When sowing in the open, wait until ground and season are ready. Rarely is anything gained by sowing before this time. The seeds rot, or the seedlings are weak. The soil must be fitted after the plants are up. Have everything ready, then make the plants grow.

Sterilizing the soil.

If the soil is infected with damping-off and other fungi, with nematodes (eel-worms) and insects, it may be sterilized. This is a common practice in greenhouses, and it should be oftener undertaken in hotbeds and outdoor seed-beds. If the soil is exposed to hard freezing, as in the open in the Central and Northern States, the nematodes are dispatched. Far South, however, they are very troublesome, as also in greenhouses and forcing-houses. One must not run the risk of infecting the garden, even for a single season, with soil or plants from the greenhouse.

The usual process of sterilization of soil is heating it with steam until a potato buried in it is thoroughly cooked. The outdoor bed is heated by inverting over it a tight metal or board box, four to five inches deep (and the size of the bed or of a part of it), banking the sides well to prevent leaking, and then turning live steam under pressure into the box. The steam is provided by a portable boiler or traction engine. It is conveyed to the sterilizing box through an iron nipple inserted in the side or end.

CHAPTER XVIII

OTHER MANAGEMENT OF THE VEGETABLE-GARDEN

Tillage is the most important item in the subsequent care of the vegetable-garden. If the land has been well fitted before the crop is put on it, tillage need be employed only for the purpose of maintaining the surface mulch. This tillage may be light, rapid and easy.

The rationale of the garden system is this: In the cool and ambitious days of spring, put the effort and the muscle into the land; work it into condition. In the long and hot days of summer, keep it in condition.

1. DOUBLE-CROPPING OR INTER-CROPPING

We must now consider the crop-scheme. To do this we must have two definitions, to clarify the situation. In the first edition of this book, 1901, they were adopted in order to clarify the subject.

Double-cropping which is the raising of more than one crop on the land in a season, is of two species: (1) succession-cropping, or the growing of one crop after another on the same land; (2) companion-cropping, or the growing of two or more crops together.

Succession-cropping.

Succession-cropping is a kind of short rotation. These are the considerations: (1) each crop in the succession should be able to mature in less time than the whole season; (2) the tillage demanded by the first crop in the series should be such as will leave the land in proper condition for the succeeding crop; (3) the crops should be so much unlike each other that they will not tend to deplete the soil by demanding similar elements, and will not carry diseases and insects from one crop to another.

It is usually preferable to grow crops of different botanical families, for by this means the fertility of the soil is not so likely to be impaired, and diseases and insects are starved in the rotation. It is well to follow root-crops with fibrous-rooted surface-feeding crops. In some cases the succession may extend over parts of two years, as when strawberries are followed by late potatoes or cabbages. In this case the strawberries are set the year before the succession-crop is grown. A crop of rhubarb or asparagus may be followed, when the crop is finally turned under, by a short-season crop, thereby allowing the cutting of the asparagus or rhubarb in its last season. It is usually best to follow a perennial crop with an annual.

When the succession-cropping extends into general farm operations, one or two entire seasons may be covered by each crop in the series. In this case we have a true rotation of crops, as that term is understood by most agricultural writers. The value of rotation in the vegetable-garden, by means of which lands are rested in clover or other sod crops, has already been discussed (Chapter XV).

Following are examples of succession-crops:

Strawberries, followed by main-crop cabbage or late potatoes.
Peas, followed by cabbage, beans, tomatoes or celery.
Onions, beans, early beets, summer squash by kale, turnip, kohlrabi, winter radish.
Spring spinach by beans and tomatoes.
Radish and bunch onions by early cabbage or celery.

Lettuce by beans and tomatoes.
Early carrots by autumn spinach, kale, turnip, winter radish.
Early potatoes, followed by fall cauliflower or turnips.
Cucumber by spinach, kale, turnip, winter radish.
Early sugar corn by second crop of same or autumn spinach, beans, tomatoes, celery.
Early cabbage, followed by late beans (for canning), or by horse-radish.
Dandelions by potatoes.
Fall-sown spinach by strawberries.
Kale, followed by potatoes or other main-season crop.

These crops can be worked into succession-cropping schemes:

Early, or incidental crop

Beans, snap,	**Mustard,**
Beet,	**Onion (from bulbs),**
Cabbage,	**Parsley,**
Carrot,	**Pea,**
Cauliflower,	**Potato,**
Cress,	**Radish,**
Kohlrabi,	**Spinach,**
Lettuce,	**Turnip.**

Late, or main crop

Beans, shell and lima,	**Muskmelon,**
Beet (mostly a farm crop),	**Okra,**
Brussels sprouts,	**Onion (from seed),**
Cabbage,	**Parsnip,**
Carrot (farm crop),	**Pepper,**
Cauliflower,	**Potato,**
Celery,	**Pumpkin,**
Corn,	**Salsify,**
Cucumber,	**Spinach (fall crop),**
Eggplant,	**Squash,**
Horse-radish,	**Sweet potato,**
Kale (fall and winter crop),	**Tomato,**
Kohlrabi (fall crop),	**Turnip and rutabaga,**
Leek,	**Watermelon.**

Companion-cropping.

In companion-cropping, or the growing of two kinds of plants on the land simultaneously, the following items are to be considered: (1) the crops should be such as will mature at widely different seasons; (2) one crop should be of distinctly less importance than the other, or be a "catch crop"; (3) the crops should be such as will profit by the

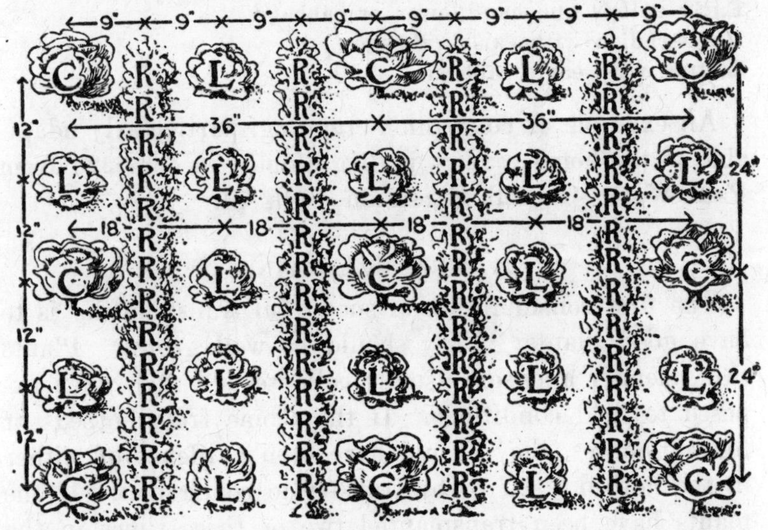

224. Companion-cropping, in which three crops are growing at the same time. The main crop is cabbage (C). The lettuce (L) and radishes (R) mature before the cabbages require all the land.

same methods of tillage and fertilizing; (4) so far as possible, they should be of different botanical families or kinds, that they may not tend to leave the soil unbalanced or to breed the same kinds of insects and fungi.

It will be seen that there is a main crop and a secondary crop. Ordinarily, the main crop occupies the middle part, or middle and later part of the season. The secondary crop matures early, leaving the ground free for the other. In some cases, the same species is grown for both crops, as when late celery is planted between the rows of early celery.

Following are examples of companion-crops:

Radishes with beets or carrots. The radishes can be sold before the beets need the room.
Corn with squashes, citron, pumpkin or beans in hills.
Early onions and cauliflower or cabbage.
Horse-radish with early cabbage.
Lettuce with early cabbage.

An example of companion-cropping, particularly adaptable to the home garden, is diagrammed in Fig. 224 from Paul Work, Cornell Extension Bull. 14.

2. TRANSPLANTING

The first consideration in successful transplanting is to have good plants. They should be well grown. Plants thin, slender and soft usually collapse or suffer when exposed to field conditions. If they come from hotbeds or forcing-houses, they should have been hardened-off either in the hotbed itself or by transfer to coldframes. If the plants have been transplanted two or three times in the seed-bed, they suffer less when put in the open field.

The second consideration is to have the land in prime condition. It should be in fine tilth and thoroughly and deeply worked. Plants live better when transplanted into newly turned land. Such land is moist. The plants quickly secure foothold.

Transplanting is more successful and is employed to a larger extent in the humid climates east of the Great Lakes than in the West. In fact, in the more arid parts of the country it is usually discouraged, and it is recommended that seeds be sown where the plants are to stand.

The perfect time to transplant is just before a rain. Just after a rain is also good, particularly if the weather comes off cloudy. Cool and cloudy days should be chosen if possible. When it is necessary to transplant in hot and dry weather, the late afternoon or evening should be chosen, that the plants may have time to straighten up in the night. When, however, the land is thoroughly prepared and the plants are well grown and not too large, there will be little difficulty in transplanting throughout the day.

If the season is very dry, the plants may be watered. It is common practice to have a boy follow with a pail and put a dipperful of water about each plant. Or, in larger operations, a tank on wheels is drawn through the fields. After the water soaks away, the dry loose earth should be drawn about the plant to provide a surface mulch and to prevent the soil from baking. In small gardens, it is practicable to shade the plants for a day or two by setting a shingle or slate on the south side of them, letting it slant over the plant.

When transplanting, the plants must be kept away from the sun when out of the ground, and they should also be kept wet. It is nearly as important to wet the tops as the roots. The roots are wet to prevent them from dying. The tops are wet to prevent transpiration or evaporation of moisture. Puddling, or dipping the roots in mud, is sometimes advised as a protection, but it is less useful with small plants than with trees, because the fine roots are matted together by the operation. When transplanting by hand, it is customary to have a boy carry the plants in a covered basket or box, and to drop them just ahead of the planters. One boy ordinarily drops for two rows of planters. The boy should not drop faster than the plants are required by the workmen.

225. Dibbers, flat and cylindrical.

Set the plants deep. Gardeners usually prefer to set them to the seed-leaf, even though they were an inch or two higher in the original seed-bed. This deep planting holds the plants in position and places the roots in the moist and cool earth. Press the earth firmly about the roots and the crown.

The best tool for opening the land is a dibber (Fig. 225), which makes a hole without removing the earth. In the working hand hold the dibber; in the other hand hold the plant; the plant is lowered into the hole made by the dibber, and both hands are then pressed tightly about the plant as the earth is closed against it. Sometimes the dibber is thrust alongside the plant and the hole filled by pressing the earth against it (Fig. 226).

226. The dibber and how to use it.

Another dibber-like tool is the "scandigie," shown in Fig. 227, adapted from Circ. 160, Calif. Exp. Sta., on lettuce, said to be "used for transplanting."

If the plants are rather large, and particularly if they have not been transplanted before, it is well to cut off a part of the foliage to hinder evaporation. One-half or one-third of the top may be twisted or cut off with good results (Fig. 228).

Of late years, transplanting machines drawn by horses have become popular for the planting of cabbages, tomatoes and other large-area crops. If the plants are well grown and of the right size, these machines work very satisfactorily. They not only expedite and lessen labor, but the plants are more likely to live than when transplanted in the ordinary way. They are supplied with a watering device. There are also various kinds of hand-transplanting devices that remove a large body of earth with the plant and drop it into a hole of similar size. These tools are useful for small areas or for amateur work, but they are not adapted to general field operations. Machines for aiding transplanting by hand have come into use, and are often very satisfactory.

227. Another form of transplanting tool.

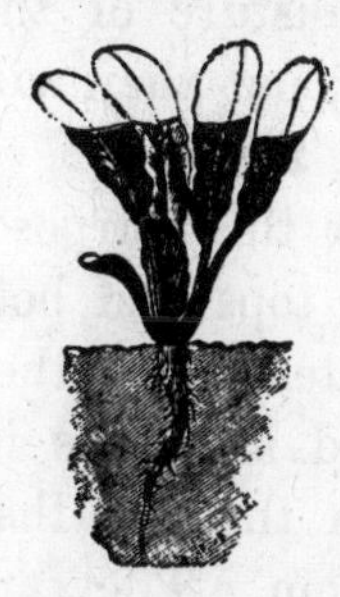

228. To show how much of the top may be removed in transplanting.

Some kinds of plants, of which melons and cucumbers are examples, do not transplant readily. It is customary to start them in boxes, pots or on the bottoms of hard sods. The plants can then be taken to the field with the earth intact, and they will not suffer in the removal. There are various kinds of transplanting boxes in the market. Some melon growers use ordinary splint pint or quart berry baskets.

Others use paper oyster buckets. A useful receptacle is shown in Fig. 229. It is a band or strip of basket-splint tacked together at the end and has neither top nor bottom. The strip is 14 inches long, and 3¾ inches wide, making a box 3¾ inches deep and about 3 inches square. The material is cut at a basket-factory. These forms are nested in the hotbed or coldframe, filled with earth, and four or five seeds planted in each. They are readily moved by running a spade, flat trowel or a shingle under them. A box makes a hill of plants. Note the discussion on page 357.

229. A "form" in which to start melons.

One of the ways to handle cucumbers and melons is to plant on sods, which are laid bottom up in the hotbed. They are cut into squares of about four inches. A little fine earth is sifted over and between them, in which the seeds are planted. With the heat and moisture of the bed these sods decay and the plants thrive; but they will hold their shape for a month or more (Fig. 230).

Old tin fruit-cans are sometimes used for this purpose. The cans are thrown into a fire, when the tops and bottoms melt off, and the sides are then fastened together with a tack or a bit of wire and are used as forms in which to grow plants. One difficulty with them is that they are too large and take up too much room. They are relatively too deep.

If the grower has a greenhouse equipment, he may use 2-inch or 3-inch pots (Fig. 230); but unless he has the pots on hand for other uses, it might not pay to buy for this particular purpose. They are easy to handle and to store, and plants thrive in them admirably; their uniformity makes them very handy.

It is customary to handle plants in flats (Fig. 230). These are shallow boxes about 3 inches deep, and of any convenient size. A box 15 x 20, or 18 x 24 inches is easily handled. The boxes may be made to order; but many gardeners make them from soap boxes, by sawing each box into several flats or sections and adding bottoms. Such a box will hold 100 plants if they are not transplanted, or one-third or one-half that number of transplanted plants. From flats a quick man can transplant 5,000 to 6,000 plants in a day if the soil is light and in good condition. With a horse transplanting machine several times this number can be set. Ten acres of cabbage plants sometimes may be set in a day by means of a horse machine. From 20,000 to 40,000 plants have been set in one day.

230. Melon plants on a sod; gardener's flat; plants in 2-inch and 3½-inch pots.

3. WEEDS

Weeds are mere incidents in good farming. They are the constants in poor farming. This is not because the good farmer spends more time killing weeds, but because he tills better and manages his land more skilfully. It is in neglected areas that weeds are most prevalent,—along the roadside, in the run-out meadow or pasture, in the barnyard or front yard, in the poorly tilled vegetable-garden. Many farmers seem to think that good farming consists in killing weeds and bugs; but the best farming consists in not having them. Of course the farmer cannot expect ever to be rid of these neighbors, but he should think more of prevention than of eradication.

A weed is a plant that is not wanted. Horse-radish may be a weed in a potato field, and potatoes may be weeds in a horse-radish field. Potatoes are weeds in potato fields when potatoes are planted too thick.

There is no royal road to weedless farming. Following are some of the means of keeping weeds in check:

1. Practice rotation; keep ahead of the weeds. Certain weeds follow certain crops; when these weeds became serious, change the crop.

2. Change the method of tillage. If a weed persists, try deeper or shallower plowing, or a different kind of harrow or cultivator, or till at different times and seasons.

3. Harrow the land frequently when it is fallow or is waiting for a crop. Harrow it, if possible, after seeding and before the plants are high enough to be broken by the implement. Potatoes, corn and other things can be harrowed after they are several inches high; and sometimes the land may be harrowed before the plants are up

4. Practice frequent tillage with light surface-working tools throughout the season. This is hard on weeds and does the remaining plants (the crop) good.

5. Pull or hoe out stray weeds that escape the wheel tools.

6. Clean the land as soon as the crop is harvested: and if the land lies open in the fall, till it occasionally. Many persons kept their premises scrupulously clean in the early season but let them run wild in the fall, and thus is the land seeded for the following year.

7. Use clean seed, particularly of crops sown broadcast, and which, therefore, do not admit of tillage.

8. Do not let the weeds go to seed on the manure piles, in the fence corners, and along the highway.

9. Avoid coarse and raw stable manure, particularly if it is suspected of harboring bad company. Commercial fertilizers may be used for a time on foul land.

10. Sheep and pigs sometimes can be employed to clean the weeds from foul and fallow land. Land infested with girasoles (Jerusalem artichokes) are readily cleaned if hogs are turned in.

11. Induce your neighbor to keep his land as clean as you keep yours.

Rank pigweeds and their ilk are a compliment to a man's soil. Land that will not grow weeds will not grow crops,—for crops are only those particular kinds of weeds that a man wants to raise. Weeds have taught us the lesson of good tillage. There is no indication that they intend to remit their efforts in our behalf.

4. INSECTS AND FUNGI

The vegetable-gardener must expect the visits of energetic bugs and furtive fungi. Many of these squatters are beyond the direct control of the cultivator. The gardener must circumvent them rather than combat them. He must avoid them by means of strategy.

Insects that feed openly on the tops of plants are usually amenable to direct treatment with poisons or other sprays. Of this class are potato-bugs and plant-lice. Those troubles that appear in the inner parts of plants or in their roots are not open to direct treatment, and in such cases the general management of the place must be relied on to keep the enemies in check. Insects and diseases are incidental or secondary facts in every garden plantation. The primary thing is to make the plants grow; the secondary thing is to keep the bugs off. One's attitude toward these invaders must be the same as that toward weeds: one must rely first on management.

In these popular writings, many other creatures are naturally included with insects, as millipedes, slugs, nematodes, sowbugs, and mites; but the general strategy and treatment are the same.

Following are some of the means by which the vegetable-gardener may hope to lessen or avert the losses from insects and diseases:

1. By means of rotation in crops and in methods of tillage. The shorter the rotation, the less is the liability to serious insect attacks. It is rare that insects and diseases appear suddenly in great numbers. They increase year by year, and in a favorable season prove very destructive. If the kinds of crops have been various, the probability is that they will not have gained a serious foothold, and that they will be held in comparative subjection. It is essential that the crops of a rotation be of such different kinds that the same kinds of insects or fungi will not thrive on them.

2. If the land becomes seriously infested with any one

pest, it is best in general to discontinue, for two or three years, the growing of the crop on which they live. This ordinarily is cheaper and quicker than to endeavor to destroy the pest by direct means. This is well illustrated in the case of the clubroot of cabbage and cauliflower. The disease may be lessened somewhat by thoroughly dressing the land with lime; but it is usually cheaper, and always more effective, to cease the growing of cabbage, cauliflower and turnips for a time, and to grow other kinds of crops on the land.

3. Make every effort to secure strong, stocky, continuous-growing plants. Even if they are attacked, they have a better chance of coming through alive. Weak and soft plants are poor for any purpose, but they are particularly unsatisfactory when they must withstand the attacks of insects and fungi.

4. Destroy seriously affected plants, particularly those attacked by fungi. If the vines are thrown on the manure pile, the probability is that the disease will be distributed the next year in the manure. If the manure is thoroughly rotted and composted, much of the danger will be averted; but even in that case it is wise not to take the risk with such serious diseases as clubroot, potato blight and rot, and the blight of melons, cucumbers and tomatoes. In autumn, all diseased plants and products should be collected and burned.

5. On infected seed-beds, use new or sterilized soil. Do not add to the seed-bed soil from a field in which diseased crops of the given kind have grown (page 413).

6. Insects and fungi can be killed. Nowadays, spraying is an economical means for many of the pests. Never-

theless, the old method of hand-picking is not gone by, and the gardener must not hesitate to resort to it on occasion. The gardener should know what insects and diseases are likely to appear on any crop and then be prepared to fight them. The time to make this preparation is before the crops are planted. In the winter, he should secure his pumps and nozzles, buy materials for the various mixtures, and inform himself as to what difficulties are likely to confront him. He is then forehanded and knows immediately what to do when the trouble comes. Every gardener should buy a good book on insects and another on fungous diseases, and then keep up to date by reading the agricultural papers and the experiment-station bulletins.

An essential point in the application of any spray or other material is timeliness. The minute the trouble appears, the antidote should be applied. The pest may be dispatched more readily at this time, and also with less expense of material and effort; and the plants will not have suffered seriously. Another important item is thoroughness. A bug will not go to get the poison: the poison must be put where the bug is. The only safe way is to put the poison on every part of the plant. One thorough spraying, which covers the plant, is worth more than a half dozen efforts when the operator merely sprinkles the tops of the leaves. Be sure that the spray is of the right kind and well made: then do not be afraid to use it.

The substances or materials employed for the destruction of the insect pests are insecticides and those for the destruction and control of fungous diseases are fungicides. The prevailing kinds may be listed, together with statements of the ingredients; how to compound and use them is the subject of many bulletins and popular articles, to which the reader is referred (also previous pages).

Insecticides

Arsenate of lead.
    4–10 lbs. to 100 gals. water.

Paris green.
    Used in place of arsenate of lead on potatoes, usually combined with bordeaux mixture.

Hellebore.
    4 oz. to 2 or 3 gals. water.
    1 lb. to 5 lbs. flour or slaked lime.

Kerosene emulsion.
    ½ lb. soap, 1 gal. water, 2 gals. kerosene; dilute with 5–7 parts water for use on dormant trees, and with 10–15 parts for plant-lice on foliage.

Carbolic acid emulsion.
    1 lb. soap, 1 gal. water, 1 pint crude carbolic acid; dilute with 30 parts water for use against root-maggots.

Tobacco.
    Nicotine is used in many forms and preparations.

Whale-oil soap.
    1 lb. soap to 5–10 gals. water.

Miscible oils.
    Preparations are on the market; for use specially against scale insects on woody plants.

Lime-sulfur.
    Several formulæ; it can be purchased in the prepared dry state; used specially against scale insects on trees.

Fumigation.
    Greenhouses and hotbeds may be fumigated with tobacco preparations, or with the deadly hydrocyanic acid gas.

Fungicides

Bordeaux mixture.
    A standard fungicide, the use of which is now well understood; it may be purchased in prepared dry form.

There are several formulæ; one of the best is 4 lbs. copper sulfate, 4 lbs. lime, 50 gals. water.

Lime-sulfur.

Both fungicide and insecticide.

Corrosive sublimate.

1 oz. to 7½ gals. water; for treating seed potatoes for scab.

Formaldehyde.

1 pint in 30 gals. water; for treating seed potatoes for scab.

This book is not a treatise on insect pests and diseases of vegetable-garden plants; yet condensed advice on the procedure in combating them is given with each of the vegetables in the regular sequence.

But some insects and similar animals are general marauders. They attack several or many kinds of plants, and therefore cannot be discussed under the particular crops without too much repetition. They are discussed here (by Crosby and Leonard, for this publication).

Cutworms and army-worms.

Cutworms are smooth, nearly naked caterpillars, 1 to 2 inches long, usually dull colored and indistinctly marked with spots and stripes. Many species have the habit of cutting off young plants at the surface of the ground or just above it. They feed mostly at night and in the day remain hidden away under stones or rubbish or in the ground. The adults are dull-colored rather heavy-bodied moths. More than a score of species has been recorded as pests of vegetables. Under certain circumstances almost any cutworm may become so abundant that it is forced to migrate for food and thus assume the army-worm habit.

The name army-worm is restricted to four or five species in which this habit is pronounced.

Control of cutworms.

The means employed for the control of cutworms varies according to the crop, the conditions under which it is grown and on the habits of the species causing the injury.

In small vegetable-gardens and greenhouses hand-picking may be practiced to advantage. Careful watch of the plants should be kept and whenever injury is noticed the soil around the base of the plants should be searched and the cutworms destroyed. Shingles or small boards laid about the beds will form attractive hiding places for the worms during the day; here they may be easily found and destroyed. When such plants as tomatoes are transplanted, they may be protected by using cardboard or tin cylinders sunk a short distance in the soil. Tin-cans with the top and bottom removed are convenient for this purpose. Greenhouses often become infested by cutworms in the rotted sod used in the beds. This may be prevented by sterilizing the soil with steam before using.

Probably the most practical, cheap and convenient method of cutworm control is the use of poisoned baits. These may be employed equally as well in the home garden, greenhouse or in the field. A bait made according to the following formula is effective against the variegated cutworm and others of similar habits:

Bran	**20 pounds,**
Paris green	**1 pound,**
Molasses	**2 quarts,**
Oranges or lemons . . .	**3 fruits,**
Water	**3½ gallons (about).**

The dry bran and paris green are thoroughly mixed in a tub or similar receptacle. The juice of the oranges or lemons is squeezed into the water; the remaining pulp and peel is chopped into fine bits and added to the water. The molasses is dissolved in the water and the bran and poison wet with it, the mixture being constantly stirred so as to dampen the mash thoroughly. Only enough water should be used just to moisten the mash, but not enough to make it sloppy.

This quantity of bait will treat about three acres. The material should be scattered broadcast evenly over the infested area at nightfall. If applied in the day, it dries out and is not then attractive to the cutworms. In the garden or greenhouse a small quantity of the bait may be placed near each plant.

Control of army-worms.

To arrest a migration of army-worms, plow a furrow across their line of march with the vertical side of the furrow towards the field to be protected. At intervals dig post holes in the bottom of the furrow as traps for the caterpillars and scatter poison bait along the edge of the field to kill those that succeed in crossing the furrow.

Wireworms.

These insects are elongate hard-shelled brownish larvæ abundant in old sod land. They eat off the smaller roots, bore into tubers and destroy germinating seed. The adults are medium-sized dull-colored snapping or click beetles. The larvæ normally feed on grass roots and thrive in old sod land.

Practice a short rotation of crops for wireworms, in which the land is not left in sod for more than two or at most three years. Do not plant vegetable crops susceptible to injury in land known to be infested. Peas and buckwheat may be used as intermediate crops between sod and vegetables. In the garden, poison baits are sometimes used for killing the wireworms. Dip small bunches of clover in paris green water and place them in the field covered by pieces of boards. Sweetened cornmeal dough poisoned with paris green may be used as a bait. The bait should be distributed after the ground is fitted out before the crop is planted.

White grubs.

These are large fat white curved grubs found in land recently in sod. The parent insect is a large brown June beetle (the familiar "June-bug"). The grubs feed on the roots of grasses and thrive in old sod land. When such land is broken up and planted to vegetable crops, the grubs concentrate their feeding, often causing great damage.

For the white grub, practice a short rotation of crops in which the land is not left in sod more than two or at the most three years. Do not plant vegetable crops on land known to be infested. As an intermediate crop between sod and vegetables, buckwheat, alfalfa, clover, and other leguminous crops may be raised. Old strawberry beds are likely to be badly infested and should be treated the same as sod land.

Grasshoppers.

Many vegetable crops are liable to injury by grasshoppers. Use the poisoned bait described for cutworms.

Red-spider (*Tetranychus telarius*).

The red-spider is not an insect; it is a minute web-spinning mite, varying in color from yellowish to greenish or reddish, that infests the underside of the leaves of many plants. The mites puncture the leaves, causing small light-colored spots. When abundant the leaves become whitish, shrivel and die.

In the greenhouse, the number of red-spiders can be reduced by spraying with clear water, using a nozzle that gives a stiff spray without drenching the beds. In the open, spray with "Black Leaf 40" tobacco extract, 3/4 pint in 100 gallons water in which 5 or 6 pounds soap have been dissolved.

Blister-beetles.

Elongate long-legged beetles of various colors that often attack vegetable crops in swarms are known as blister-beetles. As far as known the larvæ feed on grasshopper eggs. A dozen species have been reported as attacking vegetables.

Blister-beetles are difficult to control because they are injurious in the adult stage. They are resistant to poisons and move readily from plant to plant. On plants on which it is safe to use an arsenical spray, at the first appearance of the beetles spray with arsenate of lead (paste), 3 or 4 pounds in 50 gallons water or with 1 pound paris green in 50 gallons water, adding 1 pound lime to prevent burning of the foliage. In the case of choice plants it might pay to screen them with mosquito netting.

Flea-beetles.

These are small usually dark-colored leaf-beetles, that have the hind legs fitted for jumping. The larvæ usually feed on the roots of plants but some of them mine the leaves or petioles. The adults eat holes or pits in the leaves. Most species spend the winter under leaves or rubbish. Nearly a score of species of flea-beetles has been reported as attacking vegetable crops.

Keep plants well sprayed with bordeaux mixture. It acts as a deterrent, driving the flea-beetles away. It is most effective when combined with 2 pounds paris green or 4 or 5 pounds arsenate of lead (paste) in 100 gallons. In some cases it is good practice to dip plants in arsenate of lead (paste), 1 pound in 10 gallons water before transplanting. Injury to plants in the seed-bed may be prevented by screening with cheesecloth, as for cabbage root-maggot.

Greenhouse white-fly (*Aleyrodes vaporariorum*).

The adult is a minute mealy white four-winged fly. In its immature stage the insect is scale-like in form, pale greenish in color and is found on the underside of the leaves. The life cycle requires about five weeks and breeding is continuous throughout the year under greenhouse conditions. Tomatoes, cucumbers, melons, and lettuce are especially liable to attack when grown under glass. It often happens that when plants are started under glass and then transplanted in the open, they become infested while young, and the white-flies continue to develop on them out-of-doors, often causing serious injury.

For destroying white-fly on tomatoes and cucumbers grown under glass, potassium cyanide should be used at the rate of 1 ounce (or sodium cyanide, ¾ ounce) to 5,000 cubic feet of space contained in the house and the fumigation should continue all night. Fumigate only on dark dry nights when there is no wind. The house should be as dry as practicable and the temperature not above 60 degrees F. Use with great care, as the material is very poisonous. All forcing-house and hotbed plants should be wholly free of white-fly when set in the field.

Root-knot nematode or eel-worm (*Heterodera radicicola*).

In the warmer parts of the country, many vegetable crops suffer serious injury from minute worms that infest the roots, causing swellings or nodules. They are very troublesome in irrigated regions and in greenhouses. The worms can persist in moist soil for a long time. They are real worms, not the larvæ of insects.

To free fields of the root-knot nematode, rotate for two or three years with a crop not susceptible to the disease and that grows rank enough to keep out weeds that harbor the pest. Immune varieties of cowpeas, such as the Iron, followed by winter wheat or rye, are sometimes used in Florida for this purpose.

In greenhouses, renew the soil from an uninfested field or sterilize with live steam under pressure. Shallow beds may be disinfected by applying a weak solution of formaldehyde, 1 part of the 40 per cent commercial solution in 100 parts of water, using 1 to 1½ gals. for every square yard of surface.

Millipedes.

The millipedes, or "thousand-legged worms," are elongate more or less cylindrical creatures, having a distinct head and a body consisting of well-defined segments, which is not divided into a thorax and abdomen as in insects. Each segment, except the first four, bears two pairs of legs. They prefer decaying vegetable matter as food but under certain circumstances attack root-crops such as carrots, beets and potatoes; infest the heads of cauliflower, cabbage and lettuce; attack seed beans, peas; eat holes in the fruit of tomatoes and melons where they touch the ground.

No satisfactory method is devised for control under field or garden conditions; trap the millipedes under boards or slices of potato laid on the ground; in the greenhouse they may be trapped in the same way or by using lumps of dough sweetened with molasses. Lime or tobacco dust placed around the base of the plants will help to drive them away.

Slugs.

Slugs are snail-like creatures that either lack the shell entirely or have it reduced to a thin plate. They do not belong to the insect tribes. They eat holes in the leaves of lettuce, celery, seedling beans and other vegetables. They often bore into ripening tomatoes. The word "slug" is sometimes used for the soft and slimy larvæ of some kinds of insects, as the cherry-slug, and sometimes of the larvæ of the potato-beetle; but here the true snail-like slug is intended.

Methods for the control of slugs that are practicable for all crops have not been fully worked out. On crops such as field beans, where a poison can be safely used, good results may be obtained by spraying with arsenate of lead at the rate of 4 pounds in 100 gals. of water. Slugs may also be killed by using a poisoned bait made according to the formula for cutworms. This bait should be scattered in small lumps around the plants in the evening. Keep poultry away from the bait. Dusting the plants and the surrounding ground with air-slaked lime or land-plaster will have a tendency to keep slugs away. Bordeaux mixture also has a deterrent effect and on some crops may be used to advantage. In some cases the plants may be sprayed with arsenate of lead either alone or in combination with bordeaux mixture.

Of course the reader understands that insecticides and fungicides may be poisonous to human beings, and that due precautions should be taken. All poisons should be labelled and kept out of reach. The greatest care should be exercised in the use of hydrocyanic gas fumigation. All carelessness everywhere should be avoided.

CHAPTER XIX

MARKETING, STORING, DRYING

Probably half the profits in commercial vegetable-gardening depend on the marketing. Where there are ten men who can grow a product to advantage there may be only one who can sell it to advantage. Horticulturists have not even yet learned the art of advertising. They are afraid to spend money for natty packages, attractive labels, and advertisements in local papers. The bases of good marketing are at least five: (1) a good and seasonable produce; (2) uniform grades in the marketed product; (3) good packing; (4) attractive packages; (5) honesty on the part of both grower and seller. Given these qualifications, the gardener need not hesitate to push his product and to ask the buyer to pay him an extra price.

Other things being equal, the local market is most to be desired. The grower is known, and he has an opportunity to establish a reputation. He can hold his customers year by year. All the business may be within his own observation. He knows what is done with his products. There is demand for vegetables and fruits at good prices. In any city of 10,000 and upwards a special trade can be established, particularly if the city is mature. This is often denied, but it is nevertheless true. If the grower sells his products in attractive packages, with neat labels if need be, and properly sorted and arranged, and places them in the hands of an enterprising grocer who caters to the best trade, he will not need to peddle his wares. The grower for the home market must be sure to have his vegetables in season; and he will do well, also, to provide a continuous and varied supply, for thereby he can hold his customers. He must set a standard and live up to it.

In truck-farming, however, the man grows for a less personal market. His produce partakes more of the nature of staples, that have more or less generalized market quotations. The trucker must meet these general market demands and conditions, as expressed by salesmen and prices-current, rather than the wishes of individual customers. With him quantity-production is a major item.

The present remarks, therefore, may not apply to those who grow things on a large scale, but such persons usually find special means and outlets for disposing of their products: because they have found such outlets is the reason for the growth of their business.

To do the best with one's products, the grower must keep track of the market. If possible, he should visit the market. He should consult the trade papers. He should ask his dealer about the new ideas in packages and packing. Ordinarily he will be able to secure better information if he deals continuously with one reliable firm.

This is a book about vegetable-gardening, not about selling; yet some of the homelier aspects of marketing may be discussed, particularly as respects grading, packing, and common storage, inasmuch as these affairs react on the growing of the crop. The reader must not expect this book to enter the field of cooperation, market associations, transportation, refrigeration, cold storage, commission and auction systems, food preservation, and the like.

1. GRADING AND PACKING

The sorting and packing of a crop should begin in the field. The better the crop is grown, the fewer will be the culls and the less the labor of grading.

In crops which are not to be carefully sorted into sizes and packed by hand, as potatoes and many of the root crops, the vegetables may be placed directly in the package in which the product is to be taken to the market. Nothing is better for the handling of heavy products than a bushel box (Fig. 231) or the crate. Formerly baskets of various sizes were used for this purpose, but the box or crate is much better because it is cheaper, more durable and it stows better on the wagon or in the store-house. One tier of boxes may be piled on another, but this is impossible with baskets unless one resorts to expensive staging.

231. Serviceable bushel box. It is 16 inches square, and 8 inches deep.

In handling the products in the field and in the storehouse, it is important that they be kept dry and cool. Over-ripeness and decay are then prevented. They should be put on the market or in storage quickly, before they have been subjected to unfavorable conditions of weather or to accidents. Some vegetables, as onions, are not injured by being left in the sun for a few hours or even days; but, as a rule, it is better to keep the vegetables in partial shade, particularly such as remain green or soft in their marketable stage.

If one has any quantity of vegetables to handle, it is necessary to have a packing-house or shed. In this place there should be tables or counters on which the sorting or grading can be performed. If possible, this house should have a pit or cellar at one end in which vegetables can be kept temporarily or even stored for the winter. The place should be provided with pipe water, tubs or vats for washing vegetables, and devices for trimming, tying and bunching.

All vegetables not sold in bulk on the general market should be sorted and graded; and even the bulk products are now graded, as potatoes. Grading contributes not only to the appearance of the product, but also to the snugness of packing. Vegetables like melons, tomatoes and others used as table delicacies and accessories, are usually sold by the smallest specimens in the package rather than by the larger ones. If the specimens are sorted into two grades, the smaller ones usually sell as well as the mixed lot, and the larger ones sell much better. Since the grading of vegetables is a matter of mental pattern, the grade varies with every packer, and it is therefore often difficult to secure sufficient uniformity to enable one to sell his products under a trade-mark. However, if one has uniform packages and gives close attention to the details of the business, he should be able to establish a series of grades that will be associated with his name in the market.

The uniformity of grading is now much facilitated by the mechanical graders; and given markets usually demand a particular kind of grade and pack of the hand-sorted products, as melons, celery, rhubarb. A certain number of stalks of asparagus may be required in the bunch, and a certain number of tomatoes in the tray.

In the grading and the packing, certain admonitions may be stated, having in mind the market-garden type of enterprise:

1. Pack and sell only the products that are mature, well grown, free from blemish, bruises, insects and disease injury.

2. See that the vegetables are carefully cleaned, neatly and uniformly arranged or tied, and that they arrive fresh and unwilted.

3. Choose the package itself with care, to meet the demands of the particular market; see that it is clean, bright and unbroken.

4. Pack snug; see that the receptacle is full, and that all weights and capacities are full measure, and that all bunches are full count. Snug packing is particularly important if the vegetables are to be shipped any considerable distance. A large part of the vegetables in the markets is handled from ten to fifteen times from the field to the consumer. Vegetables packed tight not only bear transportation better, but they keep longer and present a more attractive appearance. In the better kinds of vegetables this firm packing is secured by placing each specimen by hand.

5. Pack the vegetables cool. They should go into the packages with a low temperature, rather than warm. They

keep longer and hold their quality better under such conditions. This is particularly true of dessert and perishable products.

6. Use relatively small packages, with all the better kinds of vegetables. Aim, so far as possible, at special and dessert trade. The warmer the season, the smaller should be the quantity.

7. Look out for ventilation. If one is shipping green stuff, as cabbage, spinach and kale, the package should be

232. A lettuce re-pack, 12 heads in a layer or tier; empty paper-lined box at right.

open to air to prevent heating, particularly if as large as barrels. It is well to use open or ventilated packages for all green vegetables in warm weather, at least for those to be shipped long distances.

8. In the finer or dessert vegetables it is well to pack in a distinctive package or to use a trade-mark or label that will distinguish one's products from others. This is essential if one is to establish an individual reputation and to hold customers from year to year. With such heavy and staple products as potatoes, beets, or cabbages, it is

usually inadvisable to attempt this kind of marketing; but even with them it can sometimes be undertaken. It is usual to associate a special package with fruits, but not with vegetables; but this condition of affairs is to be changed. The use of hampers, paper cartons, splint boxes, and other receptacles allow the making of very attractive packages. For the finer products and the best markets the receptacles are often neatly lined with paper.

9. If the grower or seller reaches an individual customer whose attention he desires to hold, he may well afford to sell under a guaranty.

233. Sweet potatoes graded and ungraded.

10. Be not misled by mere quantity results. In the end, one's success depends on the quality, regularity and dependability of the produce,—that is, on the satisfaction to the consumer.

The nature of good packages for vegetables is suggested in the pictures. Fig. 232 is adapted from Stanley S. Rogers, Calif. Circ. 160, on lettuce-growing in California: "After the lettuce is cut it is hauled directly to the packing house where it is sorted, trimmed and repacked. That which is to be shipped a considerable distance, and especially during warm weather, should always be protected from the heat; a layer of chopped ice should be placed between the bottom and the second layer of lettuce, and one on the top layer. If the pony crate is used the ice is put on the top layer only. The inside of the crates should be lined with heavy paper, which prevents the contents from drying and aids in keeping it cool. Lettuce is shipped in iced or refrigerator cars, the temperature of which should be kept as even as possible." This quotation explains the care that is taken in the grading, packing and shipping of high-class vegetables.

234. A dozen cauliflowers in a serviceable tray. See also Fig. 31.

Graded and ungraded sweet potatoes are shown in Fig. 233 (adapted from H. C. Thompson, Farmers' Bull. 970). Figs. 234 to 237 carry the suggestions still farther. Produce fit for these packages and for such care must be well grown, so that there will be the minimum of loss in the sorting. The efficiency (or the lack of it) of vegetable-gardening and fruit-growing is often measured by the produce left behind in culls and waste rather than by the first-class proportion that finds its way to market.

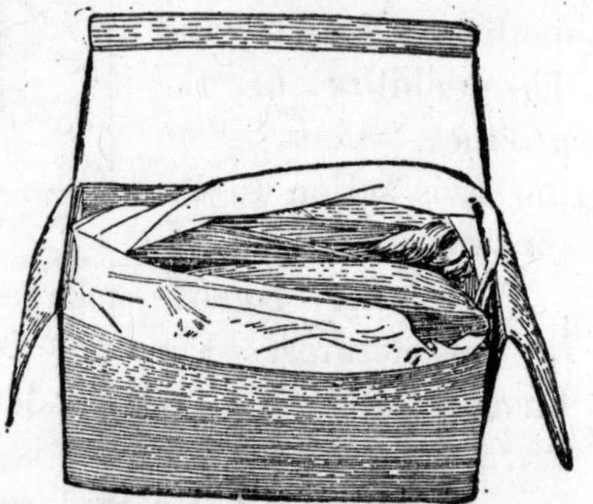

235. Sweet corn in the husk, in a paper lined carton.

In truck-growing, a much heavier and coarser package must be used; sometimes the produce shipments are in bulk. The product finds its way largely to wholesale markets, or at least to disposition in large lots. Some of the packages employed in long-distance shipments are shown in Fig. 238. The crate, splint-basket and barrel types prevail in this kind of commerce.

2. STORING

It is impossible to state principles that apply to storing all kinds of vegetables, for these products include fruits, roots and leaves. Some of them must be kept warm and some cool. Others, as onions and squashes, must be dry; still others, as cabbages and roots, must be kept moist. Each class of vegetable is a law unto itself.

236. Crate of uniform melons, just twelve grown and sorted to fit the crate.

237. A neat basket of celery.

With the exception of root and tuber crops, most vegetables are uncertain in storage unless kept in an establishment cooled by artificial means, and which, therefore, maintains uniformity of moisture and temperature. In general, it is better to sell in the fall, even at a somewhat reduced price, than to go to the expense and risk of storing. When, however, the fall market is so low as to preclude any profit, storing is a necessary recourse. Persons who have become expert in the handling of any one vegetable may store it with relative safety. If one has had no experience in the storing of those vegetables that are difficult to keep, it is generally better to put them in the hands of a person who makes a business of cold storage and pay him for his labor, investment, and experience.

In general, a low temperature is essential to the keeping of the product. It prevents over-ripening and delays the work of fungi and other disorganizing agents. Usually it is well to keep the temperature relatively near the freezing point; but there are some vegetables, as melons and sweet potatoes, that are injured by a low temperature. Products either over-ripe or markedly under-ripe usually do not keep well. It is essential to any success that the specimens be perfectly sound when put in storage, and in the proper state of maturity. No doubt some of the loss in the storing of cabbages, for example, is due to the infection of the plants with the rot fungus before the heads are put in storage. Onions

238. Vegetable containers for long-distance shipments.

seriously attacked by the smut or rust may not be expected to keep well, however good the storage.

The following essentials apply to the storing of most vegetables: (1) Protect from frost; (2) keep them cool, to prevent decay; (3) keep them relatively moist, to avoid excessive evaporation and wilting; (4) avoid a wet and stagnant evaporation, as this is likely to engender rot, particularly when the temperature is too high; (5) protect from natural heating or fermentation; (6) provide change of air, without exposing the products to such draughts that they shrivel.

Several kinds of storage are illustrated in earlier chapters, for potatoes, sweet potatoes, onions, cabbage, celery. The more general-purpose forms are shown herewith, in enough detail to suggest the essential points.

For home use, it is well to store roots and tubers in moist sand or in sphagnum moss (such as nurserymen and florists use). Beets, carrots, parsnips, and potatoes stored in this way keep plump and fresh for a twelvemonth or more, if the temperature is kept low enough to prevent sprouting. The reason for this good result is that the sand or moss prevents evaporation and maintains uniformity of conditions.

The house cellar is commonly one of the poorest places in which to store vegetables, particularly if it contains a heater for the residence. In such case it is likely to be too warm and too dry. The vegetables shrivel and tend to start into growth, or to decay quickly. Cellars that contain much vegetable-matter are likely to make the house unwholesome unless there is ample ventilation and pains is taken to pick over the vegetables from time to time and remove all unsound specimens. If the house cellar is used for the storing of vegetables, it is well to have a special vent or chimney. This may be a cheap board affair extending up the back side of the house as high as the roof. This flue carries off the foul and warm air, and thereby keeps the cellar sweet and at a relatively low temperature. In some cases an extra flue may be provided in the chimney when the house is built, and the warmth of the chimney will cause a strong draft. Fig. 239 shows a simple intake shaft for cool air and an open window for the outgoing warm air (from Cornell Reading-Course for the Farm Home, No. 113): "Warm air should be permitted to pass out at the top of the room through ventilators, and cool air from outside should be admitted to the room at the bottom. In a cellar this can be accomplished by means of a shaft leading down the wall from a window and opening near the floor. A few windows at the top of the wall constitute the system of ventilation for most farm cellars."

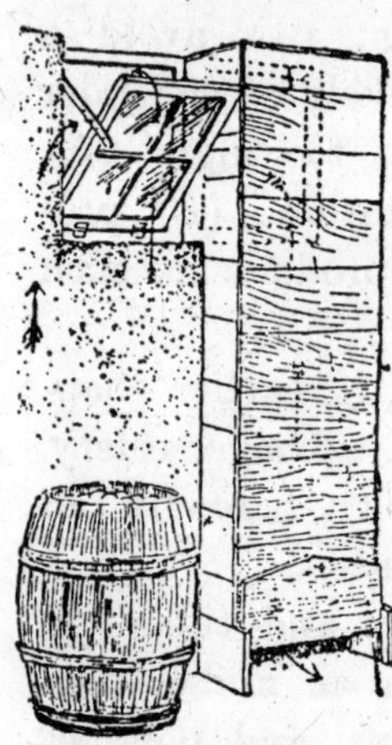

239. Intake and outlet for a house cellar.

The old-fashioned "outside cellar" usually gives better conditions for the storing of vegetables than the house cellar. This structure has been much advised of late, and many improved plans are available. It is likely to be uniform in temperature and moisture conditions. With various modifications these cellars are used largely by market-gardeners for the storing of roots, leek, celery and other products that do not require a dry air.

The outside cellar is little more than a pit sunk to the level of the ground with a gable roof covered with earth and sod so that frost cannot enter; or if the ground is likely to be moist, the pit is built partially above ground. If the cellar is to be permanent, the walls may be laid of stone or brick. If the masonry wall is lined with hollow or "lining brick," more uniform conditions are secured. It is important that provision be made for ample drainage, and also for ventilation without opening the main doors. This ventilation is usually secured by a little cupola or shaft near the center of the structure or by windows in the gables. A vestibule entrance is desirable if the climate is severe. It is preferable that the cellar have a natural earth bottom, provided the drainage, either natural or artificial, is complete.

A great difficulty with a permanent field or outside cellar is the danger of its holding so much moisture and being so "close" as to encourage the growth of fungi and thus engender decay. Investigations into the causes of the rotting of celery in storage have shown that the disease is associated largely with poor and damp houses.

Pits, or field storage.

The field cellar or pit is a temporary structure. A style much used in parts of the Northern States may be described as an example: On warm and well-drained land (preferably sand or gravel) an excavation is made one to two feet deep, usually fourteen to eighteen feet wide, and of the length required to hold the crop one has to store.

The sides or margins of the excavation are held by one or two planks placed on edge and secured by stakes driven into the ground. The pit is then covered with a gable roof made by laying boards from the margin to a ridge-pole. The ridge-pole stands three to five feet above the bottom of the pit and is held on stakes driven through the center of the pit lengthwise. Usually it is necessary to support the boards between the margin and the ridge by another run or plate held on stakes driven midway between the side and the ridge. Boards about twelve feet long are now laid from the ground to the ridge-pole, making a continuous roof. Ordinarily these boards are lapped, and the upper run is nailed lightly to hold the roof in place. The boards are not nailed very securely, however, for it may be necessary to use the boards the following

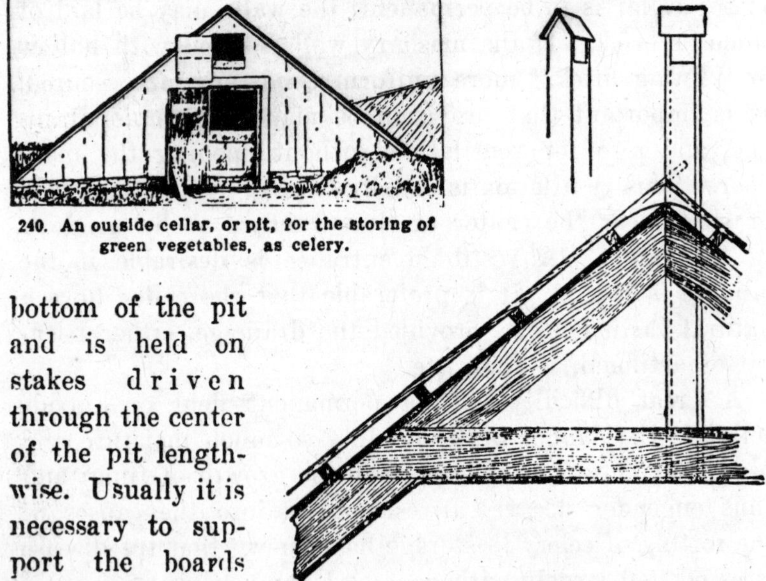

240. An outside cellar, or pit, for the storing of green vegetables, as celery.

241. Ventilating shaft, and structure of the gable.

year, and the subsequent covering will hold them in place. At intervals of ten or twelve feet, two or three boards are left without nailing to allow of an entrance, and the place is marked by a stake driven into the ground. These pits or temporary cellars are made late in autumn, and until severe freezing weather comes the protection of boards is sufficient; but as winter approaches, straw, grass or other litter is thrown over the roof, and subsequently manure or earth is added. In pits of this character, that contain a large body of air, very uniform conditions are secured. In them celery, leek and brussels sprouts, and even cabbage, may be set compactly in rows. The plants often make a root-hold in the soil, and therefore do not shrivel and are not so likely to rot as those thrown in loose.

242. Interior of a good storage pit.

243. The sill and roof construction.

Pits of this kind are very useful for the storing of late or winter celery. In them the celery grows somewhat, and it blanches by spring. If, however, it is desired to keep celery only a short time, and particularly if the crop has been blanched in the field, another kind of structure is usually more desirable. In that case, a house that has a little artificial heat is usually better.

Various patterns of storage structures.

To visualize the foregoing statements, pictures are here assembled of several forms and details of home-made or farm storage structures.

The building may be a wooden structure over a pit, as

244. Home-made storage cellar.

in Fig. 240, with a ventilating window in the gable. Usually some kind of roof ventilation is provided, perhaps in the way of such shafts or chimneys as those in Fig. 241. Some of these houses are of excellent construction, as indicated in Figs. 241 and 242. They may be ceiled to keep them warm and prevent too rapid changes in temperature, and roof windows may be provided for light. If the side walls are brick, the structure may be something like that in Fig. 243. The air-space in the roof is to be noted.

The structure may be wholly or mostly buried, either by being sunken or by having earth covered over it. Fig. 244 is a well-made outside cellar (James H. Beattie, Farmers' Bull. 879), with ventilation and drainage. Detail for the interior of an outside cellar is given by James

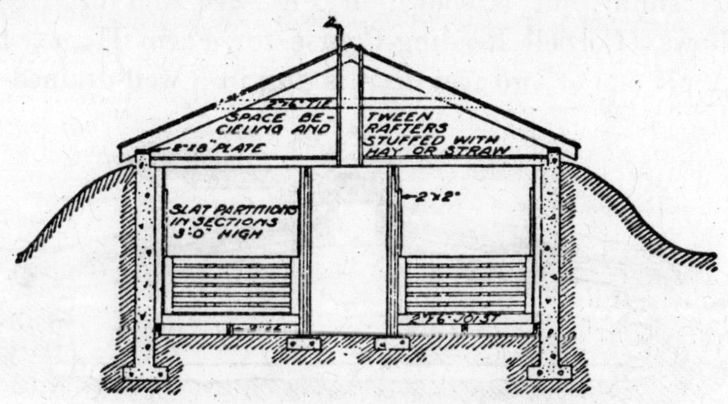

245. Detail of a partially buried cellar.

L. Strahan, Cornell Extension Bull. No. 22, shown in Fig. 245. "The interior is arranged in a double row of bins each 8 x 8 feet with a 4-foot alley through the middle. At the alley corner of each bin is a 6 x 6-inch post built up of 3 pieces of 2 x 6-inch material. The center piece is cut 6 inches short to allow for a 2 x 6-inch stringer, or ceiling support, which runs longitudinally through the cellar along the top of the posts. On this 2 x 6-inch piece rest 2 x 4-inch joists spaced 2 feet and 6 inches on centers, and these in turn support a ceiling of 1-inch unmatched boards. A 4-inch shoulder, 10 inches from the top, is constructed on the inside of the long walls to receive the ends of the 2 x 4-inch joists. This allows a space of about 15 inches between the ceiling and the roof at the point where the roof joins the wall, which can be stuffed with old rags, carpet, or burlap as an added protection against frost. Ventilation inside the cellar is provided by means of a raised slatted floor and slatted bin divisions."

A simple pit is shown in Fig. 246 and described as follows (Cornell Reading-Course for Farm Home, 113): "A pit one or two feet deep is dug in a well-drained spot,

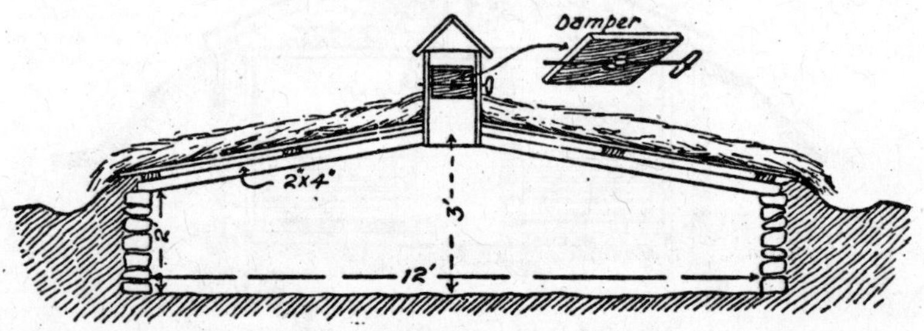

246. A simple ventilated pit.

and a foundation wall of stakes and boards, or, better, of concrete, is built around it. On this wall, rafters are erected for the support of roof boards. The roof is covered with soil and sod, or with straw and a light covering of earth, or with manure. Such a pit will last several years, especially if a rot-resistant wood, as the so-called 'pecky' cypress, is used. With the specific directions that are furnished by cement manufacturers, concrete work is within the range of any handy man, and a permanent concrete cave or pit may be built with little expense and trouble. No matter what the form of construction, one or two small ventilators should be provided at the top of the cave, and one at the bottom of the door. These should be arranged to open and close."

The burying of vegetables.

Most root crops, as beets, carrots, potatoes, are kept over winter with ease by burying them in the field. It is well to choose a warm and well-drained place. The pit is covered very lightly at first, and more covering added as the cold weather comes on. If the full covering is applied at first the products are likely to heat and decay sets in. Be sure that the beets and potatoes are not attacked by fungous diseases before they are put in the pit.

It is customary to make a small circular or rectangular excavation six inches to a foot deep and from six to eight feet across. In this the roots are piled in a tall cone. Straw or salt-hay or rather dry litter is then thrown over the pile to protect from the early frosts. As the season advances, an inch or two of earth is thrown over the straw and finally, when winter threatens to close in, the pile is covered deep enough to give full protection. Usually ten to twelve inches of earth over the straw will be sufficient, the straw itself being four to six inches thick after it is well matted down. In severe climates the earth may then be covered with a foot or two of horse manure. Apples can be buried in this way with very good results, particularly the long-keeping varieties, as Russets.

The pit may be elongated to any distance required. It is well not to make it much wider than six or eight feet, else the vegetables are likely to heat and there may be too great pressure on the lowermost tubers.

An excellent modification of the long pit is the compartment-pit. This has narrow partitions of earth every four or five feet, thus preventing the heating of the vegetables and also allowing one compartment to be emptied in winter without exposing another. A good one is shown in Fig. 247 (from Cornell Reading-Course). Usually these compartment-pits are sunk two or three feet in the earth and a partition of soil six to twelve inches is left between the excavations. Each pit is then filled until it is "rounded full" and is covered as above described. It is often difficult to make these partitions hold their shape, however, particularly in loose and sandy land. In such cases the vegetables may be heaped in several piles in a long pit and earth tramped in between the piles.

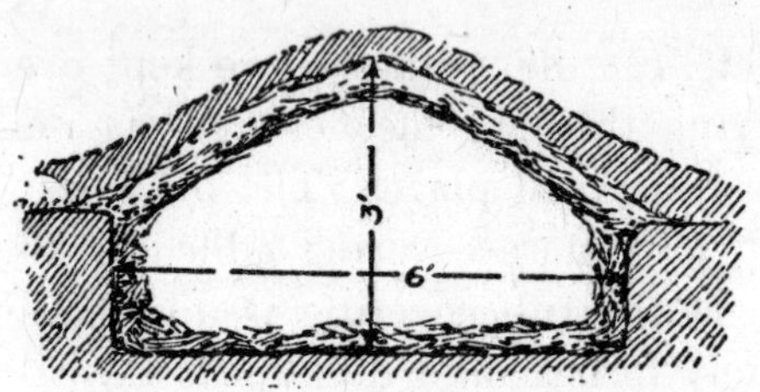

247. Cross-section of a trench.

Whatever the style of pit, it is essential that the soil be naturally well drained, and a furrow or ditch should be opened around the pit to carry off surface water.

3. DRYING

It is not the purpose of this book to discuss the preservation of vegetables; yet the importance of the subject has latterly become so great that the drying of vegetables for home use can hardly be passed over. The home canning of vegetables is better understood, and, moreover, a discussion of it would be too extensive for this place, and it is a culinary operation.

"Comparatively speaking, the evaporating plant has many advantages over the cannery," writes J. S. Caldwell in his exhaustive bulletin on "The Evaporation of Fruits and Vegetables" (Bull. 148, Wash. Exp. Sta.). "The initial cost of building and equipment necessary to handle a given volume of material is much less, the machinery is less costly and depreciates much less rapidly. The employment of a technically trained, high-salaried supervisor is not necessary."

"It must not be forgotten," Dr. Caldwell continues, "that in supplying the actual necessities of life, fruits and vegetables are as indispensable as grains and meats, and that without them it is impossible for human beings to maintain continued normal health. But fruits and vegetables retain all their nutritive value and their health-preserving powers after having been subjected to drying, which eliminates all inedible portions and converts the material into non-perishable form while reducing its weight by three-fourths to seven-eighths."

Only briefly can the subject of the drying of vegetables be opened here, and mostly by way of suggestion for the home-maker. "The nutritive value of food," writes Pearl MacDonald in Ext. Circ. 61, Pa. State Coll., "is practically unchanged by drying. In addition to the difference in flavor produced by drying, there is usually a difference in color. Green shell peas and beans remain practically unchanged in color; but apples, for example, when pared and exposed to the air are changed to a darker color due to the action of the oxygen of the air upon certain of their elements. According to the laws of nature, this is the result to be expected and everyone should recognize the fact.

Many factory-evaporated products are treated chemically to give them a lighter color, because the public demands a less highly-colored product. Such treatment, however, detracts from the natural color and flavor. * * * The amount of water in the dried fruits and vegetables is greatly reduced, which means that there is a greater concentration of food elements in dried products. Pound for pound the nutritive value is greater in dried than in fresh food. When dried foods are prepared for the table, however, the water lost by evaporation is replaced by soaking, so that the nutritive value of cooked dried material and of fresh material is virtually equal.

"Not all fruits and vegetables lend themselves to this method of preservation. * * * Of the vegetables, green shell peas, green shell beans (any of the bush and pole bean varieties such as are used for green shell beans), string beans, green shell lima beans, corn and pumpkin are the best to dry.

"The reason for drying the green shell peas and beans, green limas and string beans is to supply a greater variety for the winter diet. The family may tire of having the mature peas and beans frequently. If there is, however, a supply of canned, green dried and mature dry peas and beans, and these are properly placed in the menu, there is little danger that this type of food will become monotonous."

Various kinds of trays may be used for the drying of vegetables as for the more familiar drying of fruits. A serviceable home-made construction, for use either in the sun or on a stove is described (Pa. Circ. 61, from which Fig. 248 is adapted), as follows:

"To make this drier use strips of wood about one inch wide and one-half inch thick (lath will answer very well). Cut these and the cross-pieces to which the strips are nailed to fit the oven in which they are to be used; or any desired length if they are not to be used in oven

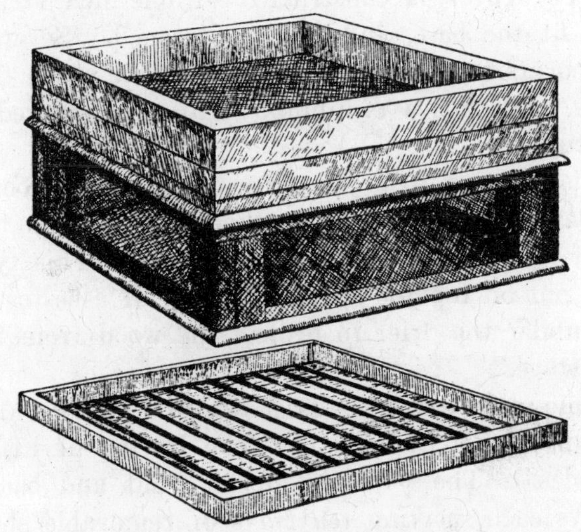

248. Home-made drier for vegetables. The drier complete with wire-bottom tray on top, which may be taken off and used separately. Beneath the top is a slat tray (shown separately) that may be used instead. If the slat tray is used, the top wire-screen tray may be inverted over it for protection from flies and dust.

drying. Around the top of the drier construct a frame of one-inch strips to prevent the material that is being dried from dropping out. To the bottom of the drier, nail wooden legs about four inches in length. This will raise the frame so as to permit a circulation of air under the drier. Wire screen or mosquito netting should be tacked around the sides of the frame to protect the food.

"Copper or other screen may be used in place of the slats. Copper screen is preferred because it will not corrode if the food material comes in direct contact with it, except of course foods that contain acid. Spreading cheesecloth over the screen will overcome trouble of this sort. The cover is constructed of one-inch strips of a size to fit the base, and is also screened. Several additional covers may be made and placed one above the other in tiers, and thus a minimum of area be covered by the apparatus.

"This type of drier may be used out of doors on a fairly level porch roof; a southern exposure is best. A tin roof is excellent for drying purposes. This type may also be put on top the stove. A piece of asbestos may be placed under the drier to protect the wood from the heat of the stove."

A home-made dry-house is shown in Fig. 249, also from the Pennsylvania Circular on "The Drying of Fruits and Vegetables." The "house" opens front and back, with five trays each, making ten trays or removable shelves to hold the produce. The construction carries its own heating facilities, in the nature of a fire-place beneath. Of course there are completer outfits manufactured for the purpose.

The drying of vegetables received special impetus in the war time, and much was written on the subject. Farmers' Bulletin 841 gives detailed advice. Its preliminary statement is as follows: "Fruits and vegetables may be dried in the home by simple processes and stored for future use. Especially when canning is not feasible, or cans and jars are too expensive, drying offers a means of saving large quantities of surplus products which go to waste each year in gardens and fruit plots. Drying also affords a way of conserving portions of food which are too small for canning.

"The drying may be done in the sun, over the kitchen stove or before an electric fan. Manufacturers have placed driers on the market. Home-made driers are satisfactory.

249. Simple dry-house.

"A good home-made drier should have the following features: (1) It should be light, easy to operate, of simple construction, inexpensive, and, as nearly as possible, non-inflammable. (2) It should permit a free circulation of air, to allow the rapid removal of the air after it has passed over the vegetables and absorbed moisture. (3) It should provide for protection of the food product against dust, insects, etc. (4) It should protect the materials from being moistened by steam, smoke, rain, or dew while drying."

"Three main ways of drying are applicable in the home manufacture of dried fruits and vegetables," according to Farmers' Bulletin 841, "namely, sun drying, drying by artificial heat, and drying by air blast. These, of course, may be combined. In general, most fruits or vegetables, to be dried quickly, must first be shredded or cut into slices, because many are too large to dry quickly or are covered with a skin, the purpose of which is to prevent drying out. When freshly cut fruits or vegetables are to be dried by means of artificial heat, they should be exposed first to gentle heat and later to the higher temperatures. If the air applied at the outset is of too high a temperature, the cut surfaces of the sliced fruits or vegetables become hard, or scorched, covering the juicy interior so that it will not dry out. Generally it is not desirable that the air temperature in drying should go above 140° to 150° F., and it is better to keep it well below this point. Insects and insect eggs are killed by exposure to heat of this temperature.

"When freshly cut fruits or vegetables are spread out they immediately begin to evaporate moisture into the air around them, and if in a closed box will very soon saturate the air with moisture. This will slow down the rate of drying and lead to the formation of molds. If a current of dry air is blown over them continually, the water in them will evaporate steadily until they are dry and crisp. Certain products, especially raspberries, should not be dried hard, because if too much moisture is removed from them they will not resume their original form when soaked in water. On the other hand, the material must be dried sufficiently or it will not keep, but will mold.

Too great stress cannot be laid upon this point. This does not mean that the product must be baked or scorched, but simply that it must be dried uniformly through and through.

"It will be found advisable also to 'condition' practically all dried vegetables and fruits. This is best done in a small way by placing the material in boxes and pouring it from one box into another once a day for three or four days, so as to mix it thoroughly and give to the whole mass an even degree of moisture. If the material is found to be too moist, it should be returned to the drying trays for a short drying."

The yield of the produce in dried material is stated by E. L. Kirkpatrick in Cornell Reading-Course for the Farm, Lesson 132. The water content of various fresh fruits and vegetables and the amount of dried produce that one hundred pounds of the fresh fruit or vegetable will yield, are shown:

	Percentage of water in the produce	Pounds dried product from one hundred pounds fresh produce
Tomatoes	**94.3**	**8.5**
Celery	**94.0**	**7.0**
Spinach	**92.0**	**7.5**
Cabbage	**91.0**	**8.0**
Carrots	**88.0**	**10.0**
Onions	**87.0**	**8.5**
Squash	**86.0**	**6.0**
Apples	**84.6**	**15.0**
Potatoes	**78.3**	**22.0**
Sweet corn	**75.4**	**21.0**

CHAPTER XX

THE HOME GARDEN

The home gardens of the country ought to be more important than the commercial gardens and the trucking areas. Perhaps they are, but as we have no statistics of them we are unable to compute their produce or to estimate their influence in the life of the people. If there are twenty-five million families in the United States, there ought to be several million home gardens. The value of these gardens in the education and discipline of children and in the raising of supplies should exceed all estimate.

The elements to be considered in the home garden are: (1) An enterprise within the means and labor supply of the family; (2) a sufficient product to supply the household; (3) continuous succession-crops; (4) ease and cheapness of cultivation; (5) maintenance of the productivity of the land year after year.

The quantity of product to be grown depends on the size of the family and its fondness for vegetables. An area 100 x 150 feet is generally sufficient to supply a family of five or six persons, not considering the winter supply of potatoes; but the area must be well tilled and handled.

The ease and efficiency of cultivation are much enhanced if all the crops are in long rows, to allow of wheel-tool tillage, either by horse or by man-power. The old practice of growing vegetables in beds usually entails more labor and expense than the crop is worth and it has had the effect of driving more than one boy from the garden. These beds always need weeding on Saturdays, holidays, circus days, and the Fourth of July.

Even if the available area is only twenty feet wide, the rows should run lengthwise the plot and be far enough apart (one to two feet for small stuff) to allow of the use of the hand wheel-hoes, many of which are very efficient. If land is available for horse-tillage, none of the rows should be less than thirty inches apart, and for large-growing things, as late cabbage, four feet is better. If the rows are long, it may be necessary to grow two or three kinds of vegetables in the same row; and in this case it is important that vegetables requiring the same general treatment and similar length of season be grown together. A row containing parsnips and salsify, or parsnips, salsify and late carrots, affords a good combination; but a row containing parsnips, cabbage and lettuce would be a faulty combination.

One part of the area should be set aside for all similar crops. For example, all root crops might be grown on one side of the plantation, all cabbage crops in the adjoining space, all tomato and eggplant crops in the center, all corn and other tall things on the opposite side. Perennial crops, as asparagus and rhubarb, and gardening structures, as hotbeds and frames, should be on the border, where they will interfere least with the plowing and tilling.

The best results in maintaining productiveness are to be secured when it is possible to practice rotation of crops, manures and tillage. Even in a small area, this rotation can be practised to a considerable extent. The area devoted to root crops this year may be given to corn or melons next year. It is particularly important to rotate if diseases and insects become serious on any one crop; and in this case, the greatest care should be taken to choose those crops, for the rotation, on which the parasites cannot thrive. For example, the club-root of the cabbage and cauliflower works on turnips. Some insects cannot be starved out in a small area, and it is then necessary to cease growing the crop for a year or two. The cabbage maggot is an example. If this pest obtains a good foothold in the home garden, cabbages and cauliflowers may be discontinued until the insect disappears; and this is often a cheaper solution of the difficulty than to attempt to destroy the insect with the bisulfide of carbon treatment. If one lives on a farm, the cabbage patch may be placed on the farther part of the estate for a year or two. When the maggot has quit the area, the cabbage patch may be made again on the old ground.

Of the home vegetable-garden, Hunn writes as follows in the "Practical Garden-Book":

"Make the soil deep, mellow and rich before the seeds are sown. Time and labor will be saved. Rake the surface frequently to keep down weeds and to prevent the soil from baking. Radish seeds sown with celery or other slow-germinating seeds will come up quickly, breaking the crust and marking the rows. About the borders of the vegetable-garden is a good place for flowers to be grown for the decoration of the house and to give to friends.

"A home vegetable-garden for a family of six would

require, exclusive of potatoes, a space not over 100 by 150 feet. Beginning at one side of the garden and running the rows the short way (having each row 100 feet long), sowings may be made, as soon as the ground is in condition to work, of the following:

Fifty feet each of parsnips and salsify.

One hundred feet of onions, 25 feet of which may be potato or set onions, the remainder black-seed for summer and fall use.

Fifty feet of early beets, 50 feet of lettuce, with which radish may be sown to break the soil and be harvested before the lettuce needs the room.

One hundred feet of early cabbage, the plants for which should be from a frame or purchased. Set the plants 18 inches to 2 feet apart.

One hundred feet of early cauliflower; culture same as for cabbage.

Four hundred and fifty feet of peas, sown as follows:

100 feet of extra early.
100 feet of intermediate.
100 feet of late.
100 feet of extra early, sown late.
50 feet of dwarf varieties.

If trellis or brush is to be avoided, frequent sowings of the dwarfs will maintain a supply.

After the soil has become warm and all danger of frost has passed, the tender vegetables may be planted, as follows:

Corn in five rows 3 feet apart, three rows to be early and intermediate, and two rows late.

Tomatoes, one row, plants 4 to 5 feet apart.

One hundred feet of string beans, early to late varieties.

Vines as follows:

10 hills of cucumbers, 6 x 6 feet.
20 hills of muskmelon, 6 x 6 feet.
6 hills of early squash, 6 x 6 feet.
10 hills of Hubbard squash, 6 x 6 feet.

One hundred feet of okra.

Twenty eggplants.

Six large clumps of rhubarb.

An asparagus bed 25 feet long and 3 feet wide.

Late cabbage, cauliflower and celery are to occupy the space made vacant by removing early crops of early and intermediate peas and string beans.

A border on one side or end will hold all herbs, such as parsley, thyme, sage, hyssop, mints.

Much of the satisfaction in the garden of one's own hands lies in the compact housing of garden tools and supplies, keeping all the outfit painted and repaired. The daily work will suggest many conveniences and aids to be made rainy days and odd times on the work-bench in the garden-house, for it is assumed that even the home garden will have its headquarters in a small neat structure built for the purpose or in a part of the barn, woodshed, garage or basement.

The garden will be well laid out, according to a plan. All of it will express good workmanship—this is part of its educational value to children. There is also a wholesome promptitude about garden work. The place will be planned with due regard both for good appearance and economy of labor. "It improves the appearance of the garden greatly and makes the work easier if all crops are planted in straight rows," writes C. E. Durst, in Circ. 198, Ill. Exp. Sta. "For this purpose one should get into the habit of using a garden line. This should be stretched tightly and the rows made just to one side of it so that the line will not be moved and the rows made crooked as a result. For making deep drills for onion sets and peas, the point of the hoe or a wheel hoe with plow attachment should be used. For shallow drills for small seeds, the tip of the hoe handle is especially good. This makes a narrow drill which is easily weeded and cultivated. The secret of making perfectly straight rows is to take a firm grip of the hoe and make good brisk draws. For vegetables planted in hills, the holes can be made quickly with the point of the hoe. When several rows of the same width are to be planted, time can be saved by using a home-made marker, as shown in Fig. 250. The runners can easily be changed for rows of different widths."

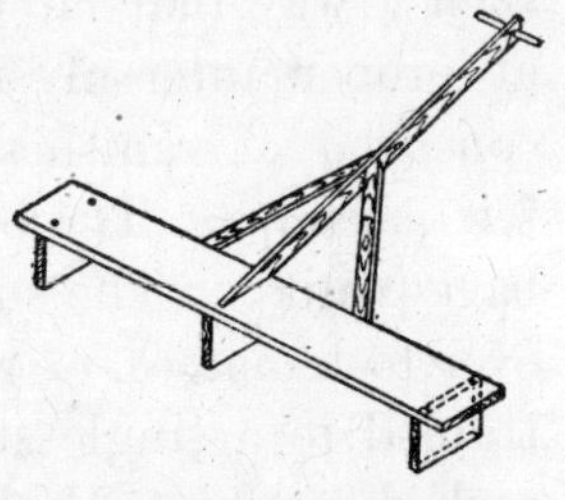

250. Hand-made marker.

The home-gardener will find it a great help to keep a simple diary of the operations, so that one year's work and rewards may be compared with those of another year. Even old hands at the business like to refresh themselves on dates for planting and to compare seasons and yields. A plan or diagram of the garden affords a good record.

There will be a tight dry place for the keeping of seeds. It is great joy to try a few novelties every year, whether vegetables or other things. It is good schooling. The seeds themselves are interesting in their fascinating shapes and markings, and in the ways they have of "coming up."

The garden shown in Fig. 251 was a city back yard 25 x 70 feet, near New York City, described in Farmers' Bull. 818: "It happened to be bounded on two sides by a board fence, and advantage was taken of this fact to plant and train grape vines. Strawberry plants were set alongside the flagstone walks and currant bushes between the walks and the fence. In the space between the bushes and the strawberries low-growing vegetables, such as bush beans, peppers, eggplants, and the like, were set out. In a space about 12 feet wide between the walks, low-growing, quick-maturing varieties of early vegetables were planted in such a way that later-maturing varieties could be put out at proper intervals between them. The early plantings consisted of radishes, early beets, lettuce, carrots, and a few parsnips. The beets gave way later to a few late cabbage plants. The sunniest portion of the yard was turned over to tomatoes, of which there were about a dozen plants trained to a single stem and set about 18 inches apart in each direction. Early and late peas were put out in the least sunny portions of the yard. Later, in the fall, spinach, kale and potato-onion sets were planted in order to provide a supply of green succulents for the winter and early spring."

A larger area, exclusively devoted to vegetables, is planned in Fig. 252, by H. C. Thompson, in Farmers' Bull. 934. "The size of the garden," the author says, "depends upon the number of persons to be supplied. One-fourth to one-half an acre is sufficient for an average family and should produce enough vegetables for use throughout the year. By close attention to the rotation of crops, the succession of crops, and interplanting, one-fourth of an acre may be made to supply a family of six. Where land is available, it is recommended that a sufficient area be set aside to allow part of the garden to be planted to a soil-improving crop each year."

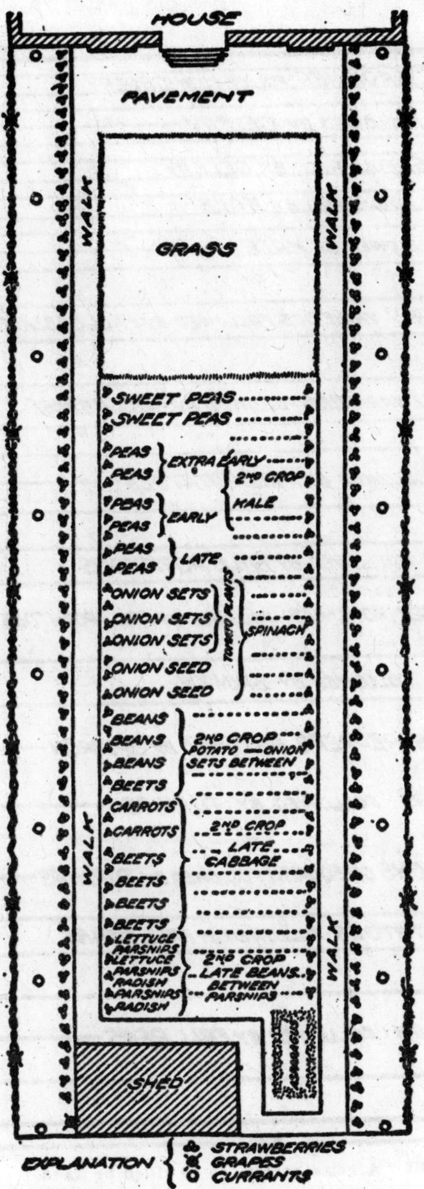

251. A back-yard garden, with vegetables and fruits.

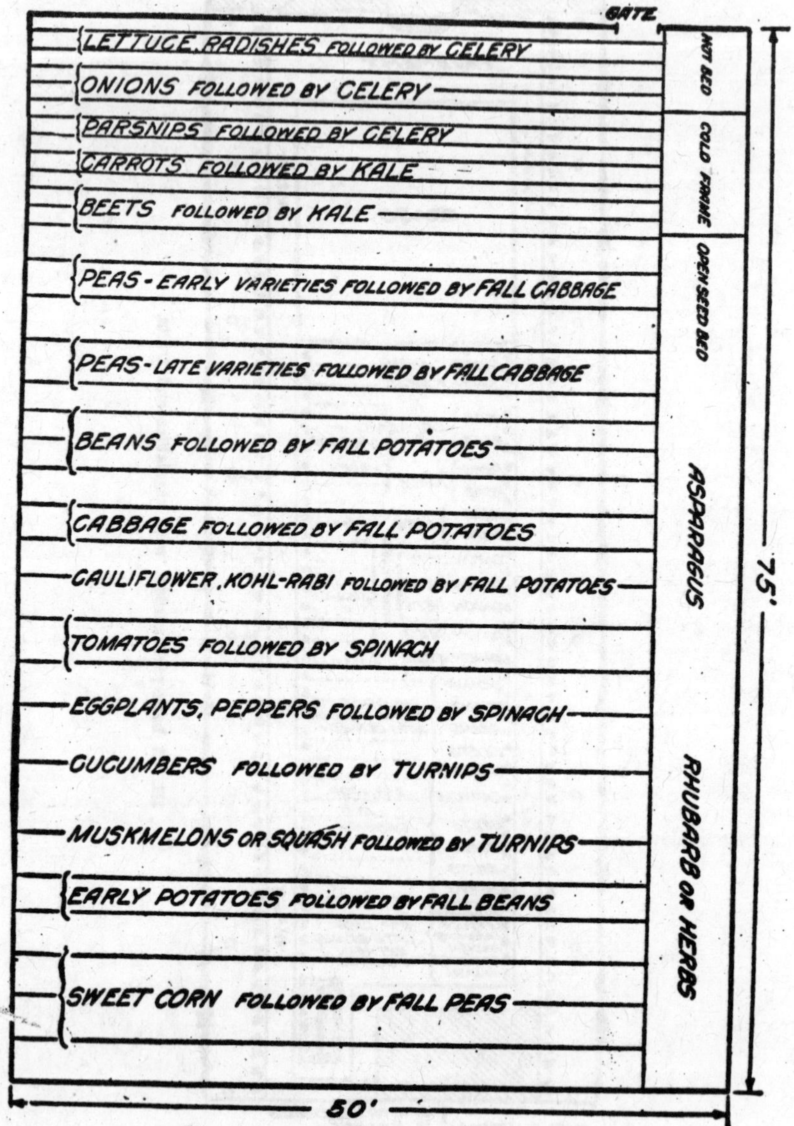

GATE

LETTUCE, RADISHES FOLLOWED BY CELERY

ONIONS FOLLOWED BY CELERY

PARSNIPS FOLLOWED BY CELERY

CARROTS FOLLOWED BY KALE

BEETS FOLLOWED BY KALE

PEAS - EARLY VARIETIES FOLLOWED BY FALL CABBAGE

PEAS - LATE VARIETIES FOLLOWED BY FALL CABBAGE

BEANS FOLLOWED BY FALL POTATOES

CABBAGE FOLLOWED BY FALL POTATOES

CAULIFLOWER, KOHL-RABI FOLLOWED BY FALL POTATOES

TOMATOES FOLLOWED BY SPINACH

EGGPLANTS, PEPPERS FOLLOWED BY SPINACH

CUCUMBERS FOLLOWED BY TURNIPS

MUSKMELONS OR SQUASH FOLLOWED BY TURNIPS

EARLY POTATOES FOLLOWED BY FALL BEANS

SWEET CORN FOLLOWED BY FALL PEAS

HOT BED

COLD FRAME

OPEN SEED BED

ASPARAGUS

RHUBARB OR HERBS

75'

50'

252. A home garden, to be tilled by hand.

Here ends the vegetable-gardening book. As it begins with plants, so it ends with the home; thereby is the personal and human interest of the book emphasized. The author enjoyed writing the book twenty years ago. Still more has he enjoyed re-writing it in his maturer years, and he has lived the subject all over again. He has had many aids not available then, for now there are numerous workers. He might have quoted endlessly from them with profit, did not the limits of the book forbid. Temptation is strong to say something of the strange vegetables now before him on another continent, for this paragraph is written far from home; but these interesting subjects must be left for another occasion. The public has been kind to the old book; the author can ask nothing better for the new one. He has tried to make it sound, but cannot hope to have escaped errors: the reader must exercise his own judgment in the use of the statements and advice. The author does not expect to re-write the book again; but if subsequent editions are needed, certain changes may be made.

INDEX